A Primer of Population Genetics and Genomics

A Primer of Population Genetics and Genomics

FOURTH EDITION

DANIEL L. HARTL

OXFORD
UNIVERSITY PRESS

Great Clarendon Street, Oxford, OX2 6DP,
United Kingdom

Oxford University Press is a department of the University of Oxford.
It furthers the University's objective of excellence in research, scholarship,
and education by publishing worldwide. Oxford is a registered trade mark of
Oxford University Press in the UK and in certain other countries

First three editions published as *A Primer of Population Genetics*.

First Edition published in 1981.

Second Edition published in 1988.

Third Edition published in 2000.

Fourth Edition published in 2020.
Impression: 2

Published in the United States of America by Oxford University Press
198 Madison Avenue, New York, NY 10016, United States of America

British Library Cataloguing in Publication Data
Data available

Library of Congress Control Number: 2019957891

ISBN 978–0–19–886229–1 (hbk.)
ISBN 978–0–19–886230–7 (pbk.)

DOI: 10.1093/oso/9780198862291.001.0001

Printed and bound by
CPI Group (UK) Ltd, Croydon, CR0 4YY

For Dana Margaret
Theodore James
Christopher Lee
& Elizabeth Adele

Preface

The goal of this book is to help instructors teach modern population genetics and genomics in light of their theoretical and empirical roots. From molecular evolution and coalescence through genome-wide association studies, transcriptome-wide association studies, ancient DNA, and studies of adaptation and speciation, modern genetics and genomics have emerged from these foundations. One motivation for writing the book is to summarize the revolution that has taken place in the field owing to low-cost, high-throughput genome sequencing coupled with vast increases in computing power to analyze the data and explore models of increasing complexity. Another is the favorable response to previous editions, which many students and their instructors have used as a brief but solid introduction to the field. The book is also suitable for self-directed learners who recognize the importance of the field and wish to know more.

Familiarity with the basic concepts of genetics and genomics has become essential in many areas of study including evolutionary biology, ecology, systematics, conservation biology and wildlife management, plant and animal breeding, human genetics, anthropology, and even medicine and public health. My own course in population genetics draws students from these fields as well as physics, applied mathematics, computer science, and history of science. Though brief, the book is intended to be sufficiently broad to appeal to readers with a diversity of interests and background.

Although the text introduces a significant number of equations, the emphasis is on explanation rather than derivation. Only elementary algebra is necessary to follow most of the material, but a familiarity with basic calculus is helpful for understanding diffusion equations, Poisson random fields, and complex threshold traits. Mathematical symbols are used consistently throughout the book; when results from the literature are quoted, I have changed the original symbols as necessary to maintain consistency.

I will be grateful for any comments or suggestions, especially those pointing out errors or descriptions that are unclear. My email address is dhartl@oeb.harvard.edu.

The companion website for this book
(www.oup.com/companion/HartlPPGG4e) features all
the figures in the book as freely downloadable JPEGs.

Acknowledgments

First and foremost, I thank the many graduate students, postdoctoral fellows, and visiting scholars who constituted my research group throughout the years. A list of their names can be found in Hartl (2019). Their hard work and creativity has advanced population genetics and helped make the field of population genomics a reality. I'm also grateful to Elena R. Lozovsky, who for many years has helped hold the group together.

Thanks also to Michael M. Desai, my co-instructor in Population Genetics and Genomics (OEB 242). His enthusiasm for the subject has been a continuing inspiration, and I have learned a great deal from his deep knowledge and insights. I'm also grateful to the authors, assessment team, and adopters of *How Life Works* for keeping me up to date on best teaching practices. Thanks also to Robin Hopkins and Sönke Johnsen for providing the gorgeous cover photograph.

I am grateful to my family, and especially to my wife, Christine Blazynski, for somehow having learned to contend with such annoying habits as my preoccupation, answering questions with a blank stare, and often spending many hours upstairs writing in my study.

Finally, I thank the editorial and production teams at Oxford University Press for their excellent work, especially Ian Sherman, Charles Bath, Nic Williams, and Kabilan Selvakumar. Thanks also to Elena R. Lozovsky and Rosa N. Capellan for help in proofreading. I myself am responsible for any mistakes or errors.

Reference

Hartl, D. L. (2019), "Q & A with Daniel L. Hartl, Recipient of the 2019 Thomas Hunt Morgan Medal," *Genetics*, 212 (2), 361–63.

Contents

CHAPTER 1

Genetic Polymorphisms

My PhD mentor once told me that he viewed population genetics as a "recondite field that will never be of great interest except to a small group of specialists" (Hartl 2011). He wasn't often wrong, but he was wrong about this. He assumed (as did nearly everyone at the time) that the ability to sequence deoxyribonucleic acid (DNA) was in the remote future, and the prospect of sequencing a genome as large as the human genome was unimaginable. Yet here we are, not so many years removed, when tens of thousands of human genomes have been sequenced, as well as tens of thousands of genomes of other organisms (mainly bacteria). The genome sequences of more than 3000 species of eukaryotes are currently available, but the Earth BioGenome Project aims to up the ante by sequencing the genomes of all 1.5 million known species of eukaryotes by the year 2028. Genome sequencing is increasingly becoming routine in clinical practice, pinpointing the genetic basis of a patient's disease and enabling personalized healthcare with the precision of prescription eyeglasses.

Large-scale genomic sequencing has had far-reaching implications for population genetics. It makes it possible to study genetics without focusing on mutant organisms that manifest visible differences, such as round versus wrinkled peas or red-eye versus or white-eye fruit flies. It also makes it possible to study genetics without doing controlled crosses. The traditional requirements for mutants and controlled crosses have become dispensable because the discovery of genetic differences among organisms or among species is no longer limited to those differences that reveal themselves by the segregation of genes in pedigrees according to the time-honored principles of genetic inheritance first described by Gregor Mendel in 1866. (For a modern translation see Abbott and Fairbanks (2016).) Large-scale genomic sequencing means that detailed genetic analysis is no longer limited to domesticated animals, cultivated plants, and the relatively small number of experimental organisms that can be cultured in the laboratory. Genetic analysis is possible in any organism. It is for this reason that the concepts and experimental approaches of population genetics have come to pervade virtually every area of modern biology.

In its broadest sense, population genetics is the study of naturally occurring genetic differences among organisms. Genetic differences that are common among organisms of the same species are called genetic **polymorphisms**, whereas genetic differences that accumulate between species constitute genetic **divergence**. We may therefore define **population genetics** as the study of polymorphism and divergence. When such studies include all or most genes in an organism, they constitute **population genomics**.

A Primer of Population Genetics and Genomics. Fourth Edition. Daniel L. Hartl, Oxford University Press (2020). © Daniel L. Hartl.
DOI: 10.1093/oso/9780198862291.003.0001

1.1 Genetic and Molecular Background

Although the genetic principles underlying population genetics are, for the most part, simple and straightforward, it may be helpful to preface the discussion with a few key definitions.

Genotype and Phenotype

Gene is a general term meaning, loosely, the physical entity transmitted from parent to offspring during the reproductive process that influences hereditary traits. The set of genes present in an individual constitutes its **genotype.** The physical or biochemical expression of the genotype is called the **phenotype.** There is a fundamental distinction between genotype and phenotype because, in general, there is not a one-to-one correspondence between genes and traits. Most complex traits are affected by many genes, including hair color, eye color, skin color, height, weight, behavior, life span, and reproductive fitness. Most traits are also influenced more or less strongly by environment. The contribution of both genotype and environment means that the same genotype can result in different phenotypes, depending on the environment; and likewise, that the same phenotype can result from more than one genotype.

Although genes alone do not determine complex phenotypes owing to environmental factors, genes do determine molecular phenotypes. Most genes specify the linear order of amino acid subunits in a **polypeptide chain**. All proteins are composed of one or more polypeptide chains. For example, in red blood cells in adults, the oxygen-carrying protein hemoglobin consists of two copies of a polypeptide chain denoted alpha-globin aggregated with two copies of a different polypeptide chain denoted beta-globin. The alpha-globin gene and the beta-globin gene specify the composition of these polypeptide chains, so these genes determine the hemoglobin phenotype.

Genes can exist in different forms or states. These alternative forms of a gene are called **alleles**. In protein-coding genes, different alleles may code for somewhat different polypeptide chains. An example is one mutant form of the beta-globin gene encoding an aberrant polypeptide chain that tends to form crystals under low oxygen tension. These cause the red blood cells to collapse into half-moon or sickle shapes, yielding the name of the associated blood disease *sickle-cell anemia*. The consequences of the chronic red cell destruction and reduced oxygen-carrying capability in sickle-cell anemia are severe. Yet the disease is maintained at a relatively high frequency in some populations because heterozygous carriers have decreased susceptibility to malaria.

Gene Expression

The essentials of gene expression in the cells of higher organisms (eukaryotes) are outlined in Figure 1.1. From a biochemical point of view, a gene corresponds to a specific sequence of constituents (called nucleotides) along a molecule of DNA. DNA is the genetic material. It is composed of four nucleotides abbreviated according to the identity of the nitrogenous **base** that each contains: A (adenine), G (guanine), T (thymine), or C (cytosine). Different sequences of nucleotides that may occur in a gene therefore, represent alleles. DNA molecules normally consist of two complementary helical strands held together by pairing between the bases: A in one strand is paired with T in the other strand across the way, and G in one strand is paired with C in the other. Each DNA strand has a *polarity* with its ends designated as 5′ or 3′ according to which carbon atom in the deoxyribose is exposed, and the intertwined strands in duplex DNA run in opposite directions.

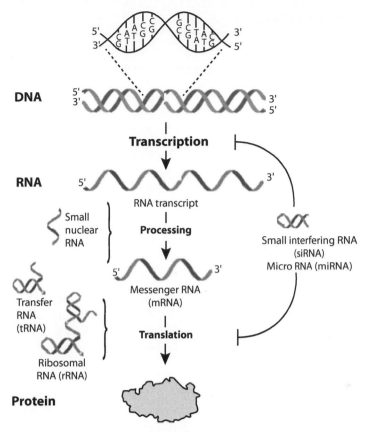

Figure 1.1 In the expression of a protein-coding gene, the sequence of bases in DNA is transcribed into a molecule of RNA that undergoes chemical modifications (processing) to produce a messenger RNA that is translated into a polypeptide chain. Processing and translation require specialized RNA molecules (indicated by the braces), many of which contain self-complementary sequences that can fold back and pair, forming hairpin or cloverleaf structures. Other small RNA molecules may inhibit transcription or translation (indicated by the T-bars). Specialized proteins are also necessary at every step in transcription, processing, and translation.

Figure 1.1 shows how DNA encodes RNA and it in turn encodes protein; however, the process requires hundreds of specialized RNA and protein molecules. The first step in gene expression is **transcription**, in which the sequence of nucleotides present in one DNA strand of a gene is faithfully copied into the nucleotides of an RNA molecule. (RNA contains the sugar ribose instead of deoxyribose.) As the RNA transcript is synthesized, each base in the DNA undergoes pairing with a complementary base in an RNA nucleotide, which is then added to the 3′ end of the growing RNA strand. The base-pairing rules are the same as those in DNA, except that in RNA nucleotides the base U (uracil) is found instead of T. The second step of gene expression is RNA **processing**, in which certain segments of the RNA transcript are removed by splicing and the ends are chemically modified. The segments that are eliminated are known as intervening sequences or **introns**. In RNA splicing, each intron is cleaved at its ends and discarded, while the ends of the flanking RNA sequences are joined together. The regions between the introns that remain in the fully processed RNA are known as **exons**. The fully processed RNA constitutes the **messenger RNA (mRNA)**, which is transported to the cytoplasm where

Table 1.1 The standard genetic code

First nucleotide in codon (5′ end)	Second nucleotide in codon				Third nucleotide in codon (3′ end)
	U	C	A	G	
U	UUU Phe/F	UCU Ser/S	UAU Tyr/Y	UGU Cys/C	U
	UUC Phe/F	UCC Ser/S	UAC Tyr/Y	UGC Cys/C	C
	UUA Leu/L	UCA Ser/S	UAA Stop	UGA Stop	
	UUG Leu/L	UCG Ser/S	UAG Stop	UGG Trp/W	G
C	CUU Leu/L	CCU Fro/P	CAU His/H	CGU Arg/R	U
	CUC Leu/L	CCC Pro/P	CAC His/H	CGC Arg/R	C
	CUA Leu/L	CCA Pro/P	CAA Gln/Q	CGA Arg/R	A
	CUG Leu/L	CCG Pro/P	CAG Gln/Q	CGG Arg/R	G
A	AUU Ile/I	ACU Thr/T	AAU Asn/N	AGU Ser/S	U
	AUC Ile/I	ACC Thr/T	AAC Asn/N	AGC Ser/S	C
	AUA Ile/I	ACA Thr/T	AAA Lys/K	AGA Arg/R	A
	AUG Met/M	ACG Thr/T	AAG Lys/K	AGG Arg/R	G
G	GUU Val/V	GCU Ala/A	GAU Asp/D	GGU Gly/G	U
	GUC Val/V	GCC Ala/A	GAC Asp/D	GGC Gly/G	C
	GUA Val/V	GCA Ala/A	GAA Glu/E	GGA Gly/G	A
	GUG Val/V	GCG Ala/A	GAG Glu/E	GGG Gly/G	G

it undergoes **translation** on ribosomes to produce the sequence of amino acids in the polypeptide encoded in the sequence of nucleotides. In the translated part of the mRNA, each adjacent group of three nucleotides constitutes a coding group or **codon**, which specifies a corresponding amino acid subunit in the polypeptide chain. The standard genetic code showing which codons specify which amino acids is given in Table 1.1. The three-letter and one-letter abbreviations are both established conventions. The codon AUG (boxed in Table 1.1) specifies methionine and also serves as the start codon for polypeptide synthesis. Any of three codons—UAA, UAG, or UGA—specifies the end, or termination, of polypeptide synthesis, upon which the completed polypeptide chain is released from the ribosome.

The totality of DNA in a cell, cellular organelle, or virus constitutes its **genome**. The term is commonly used to denote the DNA in one complete set of chromosomes in a reproductive cell such as a sperm or egg. The **chromosomes** are the microscopic threadlike bodies in the nucleus along which genes are arranged in linear order. Each human sperm or egg contains one complete set of 23 chromosomes and has a genome size of approximately 3×10^9 base pairs. The average human chromosome contains about a thousand protein-coding genes. The position of a gene along a chromosome is called the **locus** of the gene. With certain exceptions, such as the X and Y chromosomes that determine sex, each cell contains two alleles of each gene that are present at corresponding (**homologous**) loci in the maternal and paternal genomes. If the two alleles at a locus are indistinguishable in their effects on the organism, then the individual is said to be **homozygous** at the locus under consideration. And if the two alleles at a locus are distinguishable because of their sequences or because of differing effects on the organism, then the individual is said to be **heterozygous** at the locus. Typographically, genes are indicated in italics, and alleles are typically distinguished by uppercase or lowercase letters (A versus a), subscripts (A_1 versus A_2), superscripts (a^+ versus a^-), or sometimes just

+ and −. Using these symbols, homozygous individuals would be portrayed by any of these formulas: AA, aa, A_1A_1, A_2A_2, a^+a^+, a^-a^-, $+/+$, or $-/-$. As in the last two examples, the slash is sometimes used to separate alleles present in homologous chromosomes to avoid ambiguity. Heterozygous individuals would be portrayed by any of the formulas Aa, A_1A_2, a^+a^-, or $+/-$. The essence of Mendelian genetics is the **principle of segregation**, which states that each reproductive cell from a heterozygous individual contains only one of the two alleles, and that on average the reproductive cells contain the two alleles at equal frequency.

1.2 Major Types of Polymorphisms

One of the universal attributes of natural populations is phenotypic diversity. For most traits, many differing phenotypes can be found among the individuals in any population. Phenotypic diversity in many traits is impressive even with the most casual observation. Among humans, for example, there is diversity with respect to height, weight, body conformation, hair color and texture, skin color, eye color, and many other physical and psychological attributes or skills. Population genetics must deal with this phenotypic diversity, and especially with that portion of the diversity caused by differences in genotype among individuals.

DNA Polymorphisms

Genetic variation, in the form of multiple alleles of many genes, exists in most natural populations. But because most morphological and behavioral variation results from the interactions of many genes among themselves and with the environment, the simple Mendelian inheritance of variation at the molecular level is usually obscured at the phenotypic level. This paradox explains why genetic variation is usually studied at the molecular level, especially because modern methods for detecting and tracking differences in DNA among individuals are varied and powerful. Some of the most important and widely studied types of DNA polymorphisms are illustrated in Figure 1.2, which depicts a DNA duplex at high enough resolution to see the individual base pairs of the component parts of one protein-coding gene.

The **promoter** region of the gene contains sequences that control the time, tissue specificity, and level of transcription. The promoter is not transcribed but is part of the 5′ flanking region of the gene. The transcript includes introns (removed during processing into the mRNA), a 5′ untranslated region, a 3′ untranslated region, and codons to start (AUG) and stop (UAA, UAG, or UGA) polypeptide synthesis. Several types of polymorphism are illustrated.

The acronym **SNP** means a **single-nucleotide polymorphism**, one in which individuals in the population may differ in the identity of the nucleotide pair present at a particular defined site in the DNA. In the example in Figure 1.2, the SNP is in the 5′ flanking region of the gene and is a T/C polymorphism, which means that at this position, some DNA molecules in the population have a T–A base pair whereas other DNA molecules have a C–G base pair. The SNP therefore defines two "alleles," for which there could be three genotypes, namely, homozygous T–A, homozygous C–G, or heterozygous (T–A in one molecule and C–G at the corresponding position in the DNA of the homologous chromosome). The word "allele" is in quotes because the SNP need not be in a coding sequence, or even in a gene. Available data for the human genome suggest that any two randomly chosen genomes—for example, those that you inherited from your parents—are likely to differ at about three million nucleotide sites. About a million SNPs are in protein-coding genes; however, about two-thirds of the SNPs in protein-coding genes are in introns.

A **nonsynonymous** or **missense polymorphism** is a SNP present in the coding region that alters a codon to result in an amino acid replacement in the polypeptide chain. In the example in Figure 1.2, the polymorphic nucleotide substitution in the RNA is shown in boldface (**A** versus **U**), resulting from a polymorphism for the codons GAG versus GUG. The GAG codon specifies Glu (glutamic acid), whereas the GUG codon specifies Val (valine). A nonsynonymous polymorphism therefore results in an **amino acid polymorphism**. In the human genome, a Glu/Val polymorphism at amino acid position 6 in the beta-globin gene is responsible for sickle-cell anemia. On average, each of us humans is heterozygous for about 6000 amino acid polymorphisms (Chen et al. 2012).

A **synonymous polymorphism** is a SNP present in the coding region that produces a synonymous codon and so does not result in an amino acid replacement in the polypeptide chain. In Figure 1.2, the example is a GA**U** codon versus a GA**C** codon, both of which code for Asp (aspartic acid). Synonymous polymorphisms are sometimes called **silent polymorphisms** because they do not change the amino acid. Any two random human genomes differ at about 7000 synonymous nucleotide sites (Chen et al. 2012).

The existence of silent polymorphisms should not be taken to imply that organisms are indifferent to which of the synonymous codons are used for any particular amino acid. In many organisms, certain codons are preferred, especially in mRNAs that encode highly abundant proteins. Such **codon usage bias** affects speed and accuracy of translation as well as protein folding and mRNA stability (Hanson and Coller 2017).

An **indel** is an insertion/deletion polymorphism. In the example in Figure 1.2, the indel size is two base pairs and is located in the 3' flanking region of the gene. (Indels in protein-coding exons are uncommon.) In most cases of indel polymorphisms, it is unclear which sequence is the ancestral sequence; hence, it is not known whether the mutation that created the polymorphism was an insertion or a deletion. Although many indels are fewer than 10 base pairs, some are much larger. Among those in the range 1–5 kilobases (kb),

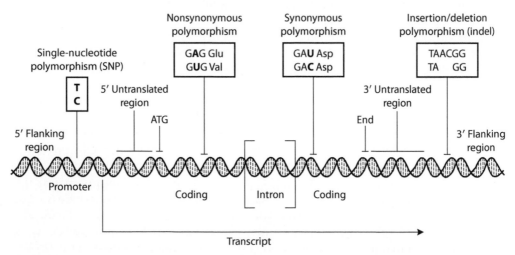

Figure 1.2 The organization of a protein-coding gene showing the major types of DNA polymorphisms that can occur with regard to individual base pairs.

most are insertions of **transposable elements**, which are specialized nucleotide sequences of various kinds that are present in the genomes of virtually all organisms and that are able to replicate and move from one position to another.

Two important types of DNA polymorphism are not shown in Figure 1.2 because they are on a larger scale than one or a few nucleotide pairs. One is a polymorphism known as a **simple tandem repeat (STR)**, in which the alleles differ according to the number of copies of a simple nucleotide sequence that are repeated in tandem along the DNA. An example of an STR polymorphism is the repeat:

$$5'-\text{CATCATCATCATCAT} \cdots \text{CATCATCATCATCAT}-3'$$
$$3'-\text{GTAGTAGTAGTAGTA} \cdots \text{GTAGTAGTAGTAGTA}-5'$$

This repeat can also be denoted as $(5'-\text{CAT}-3')_n$, where n is the number of repeats in the STR allele. Each STR has the potential for exhibiting **multiple alleles** in the population, where each allele differs in its value of n. STRs in which the repeating unit is 2–9 base pairs are often called **microsatellites**, and those in which the repeating unit is 10–60 base pairs are called **minisatellites** (or **VNTRs**, which stands for **variable number of tandem repeats**).

STRs are useful because they are present at thousands of regions scattered throughout the genome, and each has potentially a large number of alleles. Hence, each genome can, in principle, be uniquely identified according to the particular allele present at a relatively small number (ideally, 8–10) of multiallelic STR sites in different chromosomes. Individual identification by means of such repeats is one method for **DNA typing** (also called **DNA fingerprinting**).

Another type of large-scale polymorphism consists of **copy number variation (CNV)** resulting from duplications or deletions of genomic regions differing in size but averaging 200–300 kb pairs. (A *kb pair* is 1000 base pairs.) Most CNVs in the human genome are rare: the average individual is heterozygous for only 0–4 CNVs (Lupski 2015). The rarity of CNVs implies that most are deleterious, and some are known risk factors for diseases such as autism or schizophrenia. A few are beneficial, however. These include uncommon CNVs such as those of the alpha- and beta-globin genes that are associated with resistance to malaria. A few beneficial CNVs are common, such as one that increases copy number of the gene for salivary amylase in human populations. This enzyme degrades starch, and higher copy numbers of the gene are found in populations that have adapted to diets high in wheat, rice, potatoes, and corn, all of which are rich in starch (Perry et al. 2007).

Utility of DNA Polymorphisms

Why are population geneticists interested in genetic polymorphisms? The interest can be justified on any number of grounds, but the following are the reasons most often cited. Each rationale is sufficient in its own right, and most are so self-evidently important as to require no further elaboration.

- To estimate the level of genetic variation in diverse populations of organisms differing in genetic organization (prokaryotes, eukaryotes, organelles, viruses), population size, breeding structure, or life-history characters.
- To identify genetic polymorphisms that are risk factors for inherited diseases in human populations and to estimate how much each factor contributes to overall risk.
- To prudently interpret the genetic risk factors carried by any particular individual for a complex disease in light of the totality of genetic and environmental risk factors for

that disease; this issue is becoming increasingly important as *direct-to-consumer (DTC)* genetic testing and *over-the-counter (OTC)* genetic tests become more widespread.

- To use genetic differences as a means of DNA typing to be able to uniquely identify different individuals in a population for purposes of criminal investigation, personal identification, determination of genetic relatedness, tracking of viral and bacterial transmissible diseases, and so forth.
- To identify genetic polymorphisms that are associated with desirable traits in domesticated animals and cultivated plants.
- To monitor the level of genetic diversity present in key indicator species present in biological communities in habitats exposed to chemical, biological, habitat, or climatic stress.
- To understand the evolutionary origin, genetic history, global expansion, and diversification of the human population.
- To understand the wild ancestors and history of artificial selection in the origin of domesticated animals and cultivated plants.
- To analyze genetic differences between species in order to determine the ancestral history (phylogeny) of the species, and to trace the origin of morphological, behavioral, and other types of adaptations.
- To examine and understand the patterns in which different types of genetic variation (e.g., synonymous versus nonsynonymous polymorphisms) occur throughout the genome.
- To use genetic polymorphisms within subpopulations of a species as indicators of population history, patterns of migration, interbreeding, and so forth.
- To understand the evolutionary mechanisms by which genetic variation is maintained and the processes by which genetic polymorphisms within species become transformed into genetic divergence between species.

1.3 Allele and Genotype Frequencies

Alleles in natural populations usually differ in frequency from one allele to the next. The **allele frequency** of a prescribed allele among a group of individuals is defined as the proportion of all alleles at the locus that are of the prescribed type. The frequency of any prescribed allele in a sample is therefore equal to twice the number of genotypes homozygous for the allele (because each homozygote carries two copies of the allele), plus the number of genotypes heterozygous for the allele (because each heterozygote carries one copy), divided by two times the total number of individuals in the sample (because each individual carries two alleles at the locus).

In many polymorphisms both alleles can be detected in heterozygous genotypes, including polymorphisms in a SNP, an indel, an amino acid, or a simple tandem repeat. In these cases the allele frequencies can be estimated directly because each genotype can be separately identified. Consider, for example, a 32-base-pair indel found in the human chemokine receptor gene *CCR5*. This gene encodes a major macrophage co-receptor for the human immunodeficiency virus HIV-1, the principal causative agent of AIDS. Genotypes that are homozygous for the *CCR5-Δ32* deletion are strongly resistant to infection by HIV-1. The Δ32 allele is found in virtually all European populations, but the allele frequency varies (Lucotte and Mercier 1998).

Figure 1.3 shows the pattern of DNA bands observed for a fragment of the *CCR5* gene among 294 Parisians. The bands have been separated by *electrophoresis*, a procedure in

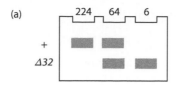

(a)

224   64   6

+

Δ32

(b)   Number of *CCR5* alleles

| + | 448 | 64 | 0 |
| Δ32 | 0 | 64 | 12 |

Figure 1.3 Separation of DNA molecules by size by means of electrophoresis. (a) Pattern of bands observed for a fragment of the *CCR5* gene among 294 individuals. (b) Number of *CCR5* + or *Δ32* alleles represented by each of the banding patterns.

which charged molecules are separated in a gel of some kind. Figure 1.3a shows that the DNA fragment missing 32 base pairs (the Δ32 deletion) is near the bottom of the gel because, being shorter than the + allele, the fragment migrates faster.

As indicated across the top of the gel, the numbers of individuals with each genotype in this sample were as follows:

$$+/+ : 224 \qquad +/\Delta 32 : 64 \qquad \Delta 32/\Delta 32 : 6$$

Expressed as proportions, these become the **genotype frequencies**:

$$+/+ : 224/294 = 0.762 \quad +/\Delta 32 : 64/294 = 0.218 \quad \Delta 32/\Delta 32 : 6/294 = 0.020$$

Figure 1.3b is a tabulation of the number of each type of allele (+ or Δ32) represented by each of the banding patterns. As estimated from this sample, the allele frequency of the + allele is:

$$\text{Estimated frequency of} + \text{allele} = (2 \times 224 + 64)/(2 \times 294) = 0.871$$

whereas that of the Δ32 allele is:

$$\text{Estimated frequency of } \Delta 32 \text{ allele} = (2 \times 6 + 64)/(2 \times 294) = 0.129$$

Because they are proportions, the sum of the genotype frequencies, as well as the sum of the allele frequencies, must equal 1. The allele frequencies just calculated are not the true allele frequencies because they are estimates based on a sample of individuals rather than the entire population. Although estimated allele frequencies are likely to deviate from the true allele frequencies owing to chance, the estimates will usually be close to the true frequencies if the sample is representative and sufficiently large. It is for this reason that allele-frequency estimates should be based on samples of l00 or more individuals whenever possible. Since the term *gene* is sometimes used as a synonym for *allele*, the expression *gene frequency* is sometimes used as a synonym for *allele frequency*; however, the term gene frequency is not recommended unless the particular allele is clear from the context.

More generally, suppose that among n individuals sampled from a population, the numbers of AA, Aa, and aa genotypes are n_{AA}, n_{Aa}, and n_{aa}, respectively. Following convention, we let p and q represent the allele frequencies of A and a, respectively, with $p+q=1$. The estimate p of the allele frequency p in the population that was sampled is:

$$p = (2n_{AA} + n_{Aa})/(2n) \tag{1.1}$$

and the estimated sampling variance is:

$$Var(p) = p\,(1-p)\,/(2n) \qquad (1.2)$$

We emphasize again that the allele frequency in a sample of individuals is only an estimate of the true allele frequency in the population as a whole. The estimates have $2n$ in the denominator because $2n$ is the sample size of the number of alleles—2 alleles in each of n individuals.

Equations 1.1 and 1.2 make use of several important concepts in statistics. Quantities used in characterizing populations are **parameters.** Although the exact values of parameters are usually unknown, their values can be estimated using samples from the population. In some cases it is necessary to distinguish parameters from their estimates typographically, typically by using unembellished symbols for parameters and circumflexes or angled brackets for the estimators. The variance of an estimate is used for judging the reliability of the estimate. Since the variance is also estimated from data in the sample, the variance in Equation 1.2 is an estimate of the true population variance. The square root of the variance of an estimate is known as the **standard error** of the estimate.

The estimate in Equation 1.2 is the sampling variance of a binomial distribution. The binomial distribution occurs in such familiar probability contexts as a series of independent flips of a coin or rolls of a die. If the sampling and estimation were repeated many times, then approximately 68 per cent of the estimates would fall within plus or minus one standard error of true value of the parameter, approximately 95 per cent would fall within two standard errors, and approximately 99.7 per cent would fall within three standard errors. These intervals are known as the 68 per cent, 95 per cent, and 99.7 per cent **confidence intervals.** The most commonly encountered confidence interval is the 95 per cent confidence interval (approximately two times the standard error). In political polling, the 95 per cent confidence interval is often called the *margin of error.*

To take a specific example, 100 independent flips of an unbiased coin would be expected to yield a frequency of heads of 0.50 with a standard error of $\sqrt{(0.5)(0.5)/100}=0.05$. Therefore, for 100 flips of an unbiased coin:

- Approximately 68 per cent of the trials would be expected to yield an observed frequency of heads in the range 0.45–0.55.
- Approximately 95 per cent of the trials would be expected to yield an observed frequency of heads in the range 0.40–0.60 (pollsters would say that the margin of error is $\pm$ 10 per cent).
- Approximately 99.7 per cent of the trials would be expected to yield an observed frequency of heads in the range 0.35–0.65.

To put the matter in somewhat different terms, in a large number of repeated trials intended to estimate the value of a parameter, 32 per cent of the trials would be expected to differ from the true value by more than one standard error, 5 per cent by more than two, and 0.3 per cent by more than three. These approximations all assume that the repeated estimates conform to the familiar, bell-shaped normal distribution.

As a population genetics example, consider again the $\Delta 32$ deletion in the *CCR5* gene. A survey of 111 French Basques from the Pyrenees gave an estimate $p = 0.018$ with a standard error of 0.009. This estimate of p is 8.2 standard errors below the average across all European populations (Lucotte and Mercier 1998), and a deviation as great or greater occurring by chance alone has a probability of 1.7×10^{-16}. Clearly, this discrepancy cannot be attributed to sampling variation. The explanation is found in the unusual population history of the Basques. Genome sequences determined from the remains of eight humans dated to 5500–3500 years ago from a cave in Spain were similar to those from early settlers in other parts of Europe who brought farming from the Middle East to Europe about 7000 years ago (Gunther et al. 2015). The ancient genomes from Spain show the closest similarity to present-day Basques, which suggests that the Basque population was founded by early migrants who interbred with local hunter-gatherers and became geographically isolated in the mountains and largely isolated genetically from other populations whose genomes later became altered by intermixing with new migrants during the spread of the Roman Empire and later the Moorish dominance in Spain. This scenario is also supported by the analysis of allele frequencies at many loci (Calafell and Bertranpetit 1994) as well as analysis of mitochondrial DNA (Bertranpetit et al. 1995).

1.4 Populations and Models

In population genetics, the word *population* does not usually refer to an entire species but rather to a group of individuals of the same species living within a sufficiently restricted geographical area so that any member can potentially mate with any other member of the opposite sex. The focus is on the local interbreeding units of possibly large, geographically structured populations, because it is within such local units that systematic changes in allele frequency occur that ultimately result in the evolution of adaptive characteristics. Such local interbreeding units are often called **local populations** or **subpopulations.** In this book we use the word *population* to mean *local population*—the actual, evolving unit of a species—unless a broader meaning is clear from the context. Local populations are also sometimes referred to as **Mendelian populations** or **demes.**

Models

Equation 1.1 specifies how allele frequencies can be estimated when each genotype can be identified. The situation is more difficult with traits determined by dominant and recessive alleles because homozygous dominant and heterozygous genotypes cannot be distinguished. In these cases, one cannot estimate the allele frequencies without making some assumptions about the relations between allele frequencies and genotype frequencies.

A **model** is an intentional simplification of a complex situation designed to eliminate extraneous detail in order to focus on the essentials. In population genetics, we must contend with factors such as population size, geographical distribution of individuals, patterns of mating, mutation, migration, and natural selection (differential survival or reproductive success). Although we wish ultimately to understand the combined effects of all these factors and more, the factors are so numerous and interact in such complex ways that they cannot usually be grasped all at once. Simpler situations are therefore devised in which a few identifiable factors are the most important and others can be neglected.

Three broad types of models can be distinguished. A *physical model* is a simplified representation of whatever is being studied. One might, for example, create paper flowers of different colors to find out which colors are more attractive to bees or other pollinators. A *conceptual model* attempts to explain how some processes act to produce others. Some conceptual models are verbal explanations. One might posit, for example, that most beneficial mutations eventually become fixed in populations. This statement seems almost a truism, but in fact it is often wrong. Most beneficial mutations have small effects on survival and reproduction, and the majority of them are lost rather quickly.

The kind of model most frequently used in population genetics is the **mathematical model**, which is a set of hypotheses that specifies the mathematical relations between measured or measurable quantities (the parameters) that characterize a population. A mathematical model may be expressed as a system of one or more equations that express how the parameters affect the relation between one or more independent variables (e.g., time) and one or more dependent variables (e.g., population size). The equations may be explicit as is the case with most mathematical models in this book, or they may be implicit in the algorithms that drive the mathematical operations in a computer program.

Mathematical models can be extremely useful. They express concisely the hypothesized quantitative relationships between parameters. They reveal which parameters are the most important in a given situation and thereby suggest critical experiments or observations. They serve as guides to the collection, organization, and interpretation of observed data. And they make quantitative predictions about characteristics of a population that can, within limits, be confirmed or shown to be false. One test of the validity of a model is whether the hypotheses on which it is based and the predictions that grow out of it are consistent with observations.

Population genetics features two complementary types of mathematical models: **exploratory models** that aim to tell you what could possibly happen, and **estimation models** that aim to tell you what actually does happen. There is a great deal of overlap because exploratory models can be used for estimation and the other way round. Models intended to be exploratory are generally simpler than models intended for estimation. Exploratory models attempt to define the main parameters that influence a population and explore how the outcome of a process (e.g., mutation, migration, or natural selection) is affected by different magnitudes of the parameters. Estimation models try to emulate real-world biology and to estimate the actual magnitudes of the parameters (mutation rates, migration rates, or selective effects). These models are often specific to a biological situation: a model for the population genetics of the pandemic covid-19 virus must be very different from a model for the population genetics of the mosquito-borne malaria parasite because of the very different biology of each situation. Estimation models overlap with exploratory models in that the effects of varying the parameters can be explored; however, their main goal is to analyze real data (in this era of genomics, sometimes massive amounts of data) to estimate the values of the parameters. In attempting to mimic biological complexity as well as to incorporate extensive data, estimation models often become so complex that they need to be analyzed computationally using numerical methods or computer simulation.

"All models are wrong," the statistician George Box famously wrote—twice in the same paper (Box 1976). Relative to keeping things simple, he comments: "Since all models are wrong the scientist cannot obtain a 'correct' one by excessive elaboration. On the contrary, following William of Occam, one should seek an economical description of natural phenomena. Just as the ability to devise simple but evocative models is the signature of the great scientist, so overelaboration and overparameterization is often the mark of mediocrity." Relative to choosing which aspects of the actual situation to include or leave out, he says: "Since all models are wrong the scientist must be alert to what is importantly wrong. It is inappropriate to be concerned about mice when there are tigers abroad."

These are words of wisdom. All models are wrong because they are intentional simplifications of the actual situations they are designed to elucidate. Many features of the actual system are intentionally left out of the model, because to include every aspect of the system would make the model too complex and unwieldy. Construction of a model always involves compromise between realism and manageability. A completely realistic model is likely to be too complex to handle mathematically, and a model that is mathematically simple may be so unrealistic as to be useless. Ideally, a model should include all essential features of the system and exclude all nonessential ones. How good or useful a model is often depends on how closely this ideal is approximated. In short, a model is a sort of metaphor or analogy. Like all analogies, it is valid only within certain limits and, when pushed beyond these limits, becomes misleading or even absurd.

Utility of Mathematical Models

Mathematical models have many uses including the following:

- They make a conceptual model precise and unambiguous. Words can be slippery things, and in some verbal arguments you may unconsciously and subtly alter the meaning or connotation of a word as the argument proceeds. Mathematical symbols are given precise definitions, and equations are unambiguous in stating the relations between these symbols.
- They allow exploration of the possible outcomes of different combinations of parameters and determine which parameters the outcomes are most sensitive to.
- They make efficient use of numerical data to estimate the values and margins of error of the parameters.
- They make quantitative or qualitative predictions about the future behavior of the dependent variables in the model.
- They suggest which further observations or measurements are most likely to result in a better understanding of the causal relations between the parameters to improve the model.
- They reveal what kinds of experiments are likely to be most informative and robust in testing the causal relations that the model assumes.
- They help define which relations between the parameters are necessary and/or sufficient for a given outcome.
- They make manifest hidden symmetries and obscure patterns that may be present in a given situation.

Discrete-Time Models

Most of the models in this book are exploratory models. They are admittedly unrealistic; however, they help to train one's intuition regarding how key processes in population genetics play with or against one another. Among the most important of these models are **discrete-time models**, which imagine that changes happen only at discrete points in time. Applied to populations, a discrete-time model assumes a very simple life history with **nonoverlapping generations** in which the individuals in each generation die as the members of the next generation are born. The model applies literally only to annual plants (Figure 1.4) and some short-lived invertebrates.

In an organism with nonoverlapping generations (often called *discrete generations*), all members of any generation are born at the same time, mature and reach sexual maturity synchronously, mate simultaneously, and die immediately after producing the new generation. The key simplification is that, at any given time, all members of the population are of the same age, and no individuals survive from one generation to the next. This model is often used in population genetics as a first approximation to populations

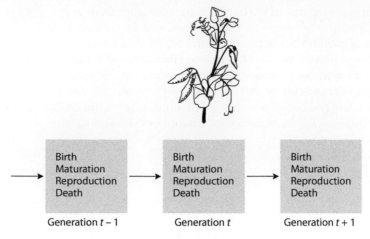

Figure 1.4 Nonoverlapping generation model.

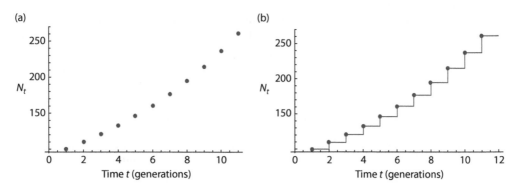

Figure 1.5 Discrete-time model of population growth. (a) Population size at times $t = 1, 2, \ldots$ when $N_0 = 100$ and $\rho = 0.10$. (b) The step function that best describes the population growth.

that have more complex life histories. Although at first glance the model seems grossly oversimplified, calculations of expected genotype frequencies based on the model are adequate for many purposes, and they are often satisfactory first approximations even for populations with long and complex life histories such as in humans.

As an example, Figure 1.5a shows population growth in an organism with discrete, nonoverlapping generations. The number of individuals in generation t (N_t) is related to the number of individuals in the previous generation (N_{t-1}) according to $N_t - N_{t-1} = \rho N_{t-1}$, where ρ may be positive, negative (provided $\rho \geq -1$), or zero. Writing this equation in slightly different form yields:

$$N_t = (1 + \rho) N_{t-1} = (1 + \rho)^2 N_{t-2} = \cdots = (1 + \rho)^t N_0 \tag{1.3}$$

In discrete-time models, the symbol Δ is often used to denote the difference in a variable between one generation and the generation before, and so we can also write $N_t - N_{t-1} = \rho N_{t-1}$ as the *difference equation*:

$$\frac{\Delta N_t}{N_{t-1}} = \frac{N_t - N_{t-1}}{N_{t-1}} = \rho \tag{1.4}$$

This equation makes manifest that ρ is the per capita increase in population size in one generation.

It is tempting to connect the dots in Figure 1.5a with a smooth curve, but the function that actually describes Equation 1.3 is the step function in Figure 1.5b. The step function correctly depicts that the population size remains constant within each generation but undergoes an instantaneous change in population size at each discrete time point.

Models in population genetics are often discrete-time models for several reasons. Although most processes in the real world happen continuously in time, measurements of such processes are usually carried out at discrete intervals. Modeling such processes as discrete in time matches the discreteness of the measurements. If the variables are changing slowly enough, a discrete model behaves essentially as a continuous process with a step function that is for all practical purposes a smooth curve. A difference equation like Equation 1.4 (also called a *recurrence equation*) is the discrete-time analogue of a differential equation, and solving difference equations is often mathematically simpler than solving differential equations, particularly in finding steady-state outcomes.

Continuous-Time Models

As the name implies, **continuous-time models** assume that a process changes at every instant in time rather than at discrete intervals. For population growth, you might imagine a large population of bacteria or some other microbe in which, at each instant in time, only a few cells are dividing. The differential equation describing population growth is the continuous analogue of Equation 1.4, namely:

$$\frac{dN(t)}{dt} \frac{1}{N(t)} = r \tag{1.5}$$

where $N(t)$ is the population's size at time t and r is called the **intrinsic rate of increase**, often denoted r_0 (pronounced "r-naught"). Equation 1.5 is an elementary differential equation, and its solution is:

$$N(t) = N(0)e^{rt} \tag{1.6}$$

A plot of $N(t)$ from Equation 1.6 is shown in Figure 1.6a.

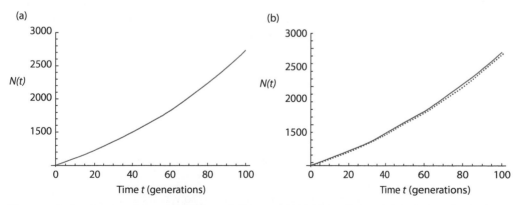

Figure 1.6 Continuous-time model of population growth. (a) Population size at time t when $N(0) = 1000$ and $r = 0.01$. (B) Continuous-time growth versus discrete growth (dots) when $N(0) = N_0 = 1000$ and $r = \rho = 0.01$.

While Equation 1.6 is mathematically a little more elegant than Equation 1.3, the timescale is ambiguous. Should time be measured in hours, days, or years? The proper timescale is one in which population growth is smooth and continuous. If the timescale is too short, then very little growth occurs in any small interval; and if it is too long, then growth is discontinuous. The proper timescale therefore depends on the life history of organism. In some bacteria, the appropriate scaling is in minutes or hours, in many insects it is in weeks or months, and in most large-bodied plants and animals it is in years. Then, too, whereas the generation time in the discrete-growth model in Equation 1.3 is clearly defined (it is the time interval at which the population size jumps from one number to another), the generation time in the continuous-growth model is undefined. As a practical matter, the generation time is usually taken as the average age at reproduction, which in bacteria can be as short as 20 minutes and in humans usually ranges from 20 to 30 years with an average of about 25 years.

When time in the continuous-growth model is measured in generations, and population growth is sufficiently slow, then Equations 1.3 and 1.6 are approximately equal over periods of 100 generations or so. Figure 1.6b shows an example for $\rho = 0.01$. The reason for the close match of the curves is that:

$$e^r = 1 + r + \frac{r^2}{2!} + \frac{r^3}{3!} + \cdots$$

When r is sufficiently small, then $e^r \approx 1 + r$ and therefore $r \approx \rho$. As t becomes large, however, the population sizes in the continuous and discrete models diverge (Figure 1.6b).

How Models are Tweaked

Populations cannot grow exponentially for very long because they soon run out of resources or space. We should be grateful. With exponential growth in a large enough flask with sufficient nutrients, a single cell of *Escherichia coli* would give rise to a clump of cells the size of an elephant in less than 24 hours. To take growth-limiting factors into account, we can tweak Equation 1.5 to the form:

$$\frac{dN(t)}{dt} = rN(t)\left(1 - \frac{N(t)}{K}\right) \tag{1.7}$$

This is called the **logistic equation** in which K is the **carrying capacity** of the environment. When $N(t)$ is much smaller than K (i.e., $N(t) << K$), then population growth is essentially exponential, but as $N(t)$ approaches K the growth rate decrease until it effectively becomes 0.

The solution to the logistic equation is:

$$N(t) = \frac{N(0)K}{N(0) + [K - N(0)]\,e^{-rt}} \tag{1.8}$$

Figure 1.7 shows an example of how population growth is damped as $N(t)$ approaches K. One can also write a discrete-time analog of Equation 1.7; however, there is no known solution that can be written in closed form (i.e., an equation using only common algebraic operations and elementary functions with a finite number of terms.) Nonlinear difference equations are usually far more difficult to solve than nonlinear differential equations, which is one reason why continuous-time models are often used in population genetics.

Perhaps paradoxically, a population geneticist would use a continuous-time model for logistic growth rather than a discrete-time model even in an annual plant in which the discrete-time model would be more realistic. The reason is that the continuous-time model

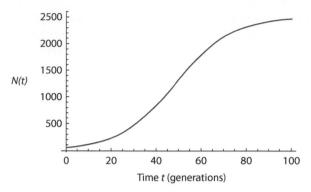

Figure 1.7 Logistic population growth. The parameters are $N(0) = 50$, $r = 0.08$, and $K = 2500$.

is easier to deal with mathematically. Population geneticists often choose a particular model primarily because of its mathematical tractability. You'll find many examples throughout this book. While there is sometimes a trade-off between mathematical convenience and reality, the hope is that what may be forfeited in relevance will be offset by the insights gained from mathematical analysis.

Problems

1.1 For a gene with two alleles A_1 and A_2 in a diploid organism:

a. Which cross with parents that differ maximally in genotype produces offspring that are genetically identical to each other?

b. Which cross with parents that are identical in genotype produces offspring that are genetically the most diverse?

1.1 ANSWER Use Mendel's principle of segregation. **a.** $A_1A_1 \times A_2A_2$ parents produce all A_1A_2 offspring. **b.** $A_1A_2 \times A_1A_2$ parents produce 1/4 A_1A_1, 1/2 A_1A_2, and 1/4 A_2A_2 offspring.

1.2 How many genotypes are possible in a diploid organism with m alleles?

1.2 ANSWER There are m possible homozygous genotypes and $m(m-1)/2$ heterozygous genotypes for a total of $m + m(m-1)/2 = m(m+1)/2$. Note that this can also be written in terms of a binomial coefficient as $\binom{m+1}{2} = \dfrac{(m+1)!}{2!(m-1)!}$.

1.3 The accompanying gel diagram shows the banding patterns observed for a two-allele polymorphism among a sample of 120 individuals in a diploid population.

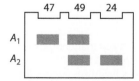

What is the estimated allele frequency p of the A_1 allele in this sample? What is the standard error of this estimate? What is the 95 per cent confidence interval around this estimate?

1.3 ANSWER The estimate of $p = (2 \times 47 + 49)/(2 \times 120) = 0.5958$. The estimated variance of the estimate equals $p(1-p)/(2n) = 0.001$ because $n = 120$. The standard error (*se*) of p is therefore $s = \sqrt{(0.001)} = 0.0317$, and the 95 per cent confidence interval is approximately $p - 2s$ to $p + 2s$ or 0.532–0.659.

1.4 The gel diagram shown here depicts the banding patterns observed for a three-allele polymorphism among a sample of 300 individuals in a diploid population.

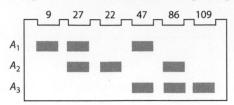

Estimate the allele frequencies p_i, p_2, and p_3 of the alleles A_1, A_2, and A_3.
1.4 ANSWER The estimates are $p_1 = (18 + 27 + 47)/600 = 0.1533$, $p_2 = (27 + 44 + 86)/600 = 0.2617$, and $p_3 = (47 + 86 + 218)/600 = 0.5850$.

1.5 An easy approximate test of whether two estimates differ from one another more than would be expected by an accident of sampling is whether their 95 per cent confidence intervals overlap. If they do not overlap, the estimates are unlikely to be so different due to chance alone. In regard to the *CCR5-Δ32* deletion, a sample of 570 individuals from Brazil yielded an estimated allele frequency of $p = 0.0544$, and a sample of 207 individuals from Denmark yielded an estimate of $p = 0.1232$. Are these allele frequencies likely to differ by chance alone?
1.5 ANSWER The Brazilian sample has a standard error of 0.0067 and that from Denmark has a standard error of 0.0162. The 95 per cent confidence intervals around the allele frequencies are 0.041–0.068 for Brazil and 0.091–0.156 for Denmark. These do not overlap; hence, the difference in allele frequency is too great for sampling error alone.

1.6 A bacterial population undergoes exponential population growth with an intrinsic rate of increase r such that the population size doubles every 50 minutes. What is the value of r in this population? What are the units of measurement of r?
1.6 ANSWER To obtain an expression for the doubling time with $N(t) = N(0)e^{rt}$, set $N(t) = 2N(0)$ and solve for t. The answer is $t = \ln(2)/r$. For $t = 50$ minutes, $r = 0.0139/\text{min}$. The units of r are reciprocal minutes.

References

Abbott, S. and Fairbanks, D. J. (2016), 'Experiments on plant hybrids by Gregor Mendel', *Genetics*, 204 (2), 407–22.

Bertranpetit, J., et al. (1995), 'Human mitochondrial DNA variation and the origin of Basques', *Ann Hum Genet*, 59 (Pt 1), 63–81.

Box, G. E. P. (1976), 'Science and statistics', *J Am Stat Assoc*, 71, 791–9.

Calafell, F. and Bertranpetit, J. (1994), 'Principal component analysis of gene frequencies and the origin of Basques', *Am J Phys Anthropol*, 93 (2), 201–15.

Chen, R., et al. (2012), 'Personal omics profiling reveals dynamic molecular and medical phenotypes', *Cell*, 148 (6), 1293–307.

Gunther, T., et al. (2015), 'Ancient genomes link early farmers from Atapuerca in Spain to modern-day Basques', *Proc Natl Acad Sci U S A*, 112 (38), 11917–22.

Hanson, G. and Coller, J. (2018), 'Codon optimality, bias and usage in translation and mRNA decay', *Nat Rev Mol Cell Biol*, 19 (1), 20–30.

Hartl, D. L. (2011), 'James F. Crow and the art of teaching and mentoring', *Genetics*, 189 (4), 1129–33.

Lucotte, G. and Mercier, G. (1998), 'Distribution of the CCR5 gene 32-bp deletion in Europe', *J Acquir Immune Defic Syndr Hum Retrovirol*, 19 (2), 174–7.

Lupski, J. R. (2015), 'Structural variation mutagenesis of the human genome: impact on disease and evolution', *Environ Mol Mutagen*, 56 (5), 419–36.

Perry, G. H., et al. (2007), 'Diet and the evolution of human amylase gene copy number variation', *Nat Genet*, 39 (10), 1256–60.

CHAPTER 2

Organization of Genetic Variation

Biological species almost always exhibit some sort of geographical structure or nonrandom pattern in the spatial distribution of organisms. Members of a species are rarely distributed homogeneously in space. There is almost always some sort of clumping or aggregation, some schooling, flocking, herding, or colony formation. Population subdivision is often caused by environmental patchiness, areas of favorable habitat intermixed with unfavorable areas. Such environmental patchiness is obvious in the case of terrestrial organisms on islands in an archipelago, but patchiness is a common feature of most habitats—freshwater lakes have shallow and deep areas, meadows have marshy and dry areas, forests have sunny and shady areas. Population subdivision can also be caused by social behavior, as when wolves form packs. Even the human population is clumped or aggregated—into towns and cities, away from deserts and mountains.

Population structure implies that allele frequencies may differ from one part of a species' range to another. For this reason, population geneticists tend to focus on local interbreeding subpopulations of organisms, often called **demes.**

2.1 Random Mating

With discrete, nonoverlapping generations, the calculation of genotype frequencies from knowledge of allele frequencies is straightforward. The genotype frequencies are determined in part by the manner in which mating pairs are formed. Under **random mating**, mating takes place at random with respect to the genotypes under consideration, as if determined by random collisions. The probability of two genotypes forming a mating pair is therefore equal to the product of their respective genotype frequencies.

Bear in mind that mating can be random with respect to some traits but nonrandom with respect to others in the same population. In human populations, for example, mating seems to be random with respect to most DNA and protein polymorphisms as well as many other characteristics, but mating is nonrandom with respect to other traits such as skin color and height. Genotype frequencies are also influenced by various evolutionary forces including mutation, migration, and natural selection. For the moment, these evolutionary forces will be assumed to be absent or at least negligibly small in magnitude. Additionally, genotype frequencies are affected by chance statistical fluctuations (**random genetic drift**) that occur in all small populations, but for now we suppose that each local population is sufficiently large that small-population effects can be neglected.

A Primer of Population Genetics and Genomics. Fourth Edition. Daniel L. Hartl, Oxford University Press (2020). © Daniel L. Hartl.
DOI: 10.1093/oso/9780198862291.003.0002

The Hardy–Weinberg Principle

The main assumptions of the standard random-mating model are:

• Diploid organism	• Equal allele frequencies
• Sexual reproduction	in the sexes
• Nonoverlapping generations	• No migration
• Random mating	• No mutation
• Large population size	• No selection

Under these assumptions, the genotype frequencies for a gene with two alleles can be deduced quite easily. We assume that genotype frequencies of AA, Aa, and aa in the parental generation are D, H, and R, respectively, where $D + H + R = 1.0$. The allele frequencies of A and a are given by:

$$p = (2D + H)/2 = D + H/2 \quad \text{and} \quad q = (2R + H)/2 = R + H/2 \tag{2.1}$$

These formulas simply express that homozygous AA genotypes carry two copies of A, homozygous aa genotypes carry two copies of a, and heterozygous Aa genotypes carry one copy of each allele. Note that $p + q = 1$, which is a consequence of the fact that the gene has only two alleles. With three genotypes, there are six possible types of mating. When mating is random, these mating types occur in proportion to the genotypic frequencies in the population. For example, the mating $AA \times AA$ occurs only when an AA male mates with an AA female, and this occurs a proportion $D \times D$ (or D^2) of the time. Similarly, an $AA \times Aa$ mating occurs when an AA female mates with an Aa male (proportion $D \times H$), or when an Aa female mates with an AA male (proportion $H \times D$)—and so the overall proportion of $AA \times Aa$ matings is $DH + HD = 2DH$. The frequencies of these and the other types of mating are given in the second column of Table 2.1.

The offspring genotypes produced by the matings are given in the last three columns of Table 2.1. The offspring frequencies follow from Mendel's law of segregation, which states that an Aa heterozygote produces an equal number of A-bearing and a-bearing gametes (**gamete** is a useful word meaning sperm or egg). Homozygous AA genotypes produce only A-bearing gametes and homozygous aa genotypes produce only a-bearing gametes. Therefore, a mating of AA with aa produces all Aa offspring, a mating of AA with Aa

Table 2.1 Demonstration of the Hardy–Weinberg principle

Mating	Frequency of mating	Offspring genotype frequencies AA	Aa	aa
$AA \times AA$	D^2	1	0	0
$AA \times Aa$	$2DH$	½	½	0
$AA \times aa$	$2DR$	0	1	0
$Aa \times Aa$	H^2	¼	½	¼
$Aa \times aa$	$2HR$	0	½	½
$aa \times aa$	R^2	0	0	1
Totals (next generation)		D'	H'	R'

where: $D' = D^2 + 2DH/2 + H^2/4 = (D + H/2)^2 = p^2$
$H' = 2DH/2 + 2DR + H^2/2 + 2HR/2 = 2(D + H/2)(R + H/2) = 2pq$
$R' = H^2/4 + 2HR/2 + R^2 = (R + H/2)^2 = q^2$

produces 1/2 *AA* and 1/2 *Aa* offspring, a mating of *Aa* with *Aa* produces 1/4 *AA*, 1/2 *Aa*, and 1/4 *aa* offspring, and so forth.

The genotype frequencies of *AA*, *Aa*, and *aa* after one generation of random mating are denoted in Table 2.1 as D', H', and R' respectively. (Population geneticists commonly use primed and unprimed symbols to denote population parameters in two successive generations.) The new genotype frequencies are calculated as the sum of the cross products shown at the bottom of the table. The new genotype frequencies P', Q', and R' simplify to:

$$AA: D' = p^2 \quad Aa: H' = 2pq \quad aa: R' = q^2 \tag{2.2}$$

a result known as the **Hardy–Weinberg principle** after Godfrey Hardy and Wilhelm Weinberg, who published their results in 1908 unaware that most researchers in the field already knew of the principle and were using it in their work.

The calculation in Table 2.1 illustrates the important principle that *random mating of individuals is equivalent to random union of gametes*. This equivalence means that we can cross-multiply the allele frequencies along the margins in the sort of square shown in Figure 2.1.

As shown at the right of the figure, the cross multiplication easily leads to the Hardy–Weinberg frequencies. The square works because it is a systematic way of going through all the possibilities of gamete combination. The probability that a sperm or egg carries *A* is p; the probability that a sperm or egg carries *a* is q. With random combination of gametes, the chance that an *A*-bearing sperm fertilizes an *A*-bearing egg is $p \times p = p^2$, which is the frequency of *AA* genotypes. The probability that an *A*-bearing sperm fertilizes an *a*-bearing egg is $p \times q = pq$, and the probability that an *a*-bearing sperm fertilizes an *A*-bearing egg is $q \times p = qp$, and so, altogether, the frequency of *Aa* heterozygotes is $pq + qp = 2pq$. Finally, the probability that an *a*-bearing sperm fertilizes an *a*-bearing egg is $q \times q = q^2$, which is the genotype frequency of *aa*. Note that $p^2 + 2pq + q^2 = (p+q)^2 = (1)^2 = 1$, thereby accounting for all of the offspring.

Constancy of Allele Frequencies

The Hardy–Weinberg principle provides the foundation for many theoretical investigations in population genetics. One of the most important implications emerges when we calculate the allele frequencies p' and q' of *A* and *a* in the next generation. Using Equation 2.1, the allele frequencies of *A* and *a* are:

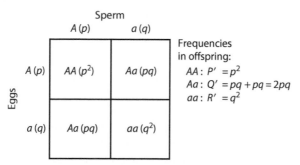

Figure 2.1 Cross-multiplication square showing Hardy–Weinberg frequencies resulting from random mating with two alleles.

$$p' = (2D' + H')/2 = (2p^2 + 2pq)/2 = p(p+q) = p$$
$$q' = (2R' + H')/2 = (2q^2 + 2pq)/2 = q(q+p) = q$$

In other words, the allele frequencies in the next generation are exactly the same as they were the generation before. This means that Mendelian inheritance, by itself, tends to keep the allele frequencies constant. Because the allele frequencies remain the same generation after generation, so also are genotype frequencies maintained in the proportions p^2, $2pq$, and q^2. These genotype frequencies constitute what is often called the **Hardy–Weinberg equilibrium (HWE)**.

The constancy of allele frequencies under Mendelian inheritance has far-reaching implications. It means that populations tend to remain the same unless altered by internal or external evolutionary forces such as mutation, migration, natural selection, or random genetic drift. More broadly, the constancy of allele frequencies means that Mendelian inheritance, by its very nature, tends to maintain genetic variation in a population. This implication has great historical significance in regard to one of the principal criticisms of Darwin's theory of natural selection that was articulated in an 1867 review of *On the Origin of Species* by the Scottish engineer Fleeming Jenkin. He pointed out that, under the then-prevailing view that hereditary factors from the parents blend together in the offspring, newly arising phenotypic variation that is adaptive would quickly be swamped because rare variants must mate with the much more common, less-adapted individuals. In his opinion, "the advantage, whatever it may be, is utterly outbalanced by numerical inferiority." Darwin regarded this criticism as among the strongest arguments against his theory of natural selection.

The natural tendency for Mendelian inheritance to maintain genetic variation refutes the argument of swamping due to numerical inferiority. Mendelian inheritance also features some strange paradoxes. For example, look in Table 2.1 at the mating $AA \times aa$, which produces all Aa offspring. In this case, parents who are genetically as different as they can be produce offspring that are genetically as similar as they can be. Look now at the mating $Aa \times Aa$, which produces offspring that are 1/4 AA, 1/2 Aa, and 1/4 aa. In this case, parents who are genetically as similar as they can be produce offspring that are genetically as different as they can be!

Chi-square Test for Hardy–Weinberg Equilibrium

The mere fact that observed genotype frequencies may happen to fit HWE cannot be taken as evidence that all of the assumptions in the model are valid. The principle is not very sensitive to small departures from the assumptions, particularly those pertaining to a very large population size with no migration, mutation, or selection. On the other hand, the relative insensitivity to departures from its assumptions gives the principle some robustness, because it implies that HWE can be valid to a first approximation even when one or more of the assumptions are violated.

The usual test for goodness of fit of observed data to HWE is a chi-square test. The test statistic is usually symbolized X^2, and under the hypothesis of HWE the X^2 has approximately a chi-square distribution. Application of the test can be illustrated with the *CCR5-Δ32* deletion polymorphism associated with resistance to infection by HIV-1. In the sample of Parisians mentioned in Chapter 1, there were 224 +/+ homozygotes, 64 +/Δ32 heterozygotes, and 6 Δ32/Δ32 homozygotes. The estimated allele frequencies p for + and q for Δ32 are $p = 0.871$ and $q = 0.129$. With HWE for these allele frequencies, the expected genotype frequencies are $(0.871)^2 = 0.758, 2(0.871)(0.129) = 0.225$, and $(0.129)^2 = 0.017$.

Multiplying each of these by the sample size (294 persons) gives the expected numbers as 222.9, 66.2, and 4.9. This conversion is necessary because the chi-square test must be based on the observed numbers, not ratios or proportions. The comparison is therefore between the observed (*obs*) and expected (*exp*) numbers:

	+/+	+/Δ32	Δ32/Δ32	
obs	224	64	6	Total = 294
exp	222.9	66.2	4.9	Total = 294

(Calculating the totals for the observed and expected numbers is a useful crosscheck.) In comparisons of this type, the value of X^2 is calculated as:

$$X^2 = \sum \frac{(obs - exp)^2}{exp} \tag{2.3}$$

where the summation sign Σ means summation over all classes of data, in this case all three genotypes. The resulting value of:

$$X^2 = \frac{(224 - 222.9)^2}{222.9} + \frac{(64 - 66.2)^2}{66.2} + \frac{(6 - 4.9)^2}{4.9} = 0.33$$

is the test statistic.

Associated with any X^2 value is a second number called the **degrees of freedom** for that X^2. In general, the number of degrees of freedom associated with a X^2 equals the number of classes of data (in this case, 3) minus 1 (because the totals must be equal), minus the number of parameters estimated from the data (in this case, 1, because the parameter p was estimated from the data), and so the number of degrees of freedom for our chi-square value is $3 - 1 - 1 = 1$. (Note: a degree of freedom is not subtracted for estimating q because of the relation $q = 1 - p$; once p has been estimated, the estimate of q is automatically fixed, hence we deduct just the one degree of freedom corresponding to p.)

The actual assessment of goodness of fit is determined from Figure 2.2. (You can also direct any internet search engine to find "chi-square P-values online.") To use the chart in Figure 2.2, first find the value of X^2 along the horizontal axis; then move vertically from this value until the proper degrees-of-freedom line is intersected; then move left horizontally from the point of intersection to the vertical axis and read the corresponding probability value. In the present case, with $X^2 = 0.33$ and one degree of freedom, the corresponding probability value is about $P = 0.57$.

The P-value has the following interpretation: it is the probability that chance alone could produce a deviation between the observed and expected values at least as great as the deviation actually obtained. Thus, if the probability is large, it means that chance alone could account for the deviation, and it strengthens our confidence in the validity of the model used to obtain the expectations—in this case HWE. On the other hand, if the probability associated with the X^2 is small, it means that chance alone is not likely to lead to a deviation as large as actually obtained, and it undermines our confidence in the validity of the model. Where exactly the cutoff should be between a "large" probability and a "small" one is, of course, not obvious, but there is an established guideline to follow. If the probability is less than 0.05, then the result is said to be **statistically significant**, and the goodness of fit is considered sufficiently poor that the model is judged invalid for the data. Alternatively, if the probability is greater than 0.05, the fit is considered sufficiently close that the model is not rejected. Because the probability in the Δ32 example is 0.57, which is considerably greater than 0.05, we have no reason to reject the hypothesis that the genotype frequencies are in HWE for this gene.

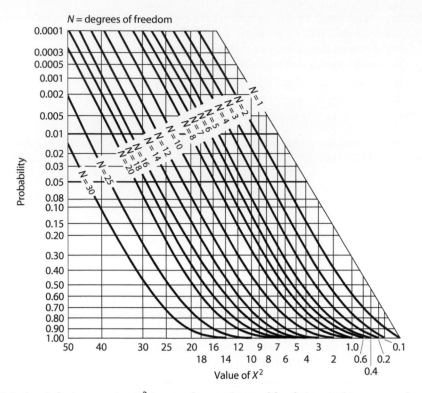

Figure 2.2 Graph for interpreting X^2 in tests for goodness of fit of observed to expected numbers. The probability value (*P*-value) for any X^2 is the probability of obtaining a fit as bad or worse than that actually observed, under the assumption that the expected numbers are correct.

In contrast to the sample of people from Paris, a sample from Rheims does not fit HWE. In this sample the observed and expected values are:

	+/+	+/Δ32	Δ32/Δ32	
obs	234	36	6	Total = 276
exp	230.1	43.8	2.1	Total = 276

and the $X^2 = 8.7$ with an associated $P = 0.003$. A *P*-value less than 0.01, as obtained in this case, is said to be **statistically highly significant**, and the hypothesis of HWE is decisively rejected. (Alternative goodness-of-fit tests are usually preferred when the smallest expected value is < 5, but here the test is for illustration only.) Why would this sample not fit HWE? In theory, any of the assumptions needed for HWE may be incorrect, but often a deficiency of heterozygotes, as observed in this case, results from combining samples from two or more subpopulations, all in HWE but differing in allele frequency. In this particular case, the observed numbers are consistent with an equal mixture of two HWE subpopulations having Δ32 allele frequencies of $q = 0.18$ and $q = 0$, which yields a X^2 of 2.1. There are also many other possibilities.

Statistical Power of the Chi-square Test for Hardy–Weinberg Equilibrium

The Hardy–Weinberg principle is one of the workhorses of population genetics. It is the starting point for exploring the interplay between mutation, migration, selection, and random drift. This may seem paradoxical because most of the assumptions on which it is based are manifestly false, such as the implicit assumption that the population is infinite in size. The main reason that the Hardy–Weinberg principle pervades much of population genetics theory is that it is relatively robust to small departures from the assumptions. To say the same thing in somewhat different terms, a chi-square test for HWE is relatively weak in rejecting the hypothesis provided that the deviations from Hardy–Weinberg frequencies are not too large.

To make these points quantitative requires a short digression into hypothesis testing, starting with some definitions. In testing any hypothesis, you must first specify the hypothesis. In statistics, the hypothesis being tested is denoted H_0 and called the **null hypothesis.** The term "null" in this context means the hypothesis that the observed data may nullify—which is to say invalidate or reject. A null hypothesis can take many forms, but in genetics typical examples are testing whether a ratio of phenotypes among progeny in a genetic cross is consistent with 3:1, or whether the genotype frequencies in a population are consistent with HWE. The null hypothesis must be specific enough to be able to calculate the probabilities of obtaining whatever results are observed, otherwise there would be no basis to compare observed with expected results and hence no grounds for rejection if the null hypothesis were false. This is why, when geneticists test whether there is an association between a single-nucleotide polymorphism (SNP) and a complex disease, such as diabetes or high blood pressure, the null hypothesis is that there is *no* association—because under the hypothesis of no association the expected results are predictable.

In testing the null hypothesis, you can make two kinds of mistakes:

1. You can reject the hypothesis when it is true. This is called a **type I error** or a **false positive.** The probability of type I error is typically represented by the Greek letter α. The value of α is chosen arbitrary and sets the **significance level** of the test. As explained earlier, α is conventionally chosen to be $\alpha = 0.05$ ("significant") or $\alpha = 0.01$ ("highly significant").
2. You can fail to reject the hypothesis when it is false. This is called a **type II error** or a **false negative.** The probability of type II error is denoted β, and β has a negative trade-off with α. Smaller values of α reduce the probability of type I error but increase β and the probability of type II error; larger values of α increase the probability of type I error but decrease β and the probability of type II error.
3. The probability of rejecting a hypothesis when it is false is given by $1 - \beta$, which is call the **power** of the statistical test. An ideal statistical test has a power of 0.90 or more, but this goal is often unrealized.

The power to reject a false hypothesis depends on the value of α, on the sample size (larger sample size increases power), and also on the extent to which the observed data deviate from the expectations of the hypothesis being tested: larger deviations more easily lead to rejection. In the context of the chi-square test for HWE for two alleles of a single gene, the test is essentially a test of whether $H^2 = 4DR$, where D, H, and R are defined as in Table 2.1. When HWE is attained, then $H^2 = 4p^2q^2$ and $4DR = 4p^2q^2$ also.

Methods for calculating the power of the chi-square test for HWE are beyond the scope of this book, but some illustrative results are shown in Figure 2.3. The curves are labeled according to the magnitude of the deviation of the frequency of heterozygotes from the expected proportion $2pq$, where p and q are the allele frequencies. For example, $\pm$ 30 per cent means that the true frequency of heterozygous genotypes could be either $2pq(1 - 0.30) = 1.4pq$ or $2pq(1 + 0.30) = 2.6pq$. When the deviation from HWE is small

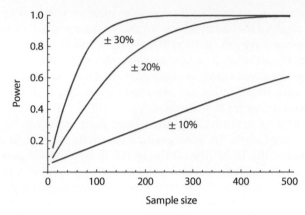

Figure 2.3 Statistical power of a chi-square test to reject a false hypothesis as a function of sample size for one gene with two alleles when $a = 0.05$. The curves are labeled according to the excess or deficiency of heterozygous genotypes relative to that expected with HWE.

($\pm$ 10 per cent), the chi-square test has little power to reject HWE even for sample sizes as great as 500, in which case the power is only 60 per cent. (Achieving 90 per cent power requires a sample size of about 1000.) The test has somewhat greater power for larger deviations; however, achieving a power of 90 per cent requires a sample size of 250 for a deviation of $\pm$ 20 per cent and a sample size of 100 for a deviation of $\pm$ 30 per cent.

Recessive Alleles Hidden in Heterozygotes

HWE helps solve a problem that arises when studying polymorphisms of genes with dominant and recessive alleles. The problem is that homozygous and heterozygous genotypes cannot be distinguished, and so the allele frequencies cannot be estimated directly. The solution is that, if one is willing to assume HWE, then the allele frequencies can be estimated anyway. The trick is to use the observed frequency of recessive homozygotes to estimate the q^2 term of the HWE. In other words, if R is the frequency of homozygous recessive genotypes found among a sample of n individuals, then the estimate of q and its sampling variance are:

$$q = \sqrt{R} \quad Var(q) = \frac{1-R}{4n} \tag{2.4}$$

The expression for the variance may look unfamiliar. This is because it is the variance of a function of a random variable rather than of the random variable itself. In this case the random variable is R and its variance is the binomial variance $R(1-R)/n$, but the function $f(R)$ whose variance we wish to estimate is $q = f(R) = \sqrt{R}$. To estimate this variance we use the so-called *delta method*, which in this case asserts that:

$$Var[f(R)] = \left(\frac{df(R)}{dR}\right)^2 Var(R) = \left(\frac{d\sqrt{R}}{dR}\right)^2 \left(\frac{R(1-R)}{n}\right) = \left(\frac{1}{2\sqrt{R}}\right)^2 \left(\frac{R(1-R)}{n}\right) = \frac{1-R}{4n}$$

To apply the formulas in Equation 2.4, consider a recessive trait examined in 100 individuals among which 25 exhibit the trait. Then $R = 25/100$ and $q = \sqrt{R} \pm \sqrt{(1-R)/(4n)} = 0.50 \pm 0.043$. The "expected" genotype frequencies are therefore 25 homozygous dominant, 50 heterozygous, and 25 homozygous recessive. There is no chance for a goodness-of-fit test to HWE because there are 0 degrees of freedom, calculated as 2 (because there are two classes of data) $-$ 1 (for the totals to match) $-$ 1 (for estimating the frequency of the recessive allele) $= 0$. The lack of any degrees of freedom is the reason why the expected frequencies fit the observed frequencies exactly.

One of the most important implications of HWE pertains to rare recessive alleles. When a recessive allele is rare, most individuals who carry the allele are heterozygous. This principle follows immediately from the genotype frequencies, because there are $2pq$ heterozygotes and q^2 recessive homozygotes. The ratio of heterozygous to homozygous recessive frequencies is $2pq/q^2 = 2p/q$. For example, if $q = 0.10$, the ratio $2p/q = 18$, which means that there are 18 times as many heterozygotes as recessive homozygotes in the population. A few other illustrative values for $(q, 2p/q)$ are (0.05, 38), (0.01, 198), (0.005, 398), and (0.001, 1998). Note that, when q is small, $2p/q \approx 2 \times (1/q)$. These examples indicate that, the smaller the frequency of the recessive allele, the greater the proportion of heterozygous carriers relative to homozygous recessives.

To take a real example, consider the recessive disease cystic fibrosis, a severe condition associated with abnormal glandular secretions caused by mutations in the gene *CFTR*. (*CFTR* stands for *cystic fibrosis transmembrane conductance regulator.*) Among Caucasians, the incidence of affected individuals is approximately 1 in 2500 newborns, yielding the allele frequency estimate $q = \sqrt{(1/2500)} = 0.02$. The frequency of heterozygotes is therefore estimated as $2(0.02)(1 - 0.02) = 0.0392$, or about 1 in 26 persons. This means that, although only 1 person in 2500 is actually affected with cystic fibrosis, about 1 person in 26 is a heterozygous carrier of the harmful mutation.

Multiple Alleles and DNA Typing

Just as HWE for two alleles can be expressed formally as the binomial square:

$$(p\,A + q\,a)^2 = p^2\,AA + 2pq\,Aa + q^2\,aa$$

HWE for multiple alleles can be expressed formally as the multinomial square:

$$(p_1A_1 + p_2A_2 + \cdots + p_kA_k)^2 \tag{2.5}$$

where p_i is the allele frequency of the allele A_i and k is the total number of alleles, hence $\Sigma p_i = 1$. Expanding Equation 2.5 into its component terms, we have:

$$\begin{aligned} \text{Genotype frequency of any homozygous genotype } A_iA_i &= p_i^2 \\ \text{Genotype frequency of any heterozygous genotype } A_iA_j &= 2p_ip_j \end{aligned} \tag{2.6}$$

A systematic way to calculate the genotype frequencies with three alleles is shown in Figure 2.4. More alleles can be handled by increasing the number of squares.

The allele frequencies of multiple alleles are usually unequal unless there is some kind of selection or other process that brings the allele frequencies into balance. For **neutral alleles**—those in which all genotypes have the same probability of survival and reproduction—the distribution of multiple alleles is given by the multivariate distribution in Equation 2.7:

$$f(x_1, \cdots, x_{k-1}) = \frac{\Gamma(k\alpha)}{[\Gamma(\alpha)]^k}(x_1 x_2 \cdots x_{k-1} x_k)^{\alpha-1} \tag{2.7}$$

This equation defines the **Dirichlet distribution**, which is a multivariate version of the beta distribution (Ewens 2004). The symbol Γ represents the gamma function, k is the

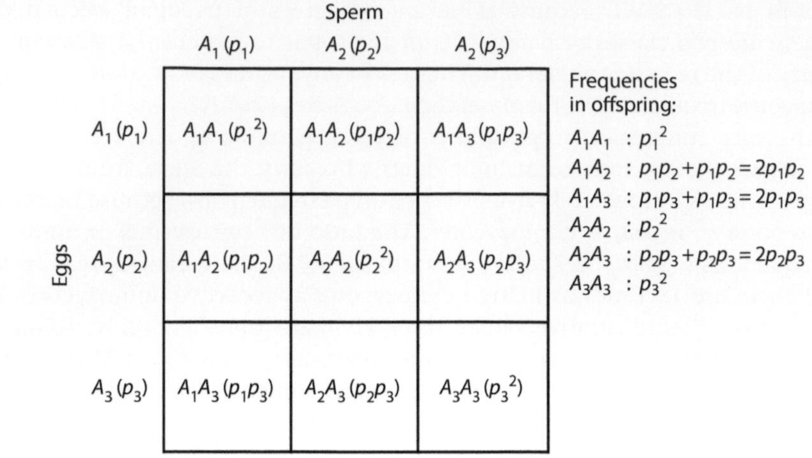

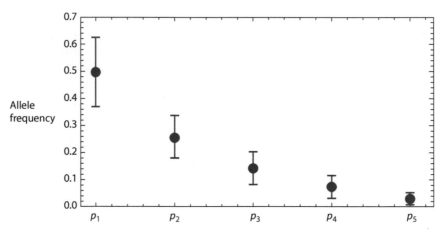

Figure 2.4 HWE for three alleles.

Figure 2.5 Mean and standard deviation of allele frequencies of a five-allele neutral polymorphism. The allele frequencies were obtained by repeated random sampling from the distribution in Equation 2.7 with parameters chosen so that the mean frequency of the major allele was about 0.5.

number of alleles, and $\alpha = \mu/(k-1)$, where μ is the probability that an allele undergoes mutation in any generation and any allele is equally likely to mutate to any other. In addition:

$$x_k = 1 - x_1 - x_2 - \cdots - x_{k-1}.$$

For neutral polymorphisms with two alleles $(k=2)$, the allele frequency of the least-frequent allele (called the **minor allele frequency** or **MAF**) is equally likely to have any value between 0 and 0.5. In other words, the MAF has a uniform distribution. Conversely, the allele frequency of the **major allele** (the most-frequent allele) is uniformly distributed between 0.5 and 1.0. These considerations imply that, even with only two alleles, the allele frequencies are likely to be quite different. The unevenness of the distribution of allele frequencies implied by Equation 2.7 is evident in Figure 2.5, which shows the mean and standard deviation of computer simulations of the allele frequencies of a polymorphism with five neutral alleles, arranged from most frequent to least frequent.

Perhaps the most important practical application of HWE for multiple alleles is that in DNA typing for purposes of individual identification, especially in matching crime-scene samples with the DNA of suspects in criminal investigations. Simple tandem repeats (STRs) are often used for DNA typing because of their multiple alleles. The potential power of the approach can be appreciated by considering an STR polymorphism with 10 neutral alleles with allele frequencies in accordance with Equation 2.7. Using parameters yielding a major-allele frequency of about 0.2, the probability of matching genotypes for a single 10-allele STR equals 0.03. This number also implies that, if the suspect is not the source of a crime-scene sample, the probability of a mismatch (and therefore exclusion of the suspect) is 97 per cent even for a single STR. For 4, 6, and 8 STR polymorphisms with 10 alleles, the match probabilities are about 10^{-6}, 10^{-9}, and 10^{-12}, respectively—and the probability of at least one mismatch to exclude an innocent suspect is virtually certain. Even though this is an artificial example, it serves to demonstrate that a simultaneous match at 6–8 highly polymorphic loci with multiple alleles makes a very strong case for identity between the suspect and the source of the criminal evidence, especially if the assumption of independence between the loci can be justified.

In practice, DNA typing raises many issues other than whether sample and suspect match. One key issue is sample handling. Has the evidence at a crime scene been properly collected and tracked through its chain of custody and treated carefully to avoid degradation or contamination? Has the biological sample from the suspect been properly collected and tracked and handled?

Another key issue is the interpretation of a DNA match, which is typically based on inferred frequencies of genotypes in a reference sample of individuals. Although the reference sample is usually of the same racial or ethnic group as the suspect, differences in allele frequency among population subgroups can result in misleading statistics. For example, a reference population of random Caucasians may be misleading for particular subgroups such as Old Order Amish or Hutterites. And what kind of reference population would be appropriate for Native Americans? A host of ancillary issues also arise. If you are asked for a DNA sample, must you comply? If you refuse, can DNA from your family members be used against you? Errors in sample handling, DNA typing, or interpretation of the results are second only to mistaken eyewitness testimony as a cause of wrongful convictions.

2.2 X-Linked Genes

Another important special case of HWE concerns genes located in the X chromosome. In mammals and many insects, females have two copies of a chromosome designated "X," whereas males have one copy of the X chromosome and one copy of a different chromosome designated "Y." The X and Y chromosomes segregate (separate) from each other during the formation of sperm, and therefore half the sperm from a male carry the X chromosome and half carry the Y chromosome. (In birds, moths, and butterflies, the sex chromosome situation is the reverse: males are XX and females XY.) Although the mammalian Y chromosome carries very few genes other than those involved in the determination of sex and male fertility, the X chromosome carries as full a complement of genes as any other chromosome. Genes in the X chromosome are called **X-linked genes**. Because males have only one X chromosome, and because the Y chromosome lacks a homolog of most genes in the X, a recessive X-linked allele present in a male will be expressed phenotypically. For X-linked genes with two alleles, therefore, there are three female genotypes (*AA, Aa*, and *aa*) but only two male genotypes (*A* and *a*).

One of the important features of random mating for X-linked genes is that conditions due to a rare recessive allele will be more common in males than in females. The reason can be seen in Figure 2.6, which shows the consequences of random mating with two X-linked alleles when the allele frequencies are equal in the sexes. Here the alleles are denoted X^A and X^a. Note that in females the genotype frequencies equal the HWE, whereas in males the genotype frequencies equal the allele frequencies. The sex difference occurs because q, which equals the proportion of males with the recessive phenotype, will always be greater than q^2, which equals the proportion of females with the recessive phenotype. For example, for the X-linked "green" type of color blindness, $q \approx 0.05$, and so the ratio of affected males to affected females is $q/q^2 = 1/q \approx 1/0.05 = 20$. For the X-linked "red" type of color blindness due to a closely linked gene, $q \approx 0.01$, and in this case the ratio of affected males to affected females is approximately $1/0.01 = 100$. These color-blindness examples show that as a recessive X-linked allele decreases in frequency, the excess of affected males over affected females increases.

In Figure 2.6 we assumed that the allele frequencies were equal in males and females. What happens when the allele frequencies differ in the sexes distinguishes X-linked from autosomal inheritance in the approach to HWE. With autosomal genes, the attainment of HWE requires two generations. The first generation is required to equalize the allele frequencies in the sexes to the average, and the second generation brings the genotype frequencies into HWE. For X-linked genes, attaining HWE requires more time because the allele frequency in males lags behind that in females owing to the fact that a male receives his X chromosome from his mother. To be specific, let $p_m(t)$ be the allele frequency of X^A in males in generation t, and let $p_f(t)$ be the corresponding allele frequency in females. For neutral alleles:

$$
\begin{aligned}
p_m(t) &= p_f(t-1) \\
p_f(t) &= [p_m(t-1) + p_f(t-1)]/2
\end{aligned}
\tag{2.8}
$$

In Equation 2.8, the first statement expresses the transmission of the X chromosome from mothers to sons, and the second statement expresses the transmission of the X chromosome from both parents to daughters. How the allele frequency changes through time is shown for the case $p_m(t) = 0.2$ and $p_f(t) = 0.8$ in Figure 2.7. In each generation, the difference in the allele frequency between the sexes is reduced by half,

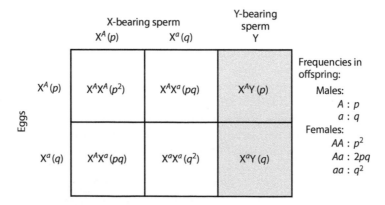

Figure 2.6 HWE for X-linked genes.

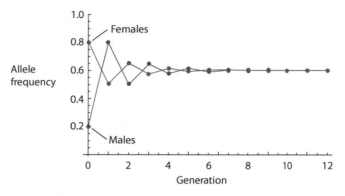

Figure 2.7 Convergence of the allele frequencies for an X-linked gene when the allele frequencies differ between the sexes.

but which sex has the greater allele frequency alternates. You can verify these principles for yourself by showing that the expressions in Equation 2.8 imply that $p_f(t) - p_m(t) = -(1/2)[p_f(t-1) - p_m(t-1)]$. Note that the allele frequency ultimately attained is not the arithmetic average as it would be for an autosomal gene, but rather $(2/3)p_f(t) + (1/3)p_m(0)$, which in this example equals 0.6; the weighted average is needed because, for an X-linked gene, 2/3 of the alleles are in females and 1/3 are in males.

The pattern shown in Figure 2.7 applies to neutral (or nearly neutral) alleles; however, if strong selection acts against males carrying a recessive X-linked allele, then the allele frequencies in males and females may differ indefinitely. Suppose, to take an extreme example, that X^a is a recessive lethal; then $p_m(t) = 1$ for any value of t, and the only source of X^a alleles is through new germ-line mutations or X^a alleles transmitted through heterozygous females.

2.3 Multiple Loci: Linkage and Linkage Disequilibrium

Statistically, HWE means that the alleles present at a locus are in random association with each other in the genotypes. It therefore may seem paradoxical that two genes, A and B, present in the same population may each obey HWE individually, yet the alleles of A and B can remain in nonrandom association in the gametes that form each generation. That this is possible is shown in Figure 2.8. The gametic types are arrayed across the top for two alleles at each of two loci, A and B. The gametic frequencies are expressed in two completely equivalent ways. One is in terms of p_{11}, p_{12}, p_{21}, and p_{22} where the first subscript refers to the allele of the A gene and the second to the allele of the B gene. The frequencies p_{11}, p_{12}, p_{21}, and p_{22} therefore correspond to the gametic frequencies of $A_1 B_1$, $A_1 B_2$, $A_2 B_1$, and $A_2 B_2$, respectively.

The other parameterization in Figure 2.8 is in terms of the allele frequencies $p_{1\bullet}$ and $p_{2\bullet}$ of the alleles A_1 and A_2 and the allele frequencies $p_{\bullet1}$ and $p_{\bullet2}$ of the alleles B_1 and B_2. In this notation, a dot in a subscript means the sum over all possible subscripts at that position; in other words:

$$p_{1\bullet} = p_{11} + p_{12} \quad p_{2\bullet} = p_{21} + p_{22}$$
$$p_{\bullet1} = p_{11} + p_{21} \quad p_{\bullet2} = p_{12} + p_{22}$$

(2.9)

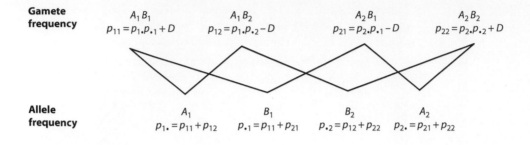

$$p_{11} + p_{12} + p_{21} + p_{22} = 1 \text{ as well as } p_{1\bullet} + p_{2\bullet} = 1 \text{ and } p_{\bullet1} + p_{\bullet2} = 1$$

Figure 2.8 Two-locus gametes and their frequencies.

The diagonal connecting lines in Figure 2.8 indicate the sums yielding the allele frequencies.

The other parameter in Figure 2.8 is designated D, and it measures the nonrandom association, called **linkage disequilibrium** or **LD**, between the loci. D is defined as:

$$D = p_{11}p_{22} - p_{12}p_{21} \tag{2.10}$$

When $D = 0$ the gametic frequencies equal the products of the relevant allele frequencies, and the loci are said to be in *linkage equilibrium*. The two types of parameterizations in Figure 2.8 are equivalent. The relation between D and the gametic frequencies can be verified by multiplying out:

$$p_{11}p_{22} - p_{12}p_{21} = \left(p_{1\bullet}p_{\bullet1} + D\right)\left(p_{2\bullet}p_{\bullet2} + D\right) - \left(p_{1\bullet}p_{\bullet2} - D\right)\left(p_{2\bullet}p_{\bullet1} - D\right)$$

which, using the identities $p_{1\bullet} + p_{2\bullet} = 1$ and $p_{\bullet1} + p_{\bullet2} = 1$, simplifies to D as per the definition.

What LD means biologically is that the frequencies of the genotype $A_1 B_1/A_2 B_2$ is not equal to the frequency of the genotype $A_1 B_2/A_2 B_1$, even though each genotype is heterozygous for both loci. With random mating the frequencies of these genotypes are $p_{11}p_{22}$ and $p_{12}p_{21}$, respectively. They are equal if and only if $D = 0$. On the other hand, each individual locus is in HWE. Random union of the gametes in Figure 2.8 results in genotype frequencies given by successive terms in the expansion of:

$$\left(p_{11}A_1B_1 + p_{12}A_1B_2 + p_{21}A_2B_1 + p_{22}A_2B_2\right)^2$$

HWE for each locus can be confirmed directly from this binomial. For example, the frequency of A_1A_1 equals $\left(p_{11} + p_{12}\right)^2 = p_{1\bullet}^2$ (the squared term of HWE), that of A_1A_2 equals $2\left(p_{11} + p_{12}\right)\left(p_{21} + p_{22}\right) = 2p_{1\bullet}p_{2\bullet}$ (the cross-product term of HWE), and so forth for the other genotypes.

Linkage equilibrium between genes is eventually attained under random mating and the assumptions for HWE. The attainment of linkage equilibrium is gradual, however, in contrast to the attainment of HWE for the alleles of a single gene, which typically requires only one or a small number of generations. The rate of approach to linkage equilibrium depends on the double heterozygous genotypes $A_1 B_1/A_2 B_2$ and $A_1 B_2/A_2 B_1$ and, in particular, on the relative frequencies of the several types of gametes that they can produce. If the genes are in different chromosomes, or are very far apart in the same chromosome, then each double heterozygous genotype produces the four possible types of gametes $A_1 B_1$, $A_2 B_2$, $A_1 B_2$, and $A_2 B_1$ in equal frequencies. In this case, as will be shown later in this chapter, the value of D decreases by half in each successive generation.

When two genes are sufficiently close together in the same chromosome, they are said to be **linked genes**, and the frequencies of the different types of gametes depends on the distance between the genes. This is because each chromosome aligns side-by-side with its partner chromosome during the formation of gametes and can undergo a sort of breakage and reunion resulting in an exchange of parts between the partner chromosomes. The distance between the genes determines the likelihood of an exchange. The result of genetic exchange is called *recombination*, and it is measured by the **frequency of recombination**, defined as the proportion of gametes that carry a combination of alleles not present in either parental chromosome. We will denote the frequency of recombination using the symbol c (for *crossover*, another name for an exchange event between homologous chromosomes). The reason we're avoiding the more conventional symbol r is that, in the context of LD, the symbol r has been preempted, as we'll see later.

Recombination is illustrated in Figure 2.9. The parental double heterozygote $A_1 B_1/A_2 B_2$ is shown at the top. In this case the nonrecombinant (parental) chromosomes are $A_1 B_1$ and $A_2 B_2$, whereas the recombinant chromosomes are $A_1 B_2$ and $A_2 B_1$. The frequency of each type of chromosome is indicated, and the overall frequency of recombinant chromosomes is c. The situation is the same in the double heterozygote $A_1 B_2/A_2 B_1$ shown at the bottom, except that the nonrecombinant and recombinant chromosomes are reversed. The minimum frequency of recombination is $c = 0$, when the genes are completely linked. The maximum frequency of recombination is $c = 0.5$, when the genes are either far apart in the same chromosome or present in different chromosomes. The reason for the maximum at 0.5 is that each recombination event yields two recombinant products and two nonrecombinant products, hence the frequency of recombination from each event equals $2/4 = 0.5$. Furthermore, the more exchanges take place between two genes, the more likely that one exchange will undo the result of a previous exchange, until eventually the alleles in the parental chromosomes become randomly combined in the gametes. This means that the gametic types are equally frequent, or in other words, that $c = 0.5$.

The frequency of recombination between genes determines the rate of approach to linkage equilibrium. To see how, consider a chromosome carrying $A_1 B_1$. This chromosome could have only two possible origins relative to the chromosomes in the previous generation. It could be a recombinant chromosome with probability c, or it could be a

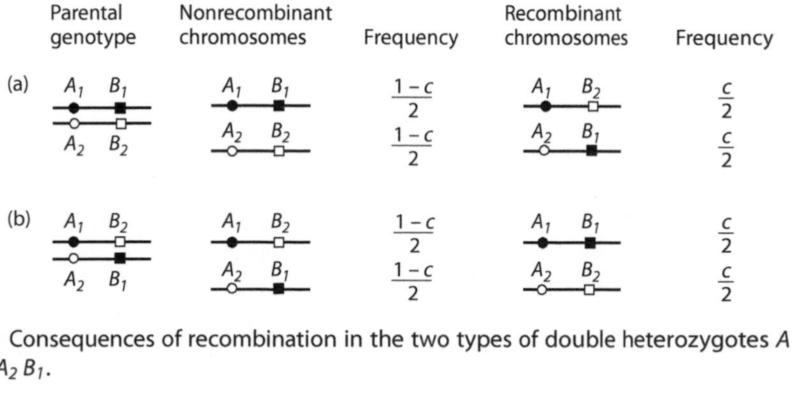

Figure 2.9 Consequences of recombination in the two types of double heterozygotes $A_1 B_1/A_2 B_2$ and $A_1 B_2/A_2 B_1$.

nonrecombinant chromosome with probability $1 - c$. If it is a recombinant chromosome (probability c), it must have come from a parent of genotype $A_1 -/- B_1$, where in this case the dash means that it is unnecessary to specify the particular allele at the locus. The frequency of this genotype is $p_{1\bullet}p_{\bullet1}$ because of the random mating. On the other hand, if it is a nonrecombinant chromosome (probability $1 - c$), then its progenitor chromosome in the previous generation must also have been $A_1 B_1$, and the frequency of this type of chromosome is p_{11}. Putting these possibilities together, and letting p_{11}' be the frequency of $A_1 B_1$ chromosomes in the present generation:

$$p_{11}' = cp_{1\bullet}p_{\bullet1} + (1-c)p_{11}$$

Subtracting $p_{1\bullet}p_{\bullet1}$ from both sides leads to:

$$p_{11}' - p_{1\bullet}p_{\bullet1} = (1-c)\left(p_{11} - p_{1\bullet}p_{\bullet1}\right)$$

However, the parameterizations in Figure 2.8 indicate that $p_{11} - p_{1\bullet}p_{\bullet1} = D$. This means that in each generation D is only $1 - c$ times as large as in the previous generation, Hence, letting D_t be the value of D in generation t:

$$D_t = D_{t-1}(1-c) = D_{t-2}(1-c)^2 = D_{t-3}(1-c)^3 = \cdots = D_0(1-c)^t \qquad (2.11)$$

where D_0 is the value of D in the initial or founder population. Because $1 - c \leq 1$, then $(1-c)^t$ goes to zero as t becomes large, but how rapidly it does depends on c—the closer to zero, the slower the rate. Recall here that $c = 0.5$ corresponds either to genes far apart in the same chromosome or to genes in different chromosomes; it implies that the disequilibrium decreases by half in each generation.

One limitation of D as a measure of LD is that its possible values depend on the allele frequencies; hence it is difficult to compare values of D from one pair of loci to the next. For this reason, D is sometimes expressed as a ratio relative to its maximum value (if D is positive) or its minimum value (if D is negative), which restricts the range from 0 to 1. This normalized value of D is denoted D', and so we can write:

$$D' = \begin{cases} D/D_{max} & \text{if } D \geq 0 \\ D/D_{min} & \text{if } D < 0 \end{cases} \qquad (2.12)$$

The maximum and minimum values of D can be deduced from Figure 2.8 by noting that the frequencies of all four gametic frequencies must be nonnegative. This consideration implies that:

$$D_{max} = \text{the smaller of } p_{1\bullet}p_{\bullet2} \text{ and } p_{2\bullet}p_{\bullet1}$$
$$D_{min} = \text{the larger of } -p_{1\bullet}p_{\bullet1} \text{ and } -p_{2\bullet}p_{\bullet2} \qquad (2.13)$$

To illustrate the use of Equations 2.9–2.13, we consider two closely linked SNPs in human chromosome 4. One SNP is an A to G substitution in the gene for glycophorin A, which results in an amino acid replacement of serine to leucine associated with the M type of the MN blood group. The other SNP is a T to C substitution in the gene for glycophorin B, which results in the amino acid replacement methionine to threonine that is associated with the S type of the Ss blood group. In a sample of 1000 British people, the genotype counts for the A-versus-G SNP were 213 *A/A*, 489 *A/G*, and 298 *G/G*; and the genotype counts for the T-versus-C SNP were 483 *T/T*, 418 *T/C*, and 99 *C/C*. You can verify for yourself that the chi-square values for goodness of fit to HWE are 0.22 and 0.38, respectively. These correspond to *P*-values of 0.64 and 0.54, and so each individual SNP has genotype frequencies in agreement with HWE.

Now let's consider both SNPs together. In this sample, the genotypes of the 2000 chromosomes examined were:

$$A \; T \; 773 \quad A \; C \; 142 \quad G \; T \; 611 \quad G \; C \; 474$$

From these data the estimated gametic frequencies are:

$$p_{11} = 0.386 \quad p_{12} = 0.071 \quad p_{21} = 0.306 \quad p_{22} = 0.237$$

and the estimated allele frequencies of the SNPs are:

$$p_{1\bullet} = 0.458 \, (A \text{ nucleotide}) \quad p_{2\bullet} = 0.542 \, (G \text{ nucleotide})$$
$$p_{\bullet 1} = 0.692 \, (T \text{ nucleotide}) \quad p_{\bullet 2} = 0.308 \, (C \text{ nucleotide})$$

The estimate of D equals:

$$D = p_{11}p_{22} - p_{12}p_{21} = 0.070$$

Is this value significantly different from 0? To find out, one calculates the expected numbers of gametes under the assumption of linkage equilibrium, and these are:

$$A \; T \; p_{1\bullet}p_{\bullet 1} \times 2000 = 633.2 \quad A \; C \; p_{1\bullet}p_{\bullet 2} \times 2000 = 281.8$$
$$G \; T \; p_{2\bullet}p_{\bullet 1} \times 2000 = 750.8 \quad G \; C \; p_{2\bullet}p_{\bullet 2} \times 2000 = 334.2$$

The chi-square for goodness of fit is 184.8 with one degree of freedom. This value is off the chart in Figure 2.2, so we may conclude that P is very much less than 0.0001. This means that chance alone would produce a fit as poor or poorer substantially less than one time in 10,000, so the hypothesis that the loci are in linkage equilibrium can confidently be rejected.

Although D is statistically highly significant, is it large value or a small value? To assess this issue, we calculate D as a proportion of its maximum possible value according to Equations 2.12 and 2.13. (If D were negative, we would use the minimum possible value.) The maximum is given by the smaller of $p_{1\bullet}p_{\bullet 2} = 0.141$ and $p_{2\bullet}p_{\bullet 1} = 0.375$, and so $D_{max} = 0.141$. The ratio D' is the value of D/D_{max}, and hence $D' = 0.070/0.141 = 0.496$. We can therefore conclude that D is about 50 per cent of its maximum possible value, given the allele frequencies.

Linkage Disequilibrium, Genetic Associations, and the Problem of Multiple Comparisons

In principle, a statistically significant value of X^2 leads to rejection of the hypothesis that $D = 0$; however, in applications to genome sequences, the P-value must be adjusted for the hundreds or thousands of pairwise tests for LD because for k SNPs there are $k(k-1)/2$ possible pairwise associations, and so the number of pairwise tests for LD grows as k^2. For as few as 100 SNPs there are nearly 5000 tests, and at the conventional significance level of 5 per cent there would be about 250 "significant" associations resulting from chance alone.

Rejecting a null hypothesis when it is true is a *false-positive* outcome, and the number of false positives grows with the number of multiple comparisons. The remedy is to adjust the significance level to decrease the P-value required for rejection, and while there are many ways to do this there is no perfect solution. One method is to reduce the probability of at least one false-positive outcome to a predetermined level, which is known as the **family-wise error rate** and symbolized α_{FWER}. Suppose, for example, that you are carrying out m independent multiple comparisons and wish to reduce the probability of at least one false positive to $\alpha_{FWER} = 0.05$. If each individual test has a significance level of α_i (equal across each of the $i = 1, 2, \ldots, m$ tests), then the probability of not rejecting any of m true null hypothesis equals $(1 - \alpha_i)^m$, and therefore you should choose α_i so that the probability of at least one false positive equals $1 - (1 - \alpha_i)^m = \alpha_{FWER}$. If α_{FWER} is small and m large, then $\alpha_i \approx \alpha_{FWER}/m$ for each of the m tests. This is the **Bonferroni correction** for multiple tests. It is appropriate when each test is independent of the others, but it can be excessively conservative if the test statistics are correlated. (This is intuitive: if the tests were perfectly correlated, then the m tests are really only one test in disguise, and so we should set $m = 1$.) Unfortunately for the Bonferroni correction, multiple comparisons carried out in genetics and genomics are often correlated in complex ways. The interdependence of the statistical tests means that the Bonferroni correction will result in failing to reject some hypotheses that are, in fact, false (type II error, or false negatives). To say the same thing in another way, application of the Bonferroni correction to hypothesis tests that are not independent erodes the power to flag null hypotheses that are false.

Another approach is to control the overall proportion of false positives across all the tests, which is known as the **false discovery rate** and symbolized α_{FDR}. In one method for implementing this approach, the first step is to arrange all of the m observed P-values from smallest (P_1) to largest (P_m) and associate the null hypothesis corresponding to P_i as H_0^i. Then, starting with P_1, find the largest integer k such that $P_k \leq k\alpha_{FDR}/m$. Having found k, reject all H_0^i for all $i \leq k$. The rejected null hypotheses are regarded as the "discoveries." This correction for multiple comparisons is known as the **Benjamini–Hochberg procedure** (Benjamini and Hochberg 1995). When all of the null hypotheses are true and the tests are independent, then controlling FWER is equivalent to controlling FDR. If the tests are correlated, or some proportion of the null hypotheses are false, then controlling for FDR is less stringent than controlling for FWER. In other words, controlling for FDR increases the probability of type I error for the sake of greater statistical power.

The distinction between FWER and FDR emphasizes the importance of striking a proper balance between the false-positive rate (error in single test) and the false-discovery rate (errors across all tests), so as not to miss too many true associations while at the same time minimizing the number of false associations. For a more thorough discussion of this subject, and alternative methods to adjust the significance level for FWER or FDR, see Austin et al. (2014), Sham and Purcell (2014), Storey and Tibshirani (2003), and Wilson (2019) and references therein. As noted, there is no perfect statistical solution, and the strongest evidence for a true association is replication in an independent sample.

2.4 Linkage Disequilibrium in Natural Populations

How widespread is LD in natural populations? The answer depends on the type of organism, the distance between SNPs, the likelihood of recombination in the life cycle, the population size, and the ancestral history of the particular subpopulation. For example, bacteria that rarely undergo recombination can show significant LD across large parts of the genome, but those that undergo frequent recombination exhibit LD only for SNPs in nearby genes. In sexual eukaryotes, organisms that regularly undergo inbreeding (mating between relatives) have reduced opportunities for recombination and exhibit enhanced LD as compared with related species that are outcrossers (non-inbreeders). Genes that are in or near chromosomal inversions often show substantial LD because, when the inversion is heterozygous, crossovers within the inverted region result in products with duplications or deletions and are not recovered. The consequent reduced rate of recombination allows

LD to build up and be maintained as new mutations occur over time. Inversions and the resulting LD are frequent in certain species of *Drosophila* (Corbett-Detig and Hartl 2012; Wallace et al. 2013). Another source of LD is natural selection, provided that there is a sufficient selective advantage of some gametic types over others to overcome the natural tendency for *D* to go to zero (Corbett-Detig et al. 2013).

LD in a human population with Northern and Western European ancestry is illustrated in Figure 2.10. The solid line depicts the observed mean values of *D'* as a function of distance between SNPs in kilobase pairs. The dots represent a typical random sample, simulated based on the observed means and variances in Reich et al. (2001). Note the extent to which the points are scattered around the line. This kind of scatter is typical of LD data because the value of *D'* between SNPs depends on the vagaries and unpredictability of recombination.

The results in Figure 2.10 are typical of non-African populations in showing a reasonable amount of LD across regions smaller than about 60–80 kb (McEvoy et al. 2011). The situation is different in sub-Saharan African populations, in which LD is only about half as great and extends only about half the distance (McEvoy et al. 2011). The simplest explanation for the discrepancy is that anatomically modern humans outside of Africa arose from a subset of African genomes, resulting in a **bottleneck** (constriction) in population size in which certain combinations of alleles were overrepresented leading to the increased LD. For the data in Figure 2.10, computer simulations suggest that such a bottleneck in population size occurred about 25,000 years ago (Reich et al. 2001). This would have been among the most recent of several migrations of anatomically modern humans out of Africa, some of whom were displaced by even earlier migrants such as *Homo erectus*. The earliest migration of modern humans out of Africa that contributed to the present-day human genome is estimated to have been about 65,000 years ago (McEvoy et al. 2011), but some evidence suggests 100,000 years (Lopez et al. 2015).

Linkage Disequilibrium as a Correlation Between Alleles of Different Genes in Gametes

The LD *D* is related to the correlation coefficient between the *A* and *B* alleles present in gametes through the following equation:

$$r = \frac{D}{\sqrt{p_{1\bullet}p_{2\bullet}p_{\bullet1}p_{\bullet2}}} \tag{2.14}$$

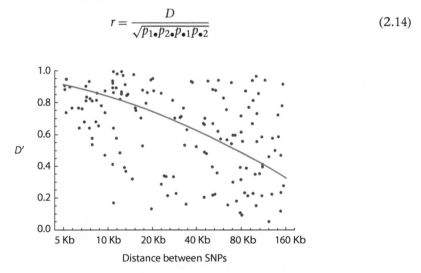

Figure 2.10 Linkage disequilibrium as measured by *D'* against distance between SNPs in a simulated sample matching the means and variances observed in residents of Utah (Reich et al. 2001).

Table 2.2 The quantity r is the correlation between the A and B alleles in gametes

Gamete	Frequency	Value of A allele	Value of B allele	Product of A and B values
$A_1 B_1$	p_{11}	1	1	1
$A_1 B_2$	p_{12}	1	0	0
$A_2 B_1$	p_{21}	0	1	0
$A_2 B_2$	p_{22}	0	0	0

Mean $\qquad p_{11}+p_{12}=p_{1\bullet} \qquad p_{11}+p_{21}=p_{\bullet 1} \qquad p_{11}$

Variance (A gene) $= p_{1\bullet} - p_{1\bullet}^2 = p_{1\bullet}(1-p_{1\bullet}) = p_{1\bullet}p_{2\bullet}$

Variance (B gene) $= p_{\bullet 1} - p_{\bullet 1}^2 = p_{\bullet 1}(1-p_{\bullet 1}) = p_{\bullet 1}p_{\bullet 2}$

Covariance (A and B) $= p_{11} - p_{1\bullet}p_{\bullet 1} = D$

Correlation coefficient $r = D/\sqrt{p_{1\bullet}p_{2\bullet}p_{\bullet 1}p_{\bullet 2}}$

The correlation r can range -1 to $+1$. The logic is outlined in Table 2.2, where A and B are treated as binary random variables with A_1 and B_1 assigned the value 1 and A_2 and B_2 the value 0. Because it has some convenient mathematical properties, the square of r given by:

$$r^2 = \frac{D^2}{p_{1\bullet}p_{2\bullet}p_{\bullet 1}p_{\bullet 2}} \qquad (2.15)$$

has become another widely used measure of LD.

Both r and D' (Equation 2.12) are measures of LD, the only difference in being how D (Equation 2.10) is normalized. While the measures are distinct they are correlated. Averaged across thousands of random, uniformly distributed quartets of gametic frequencies, the correlation between r^2 and D' and is roughly 0.75.

The measure r^2 has a number of convenient features, one of which is that the chi-square value (X^2) for goodness of fit to the hypothesis that $D = 0$ is given by:

$$X^2 = r^2 n \qquad (2.16)$$

where n is the number of chromosomes in the sample. In the glycophorin example $r^2 = 0.0924$, and when $n = 2000$, $r^2 n = 184.8$ in agreement with the chi-square for goodness of fit calculated more laboriously earlier.

D' and r^2 are somewhat complementary measures of LD. Although the value of D' depends mainly on the rate of recombination, that of r^2 also depends on the ancestral history. To understand why, imagine $c \approx 0$ between the A and B genes in a population with haplotypes $A_1 B_1$, $A_1 B_2$, $A_2 B_1$, and $A_2 B_2$. (The term **haplotype** refers to the alleles present together in a region of chromosome.) Because $c \approx 0$, the value of D'—whatever it may be—will change very slowly. On the other hand, the value of r^2 will depend on when the $B_1 \rightarrow B_2$ mutation took place. To be concrete, suppose that the $B_1 \rightarrow B_2$ occurred in the $A_1 B_1$ haplotype. If the mutation occurred early, then most of the haplotypes carrying A_1 will also carry B_2, and the predominant haplotypes will be $A_1 B_2$ and $A_2 B_1$ with smaller frequencies of $A_1 B_1$ and $A_2 B_2$ because of recombination. In this case r^2 will be close to 1. In contrast, if the $B_1 \rightarrow B_2$ mutation occurred late, then the predominant haplotypes will be $A_1 B_1$ and $A_2 B_1$ with smaller frequencies of $A_1 B_2$ and $A_2 B_2$. In this case r^2 will be close to 0. This difference explains why D' and r^2 are complementary measures of LD and also why r^2 can assume a range of values for any given value of D'. For the Utah population in Figure 2.10, r^2 is significant for a distance between SNPs of about 40 kb (McEvoy et al. 2011).

The measure r^2 is also convenient because its expected value in natural populations can be calculated. Because r^2 depends on D^2 (Equation 2.15), and D tends to 0 (Equation 2.11), one might think that $E(r^2) = 0$. But this is assumes an infinite population size. In a finite population, the gametic frequencies are subject to random drift and may change haphazardly according to chance. Therefore, by chance alone, the two types of double heterozygotes $A_1 B_1/A_2 B_2$ and $A_1 B_2/A_2 B_1$ may come to have different frequencies, and in this way random drift can create LD and counteract the diminishing effects of recombination expressed in Equation 2.11.

When the frequency of recombination between SNPs is c in a population of size N, and the minor allele frequencies are ≥ 0.05, then the expected value of r^2 for an infinite sample size is approximately:

$$E\left(r^2\right) \approx \frac{10+4Nc}{(2+4Nc)(11+4Nc)} \tag{2.17}$$

When $4Nc$ is large, then $E(r^2) \approx 1/(4Nc)$ (Charlesworth and Charlesworth 2012; Song and Song 2007). Approximations for finite sample size as well as other approaches for estimating $4Nc$ are discussed in Hahn (2019).

Figure 2.11 shows $E(r^2)$ from Equation 2.17 along with a plot of $1/(4Nc)$. When $c = 0$, then $E(r^2) = 0.45$ (i.e., 10/22), but the value drops very quickly as $4Nc$ increases. By the time $4Nc \approx 30$ the approximation $E(r^2) = 1/(4Nc)$ is already very good. We should emphasize that, in these equations, N is the so-called **effective population size**, not the actual or census population size (Charlesworth 2009). Roughly speaking, the effective population size is the number of breeding individuals in a population. For present purposes you can think of the effective population size as the size of an idealized population with the same random variation in haplotype frequencies due to genetic drift as the actual population in question.

Linkage Disequilibrium Due to Population Admixture

Before leaving the subject of LD we should point out that LD can also arise from admixture of populations with differing gametic frequencies. Just as an admixture of two HWE subpopulations differing in allele frequencies can result in a departure from HWE in the direction of too few heterozygotes in the mixed population, an admixture of two populations differing in gametic frequencies can result in LD in the mixed population, even if the unmixed populations show no LD.

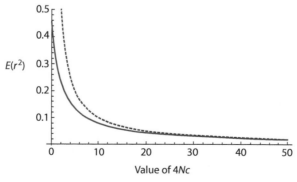

Figure 2.11 r^2 as a function of $4Nc$ from Equation 2.17. The dashed blue line corresponds to $E(r^2) \approx 1/(4Nc)$.

The effect of population admixture is best illustrated by example. Shown here are the gametic frequencies in two different populations distinguished by using the symbols p and q:

$$p_{11} = 0.0475 \quad q_{11} = 0.0475$$
$$p_{12} = 0.9025 \quad q_{12} = 0.0025$$
$$p_{21} = 0.0025 \quad q_{21} = 0.9025$$
$$p_{22} = 0.0475 \quad q_{22} = 0.0475$$

In both populations $D = 0$. In an equal admixture of these populations, denoted by the symbol m, the gametic frequencies are the averages:

$$m_{11} = 0.0475$$
$$m_{12} = 0.4525$$
$$m_{21} = 0.4525$$
$$m_{22} = 0.0475$$

The artefactual (but nevertheless real) LD in the admixed population has $D = -0.2025$, $D' = 0.81$, and $r^2 = 0.66$. This is an extreme example contrived to make the point. Most admixed populations would not exhibit LD to nearly this extent, but the example is admonitory. Note the absence of any stipulation that the genes are in the same chromosome: LD can result from population admixture even for genes in different chromosomes.

Wahlund's Principle

Population admixture results not only in LD between genes but also fewer heterozygous genotypes for each gene than expected with random mating. In the example just discussed, the allele frequency of A_1 in the two populations is $p_{1\bullet} = p_{11} + p_{12} = 0.95$ in population 1 and $p_{2\bullet} = q_{21} + q_{22} = 0.05$ in population 2, and the allele frequency of A_2 in the two populations is $q_{1\bullet} = q_{11} + q_{12} = 0.05$ in population 1 and $q_{2\bullet} = q_{21} + q_{22} = 0.95$ in population 2. With random mating in each subpopulation, a sample of an equal number of individuals from each subpopulation includes a frequency of $A_1 A_2$ heterozygous genotypes of $[2(0.95)(0.05) + 2(0.05)(0.95)]/2 = 0.095$ (the average heterozygosity in the two subpopulations). On the other hand, the average allele frequency of A_1 in the sample equals $(0.95 + 0.05)/2 = 0.50$ and that of A_2 equals $(0.05 + 0.95) = 0.50$. The expected frequency of heterozygotes with random mating is therefore $2(0.50)(0.50) = 0.500$, which is far larger than the 0.095 actually observed in the sample. (In this particular example, the deficiency in the frequency of $B_1 B_2$ heterozygotes is the same as for the $A_1 A_2$ heterozygotes, namely 0.095 versus 0.500, which you can easily verify for yourself.)

The principle that a mixed sample from subpopulations in a subdivided population has a deficiency of heterozygotes relative to that expected with random mating is known as **Wahlund's principle**. This principle has far-reaching implications in population genetics, and we will discuss it in greater detail in Chapter 3.

Problems

2.1 J. B. S. Haldane, one of the pioneers of population genetics, preferred using $u/(1+u)$ and $1/(1+u)$ to denote the allele frequencies instead of p and q as others preferred. What are the values of $u/(1+u)$ and $1/(1+u)$ in terms of p and q? What are the Hardy–Weinberg genotype frequencies in terms of u?

2.1 ANSWER Solve $u/(1+u)=p$ or $1/(1+u)=q$ to obtain $u=p/q$. In terms of u, the Hardy–Weinberg frequencies p^2, $2pq$, and q^2 are $u^2/(1+u)^2$, $2u/(1+u)^2$, and $1/(1+u)^2$, respectively.

2.2 The gel diagram shown here depicts the banding pattern of DNA fragments corresponding to two alleles, A_1 and A_2, in a diploid organism. The number above each lane is the number of individuals whose genomic DNA shows that banding pattern. Use a chi-square test to ascertain whether the observed genotype frequencies are consistent with HWE.

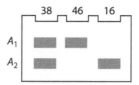

2.2 ANSWER The total number of individuals tested equals 100, and the allele frequencies are estimated as $p=(2\times 46+38)/200=0.65$ and $q=0.35$. The expected numbers are, from left to right, 45.50, 42.25, and 12.25. The chi-square equals 2.717 with one degree of freedom. The P-value is 0.099, and because $P>0.05$, these data should be judged to be consistent with HWE.

2.3 Use the frequencies of matings and their offspring in Table 2.1 to show that, when A is dominant to a and the population is in HWE, then the frequency of aa offspring from dominant $\times$ dominant matings is $[q/(1+q)]^2$ and that from dominant $\times$ recessive matings is $q/(1+q)$.

2.3 ANSWER With random mating, the frequency of dominant $\times$ dominant matings equals $(p^2)^2+2(p^2)(2pq)+(2pq)^2=p^2(p^2+4pq+4q^2)=p^2\left[(1-q)^2+4(1-q)q+4q^2\right]=p^2(1+q)^2$. Among these matings the frequency of aa offspring equals $(2pq)^2/4=p^2q^2$, and so the proportion of aa offspring from dominant $\times$ dominant matings equals $p^2q^2/\left[p^2(1+q)^2\right]=[q/(1+q)]^2$. The proportion of dominant $\times$ recessive matings equals $2p^2q^2+2(2pq)q^2=2pq^2(p+2q)=2pq^2(1+q)$. Among these, the proportion of aa offspring equals $2(2pq)q^2/2=2pq^3$. The ratio of aa offspring from dominant $\times$ recessive matings is therefore $2pq^3/[2pq^2(1+q)]=q/(1+q)$. These ratios are called Snyder's ratios after Laurence H. Snyder, a pioneer in human genetics. It is considered evidence of simple Mendelian recessive inheritance if a phenotypic trait appears at frequency x among the offspring of matings when one parent exhibits the phenotype and frequency x^2 among the offspring of matings when neither parent exhibits the phenotype, and we've just shown that $x=q/(1+q)$.

2.4 Consider a rare recessive allele at frequency q in a randomly mating diploid population with discrete, nonoverlapping generations. What is the probability that the allele remains heterozygous for t consecutive generations before meeting another deleterious recessive and becoming homozygous? What is the mean value of t?

2.4 ANSWER Intuitively, it seems reasonable that if a rare event has probability q, then the average time between such events would be approximately $1/q$. The logic is that, with random mating, the probability that the allele remains heterozygous in any one generation equals $p=1-q$, and the probability that it becomes homozygous is q. Hence the probability that it remains heterozygous for exactly t generations and then becomes homozygous is $p^t q$. The mean of this distribution equals the infinite sum $\sum_{i=0}^{\infty} i p^i q = q\left(\frac{p}{(1-p)^2}\right) = p/q$. (The variance in t equals p/q^2.)

2.5 For two biallelic loci, show that $D_{max} = min\left(p_{1\bullet}p_{\bullet2}, p_{2\bullet}p_{\bullet1}\right)$ and $D_{min} = max\left(-p_{1\bullet}p_{\bullet1}, -p_{2\bullet}p_{\bullet2}\right)$.

2.5 ANSWER First note that $p_{11} = p_{1\bullet}p_{\bullet1} + D$, $p_{12} = p_{1\bullet}p_{\bullet2} - D$, $p_{21} = p_{2\bullet}p_{\bullet1} - D$, and $p_{22} = p_{2\bullet}p_{\bullet1} + D$. The requirement that $p_{11} \geq 0$ and $p_{22} \geq 0$ implies that $p_{1\bullet}p_{\bullet1} + D \geq 0$ and $p_{2\bullet}p_{\bullet2} + D \geq 0$. In other words, $D \geq -p_{1\bullet}p_{\bullet1}$ as well as $D \geq -p_{2\bullet}p_{\bullet2}$, which means that $D_{min} = max\left(-p_{1\bullet}p_{\bullet1}, -p_{2\bullet}p_{\bullet2}\right)$. Likewise, $p_{12} \geq 0$ and $p_{21} \geq 0$ implies that $p_{1\bullet}p_{\bullet2} - D \geq 0$ and $p_{2\bullet}p_{\bullet1} - D \geq 0$. In other words, $D \leq p_{1\bullet}p_{\bullet2}$ as well as $D \leq p_{2\bullet}p_{\bullet1}$, which means that $D_{max} = min\left(p_{1\bullet}p_{\bullet2}, p_{2\bullet}p_{\bullet1}\right)$.

2.6 The gametic frequencies for a pair of genes in a population are $p_{11} = 0.03$, $p_{12} = 0.27$, $p_{21} = 0.57$, and $p_{22} = 0.13$. Calculate D, D', and r^2 for these genes.

2.6 ANSWER $D = p_{11}p_{22} - p_{12}p_{21} = -0.15$. The allele frequencies are $p_{1\bullet} = p_{11} + p_{12} = 0.30$, $p_{2\bullet} = 0.70$, $p_{\bullet1} = p_{11} + p_{21} = 0.6$, and $p_{\bullet2} = 0.4$. $D_{min} = max\left(-p_{1\bullet}p_{\bullet1}, -p_{2\bullet}p_{\bullet2}\right) = max(-0.18, -0.28)$, hence $D' = -0.15/ -0.18 = 0.83$. $r^2 = D^2 / \left(p_{1\bullet}p_{2\bullet}p_{\bullet1}p_{\bullet2}\right) = 0.45$.

2.7 Fertilized females of *Drosophila melanogaster* were sampled in a natural population in Raleigh, NC, USA, and used to found a large experimental population. After about 5 months (ten generations), 489 third chromosomes in the population were analyzed for fast (F) versus slow (S) electrophoretic variants of the enzymes esterase-6 (alleles *E6-F* and *E6-S*), esterase-C (alleles *EC-F* and *EC-S*), and octanol dehydrogenase (alleles *Odh-F* and *Odh-S*). The results were as follows:

E6-F	EC-F	Odh-F	152	E6-S	EC-F	Odh-F	264
E6-F	EC-F	Odh-S	7	E6-S	EC-F	Odh-S	13
E6-F	EC-S	Odh-F	15	E6-S	EC-S	Odh-F	29
E6-F	EC-S	Odh-S	1	E6-S	EC-S	Odh-S	8

The order of the genes in the third chromosome is *E6–EC–Odh*. The frequency of recombination between *E6* and *EC* is 0.122, and that between *EC* and *Odh* is 0.002. (That is, *E6* and *EC* are rather loosely linked whereas *EC* and *Odh* are tightly linked.)

a. Carry out an analysis to determine whether there is significant LD between any of the loci. (Note: in a chi-square test for LD, the value of the chi-square is given by $r^2 n$.)

b. If there is significant LD, express the value of D relative to its theoretical maximum or minimum.

2.7 ANSWER

a. The total sample size tested in the experimental population is $n = 489$. For the pair of genes *E6* and *Ec*, the gametic frequencies are $p_{11} = 0.3251$, $p_{12} = 0.0327$, $p_{21} = 0.5665$, and $p_{22} = 0.0757$. The disequilibrium parameter $D = 0.0061$. The F and S allele frequencies for the two genes are $p_{1\bullet} = p_{11} + p_{12} = 0.3579$, $p_{2\bullet} = p_{21} + p_{22} = 0.6421$, $p_{\bullet1} = p_{11} + p_{21} = 0.8916$, and $p_{\bullet2} = p_{12} + p_{22} = 0.1084$. $r^2 = 0.001658$ and chi-square $= r^2 n = 0.8108$. The P-value equals 0.368, hence these two genes show no evidence of LD. For the pair of genes *E6* and *Odh*, $p_{11} = 0.3415$, $p_{12} = 0.0164$, $p_{21} = 0.5992$, and $p_{22} = 0.0429$. $D = 0.0049$. The F and S allele frequencies are $p_{1\bullet} = 0.3579$, $p_{2\bullet} = 0.6421$, $p_{\bullet1} = 0.9407$, and $p_{\bullet2} = 0.0593$. $r^2 = 0.001845$ and chi-square $= 0.9023$. The P-value equals 0.342, hence these two genes also show no evidence of LD.

On the other hand, for the pair of genes EC and Odh, $p_{11} = 0.8507$, $p_{12} = 0.0409$, $p_{21} = 0.0900$, and $p_{22} = 0.0184$. $D = 0.0120$. The F and S allele frequencies are $p_{1\bullet} = 0.8916$, $p_{2\bullet} = 0.1084$, $p_{\bullet 1} = 0.9407$, and $p_{\bullet 2} = 0.0593$. $r^2 = 0.026609$ and chi-square $= 13.012$. The P-value equals 0.00031, hence these two genes show strong evidence of being in LD.

b. For the pair of genes EC and Odh, D_{max} is the smaller of $p_{1\bullet}p_{\bullet 2}$ and $p_{2\bullet}p_{\bullet 1}$, which is the smaller of 0.0529 and 0.1020, or 0.0529. Therefore $D' = D/D_{max} = 0.227$, which is to say that D is about 22.7 per cent of is maximum possible value, given the allele frequencies.

2.8 Population admixture results in samples in which there is a deficiency of heterozygotes relative to that expected with HWE. Population admixture can also produce LD. Consider, for example, a pair of biallelic genes in two populations with gametic frequencies $p_{11} = 0.16$, $p_{12} = 0.64$, $p_{21} = 0.04$, and $p_{22} = 0.16$ in one population and $q_{11} = 0.16$, $q_{12} = 0.04$, $q_{21} = 0.64$, and $q_{22} = 0.16$ in another population. Note that $p_{11}p_{22} - p_{12}p_{21} = 0$ and $q_{11}q_{22} - q_{12}q_{21} = 0$, hence both population are in linkage equilibrium. Suppose you sample an equal number of individuals from each subpopulation.

a. What are the gametic frequencies m_{11}, m_{12}, m_{21}, and m_{22} in the admixed population?
b. What is the value of $D = m_{11}m_{22} - m_{12}m_{21}$ in the admixed population?
c. For the admixed population calculate D' and r^2.

2.8 ANSWER a. The gametic frequencies in the admixed population are the averages in the two subpopulations, hence $m_{11} = (p_{11} + q_{11})/2 = 0.0475$, $m_{12} = 0.4525$, $m_{21} = 0.4525$, and $m_{22} = 0.0475$. **b.** $D = m_{11}m_{22} - m_{12}m_{21} = -0.2025$. **c.** The allele frequencies in the admixed population are $m_{1\bullet} = 0.5$, $m_{2\bullet} = 0.5$, $m_{\bullet 1} = 0.5$, and $m_{\bullet 2} = 0.5$. Because D is negative, it should be normalized by $D_{min} = max(-m_{1\bullet}m_{\bullet 1}, -m_{2\bullet}m_{\bullet 2}) = -0.25$, hence $D' = -0.2025/ -0.25 = 0.81$. The value of $r^2 = D^2/(m_{1\bullet}m_{2\bullet}m_{\bullet 1}m_{\bullet 2}) = 0.6561$.

References

Austin, S. R., Dialsingh, I., and Altman, N. (2014), 'Multiple hypothesis testing: a review', *J. Indian Soc Agric Stat*, 68 (2), 303–14.

Benjamini, Y. and Hochberg, Y. (1995), 'Controlling the false discovery rate: a practical and powerful approach to multiple testing', *J R Stat Soc B*, 57 (1), 289–300.

Charlesworth, B. (2009), 'Fundamental concepts in genetics: effective population size and patterns of molecular evolution and variation', *Nat Rev Genet*, 10 (3), 195–205.

Charlesworth, B. and Charlesworth, D. (2012), *Elements of Evolutionary Genetics* (Greenwood Village, CO: Roberts and Company).

Corbett-Detig, R. B. and Hartl, D. L. (2012), 'Population genomics of inversion polymorphisms in Drosophila melanogaster', *PLoS Genet*, 8 (12), e1003056.

Corbett-Detig, R. B., et al. (2013), 'Genetic incompatibilities are widespread within species', *Nature*, 504 (7478), 135–7.

Ewens, W. J. (2004), *Mathematical Population Genetics* (2 edn.; New York: Springer-Verlag).

Hahn, M. W. (2019), *Molecular Population Genetics* (New York: Oxford University Press).

Lopez, S., van Dorp, L., and Hellenthal, G. (2015), 'Human dispersal out of Africa: a lasting debate', *Evol Bioinform Online*, 11 (Suppl 2), 57–68.

McEvoy, B. P., et al. (2011), 'Human population dispersal "Out of Africa" estimated from linkage disequilibrium and allele frequencies of SNPs', *Genome Res*, 21 (6), 821–9.

Reich, D. E., et al. (2001), 'Linkage disequilibrium in the human genome', *Nature*, 411 (6834), 199–204.

Sham, P. C. and Purcell, S. M. (2014), 'Statistical power and significance testing in large-scale genetic studies', *Nat Rev Genet*, 15 (5), 335–46.

Song, Y. S. and Song, J. S. (2007), 'Analytic computation of the expectation of the linkage disequilibrium coefficient r2', *Theor Popul Biol*, 71 (1), 49–60.

Storey, J. D. and Tibshirani, R. (2003), 'Statistical significance for genomewide studies', *Proc Natl Acad Sci U S A*, 100 (16), 9440–5.

Wallace, A. G., Detweiler, D., and Schaeffer, S. W. (2013), 'Molecular population genetics of inversion breakpoint regions in Drosophila pseudoobscura', *G3 (Bethesda)*, 3 (7), 1151–63.

Wilson, D. J. (2019), 'The harmonic mean p-value for combining dependent tests', *Proc Natl Acad Sci U S A*, 116 (4), 1195–200.

CHAPTER 3

Inbreeding and Population Structure

Just as random mating organizes alleles into genotypes according to Hardy–Weinberg proportions without changing the allele frequencies, so does **inbreeding**, which is mating between relatives. The main effect of inbreeding is to increase the frequency of homozygous genotypes in a population, relative to the frequency that would be expected with random mating. Unlike random mating, which may affect some genes but not others, inbreeding affects all genes in the genome. In human populations, the closest degree of inbreeding that commonly occurs in most societies is first-cousin mating, but many plants regularly undergo such close inbreeding as self-fertilization.

3.1 Genotype Frequencies with Inbreeding

The increase in frequency of homozygous genotypes at the expense of heterozygous genotypes can most easily be seen in the case of repeated self-fertilization. This is a scenario first examined in detail by Mendel in 1865 in his famous "Experiments in Plant Hybridization" (Abbott and Fairbanks 2016). His example started with a self-fertilizing population of plants consisting of 1/4 AA, 1/2 Aa, and 1/4 aa genotypes. These genotype frequencies are initially in Hardy–Weinberg equilibrium (HWE). Because each plant undergoes self-fertilization, the AA and aa genotypes produce only AA and aa offspring, respectively, whereas the Aa genotypes produce 1/4 AA, 1/2 Aa, and 1/4 aa offspring. After one generation of self-fertilization, therefore, the genotype frequencies of AA, Aa, and aa are:

$$\left(\frac{1}{4}\right)(1) + \left(\frac{1}{2}\right)\left(\frac{1}{4}\right) = \left(\frac{3}{8}\right) AA \quad \left(\frac{1}{2}\right)\left(\frac{1}{2}\right) = \left(\frac{1}{4}\right) = \left(\frac{2}{8}\right) Aa$$

$$\left(\frac{1}{4}\right)(1) + \left(\frac{1}{2}\right)\left(\frac{1}{4}\right) = \left(\frac{3}{8}\right) aa$$

After the second generation of self-fertilization the frequency of heterozygous genotypes is halved again, and so forth for successive generations. After one, two, three, and four generations of self-fertilization the genotype frequencies of AA, Aa, and aa genotypes are given by:

3/8	2/8	3/8
7/16	2/16	7/16
15/32	2/32	15/32
31/64	2/64	31/64

A Primer of Population Genetics and Genomics. Fourth Edition. Daniel L. Hartl, Oxford University Press (2020). © Daniel L. Hartl.
DOI: 10.1093/oso/9780198862291.003.0003

These genotype frequencies are no longer in HWE because there is a deficiency of heterozygous genotypes and an excess of homozygous genotypes. Although self-fertilization is the most extreme form of inbreeding, all forms of inbreeding result in a reduction in heterozygosity. This suggests that one can measure the amount of inbreeding by comparing the actual proportion of heterozygous genotypes in the population with the proportion of heterozygous genotypes that would occur with random mating. To be concrete, consider a gene with two alleles, A and a, at respective frequencies p and q with $p + q = 1$. Suppose that the actual frequency of heterozygous genotypes in a population at the present time is denoted H. If the population were in HWE for the gene, the frequency of heterozygous genotypes would be $2pq$. We will denote this baseline value as H_0, hence $H_0 = 2pq$. (The symbol H_0 in this context is a numerical quantity, not a null hypothesis.) The effects of inbreeding can be defined in terms of the ratio $(H_0 - H)/H_0$, which is usually denoted in population genetics by the symbol F and called the **inbreeding coefficient**. Thus:

$$F = \frac{H_0 - H}{H_0} \tag{3.1}$$

Biologically, F measures the reduction in heterozygosity relative to that expected in a random mating population with the same allele frequencies. Because $H_0 = 2pq$, the actual frequency of heterozygous genotypes in the inbred population can be written in terms of F as:

$$H = H_0 - H_0 F = 2pq - 2pqF \tag{3.2}$$

The frequency of AA homozygous genotypes in the inbred population can also be expressed in terms of F. Suppose that the proportion of AA genotypes is actually D. Because the allele frequency of A is p, we must have, by Equation 1.3, that $D + H/2 = p$. But $H = 2pq(1 - F)$. Therefore,

$$D = p - \frac{(2pq - 2pqF)}{2} = p^2 + pqF \tag{3.3}$$

Likewise, the frequency of aa genotypes R is given by:

$$R = q - \frac{(2pq - 2pqF)}{2} = q^2 + pqF \tag{3.4}$$

A little algebraic manipulation of Equations 1.20–1.22 enables the genotype frequencies with inbreeding to be written in the alternative form:

$$\begin{aligned} &AA : p^2(1 - F) + pF \\ &Aa : 2pq(1 - F) \\ &aa : q^2(1 - F) + qF \end{aligned} \tag{3.5}$$

This formulation shows that the genotype frequencies equal the HWE frequencies multiplied by the factor $1 - F$, plus a correction term for the homozygous genotypes multiplied by the factor F. When $F = 0$ (no inbreeding), the genotype frequencies are the HWE. When $F = 1$ (complete inbreeding), the population consists entirely of AA and aa homozygous genotypes in the frequencies p and q, respectively.

The reorganization of genotype frequencies with inbreeding is illustrated graphically in Figure 3.1. To obtain the genotype frequencies, multiply $1 - F$ times the HWE frequencies in the rectangles on the left, and add F times the allele frequencies in the rectangles on the right. This calculation corresponds to Equation 1.23. The main effect of inbreeding is that some heterozygous genotypes disappear from the population and are replaced

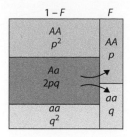

Figure 3.1 Genotype frequencies with inbreeding. The effect of inbreeding is to eliminate some heterozygous genotypes and to replace them with homozygous genotypes.

with homozygous genotypes. The arrows indicate the increase in homozygotes. Why an increase occurs is explained in the next section.

3.2 The Inbreeding Coefficient

The inbreeding coefficient F has an alternative interpretation in terms of probability. For any pair of alleles present in a single inbred individual, the alleles are called **identical by descent (IBD)** if they are both derived by DNA replication of a single allele present in some ancestral population. The probability interpretation of F is that, for a pair of alleles present in an inbred individual, F is the probability that the alleles are IBD. When expressed in terms of the probability of IBD, the inbreeding coefficient is clearly a relative concept because we have not specified the arbitrary ancestral population in which we define all pairs of alleles as being not IBD. In other words, we regard the ancestral population as having no inbreeding ($F = 0$). Relative to this ancestral population, the inbreeding coefficient of an individual in the present population is the probability that the two alleles of a gene in the individual arose by replication of a single allele more recently than the time at which the ancestral population existed. The ancestral population need not be remote in time from the present one. The ancestral population typically refers to the population existing just a few generations previous to the present one, and F in the present population then measures inbreeding that has occurred in the span of these few generations. Because the span of time involved is usually short, the possibility of mutation can safely be ignored.

If the two alleles in an inbred individual are IBD, the genotype at the locus is said to be **autozygous**. If they are not IBD, the genotype is said to be **allozygous**. The distinction is important because an autozygous genotype must be homozygous, since the alleles are IBD and we assume no mutation. On the other hand, allozygous genotypes can be either homozygous or heterozygous. Figure 3.2 illustrates how the concepts of autozygosity and allozygosity are related to those of homozygosity and heterozygosity. The essential point is that two alleles can be **identical by state (IBS)**, which means that they have the same sequence of nucleotides along the DNA, without necessarily being IBD.

The concept of IBD pertains to the ancestral origin of an allele and not to its chemical makeup. Although alleles that are IBS may have originated from a common ancestral allele at some time in the remote past, they are IBD only if they derive from a common ancestral allele after some arbitrarily designated time in the population's more recent history. In a pedigree, for example, IBD means that the alleles originated by replication of a common ancestral allele more recently than the earliest ancestors in the pedigree. As shown in

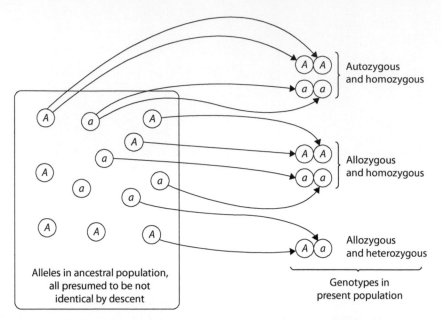

Figure 3.2 In an autozygous individual, homologous alleles are derived from replication of a single DNA sequence in an ancestor and are therefore IBD. In an allozygous individual, homologous alleles are not IBD.

Figure 3.2, two distinct alleles that are IBS and chemically identical (two A alleles or two a alleles, for example) may come together in an individual and thereby make the individual homozygous, the alleles in the remote ancestral population are, by definition, not IBD, and so the individual is allozygous. Similarly, although a heterozygous individual must be allozygous (ignoring mutation), a homozygous individual may be either autozygous (alleles IBD) or allozygous (alleles IBS but not IBD) for the reasons shown in Figure 3.2.

To show that the inbreeding coefficient defined in terms of IBD is equivalent to that defined in Equation 3.1 in terms of heterozygosity, we need only consider the implications of the probability definition for an entire population. Imagine, therefore, a population in which individuals have average inbreeding coefficient F. Focus on one individual, and consider the alleles of any gene in the individual. Either of two things must be true: the alleles must be either allozygous (probability $1 - F$) or autozygous (probability F). If the alleles are allozygous, then the probability that the individual has any particular genotype is the same as the probability of that genotype with HWE, because, by chance, this particular gene has escaped the effect of inbreeding. On the other hand, if the alleles are autozygous, then the individual must be homozygous, and the probability that the individual is homozygous for any particular allele is equal to the frequency of that allele in the population as a whole. (Because the alleles in question are autozygous, knowing which allele is present in one chromosome immediately tells you that an identical allele is in the homologous chromosome.) In symbols, the probability that an individual has genotype AA is $p^2(1 - F)$ [when the alleles are allozygous] $+ pF$ [when the alleles are autozygous]. Similarly, the probability that the individual has genotype aa is $q^2(1 - F) + qF$. Heterozygous Aa genotypes then occur in the frequency $2pq(1 - F)$ because alleles that are heterozygous must be allozygous. The genotype frequencies with inbreeding are summarized in Table 3.1. Note that the genotype frequencies are exactly

Table 3.1 Genotype frequencies with inbreeding

Genotype	With inbreeding coefficient F		Frequency in population	With $F = 0$ (random mating)	With $F = 1$ (complete inbreeding)
AA	$p^2(1-F)$	$+$	pF	p^2	p
Aa	$2pq(1-F)$			$2pq$	0
aa	$q^2(1-F)$	$+$	qF	q^2	q
	Allozygous genes		Allozygous genes		

the same as those given in Equation 3.5. The similarity shows that the IBD definition of F and the heterozygosity definition of F, although superficially quite different, are actually equivalent.

Still another definition of the F is as the correlation in state between uniting gametes. This definition can be understood with reference to Table 3.1 by associating the state of uniting gametes with binary random variables X and Y that take on a value of 1 when the gamete carries A and 0 when it carries a. Then $E(X) = E(Y) = p$ and the covariance between uniting gametes equals $E(XY) - E(X)E(Y) = p^2(1-F) + pF - p^2 = pqF$. Also, the variance of X is $Var(X) = E(X^2) - [E(X)]^2 = p - p^2 = pq$ and likewise $Var(Y) = pq$. The correlation coefficient between uniting gametes is therefore:

$$\frac{Cov(XY)}{\sqrt{Var(X)Var(Y)}} = \frac{pqF}{pq} = F \tag{3.6}$$

This is the original definition of F due to Wright (1922), who preferred it because it allows for the theoretical possibility that $F < 0$. The definition of F in terms of probability of IBD came much later (Malécot 1948) and while this definition is more intuitive, it constrains F to be ≥ 0.

Inbreeding Depression and Heterosis

In species that usually breed by outcrossing with unrelated individuals, close inbreeding is generally harmful. These harmful effects are referred to as **inbreeding depression** (Charlesworth and Willis 2009; Paige 2010). That inbreeding can result in weakness and small size has been known for hundreds of years (Crow 1998). An example of inbreeding depression in one population of cultivated corn (maize) is shown in Figure 3.3. The linear decrease in yield is what one would expect if most of the response were due to rare recessive alleles that become homozygous as a result of inbreeding. The generally harmful effects of inbreeding have important implications for animal breeding, including that of "purebred" dogs and cats, many of which suffer the effects of being homozygous for deleterious recessive alleles. Inbreeding depression is also important for conservation genetics in the preservation of endangered species, because small population size inevitably results in mating between relatives.

The flip side of inbreeding depression is **heterosis**, or hybrid vigor, which refers to the increased size and vigor of hybrids between inbred lines. The reason for hybrid vigor is that, while deleterious recessive mutations occur in all lines of corn, the particular genes that undergo mutation differ from one line to the next. When these populations are inbred, they all show inbreeding depression analogous to that in Figure 3.3, but the

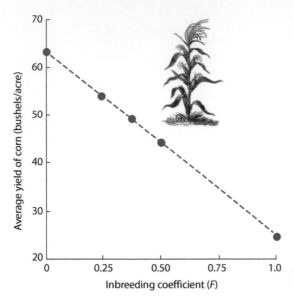

Figure 3.3 Inbreeding depression in corn (*Zea mays*). (Data from Neal (1935).)

recessive alleles that account for the effects differ among the inbred lines. When the inbred lines are crossed, the effects of the harmful recessive alleles contributed to the hybrid by each inbred line are concealed by the dominant alleles contributed by the other inbred line. The resulting hybrids are therefore heterozygous for all or most of the deleterious recessives, which accounts for the hybrid vigor and performance often exceeding that of the average of the original populations before inbreeding. Moreover, the hybrids from a cross between inbred lines are genetically identical, and hence they show more uniformity in germination time, growth rate, height, and other phenotypes that are important in agriculture. Hybrid corn was originally developed early in the twentieth century, but it was not until about 1930 that seed from hybrid corn became commercially available in large quantities. Today more than 95 per cent of corn acreage in the United States is planted with hybrids, and over the period 1960–2020 the yield per hectare has increased by more than 250 per cent.

An alternative to the hypothesis of simple dominance for heterosis is the hypothesis of **overdominance**, which posits that heterozygous genotypes typically have greater fitness and superior performance than homozygous genotypes. The contribution of overdominance to heterosis has been a subject of debate and controversy from almost the beginning of modern genetics, but after many decades of experiments and observations it seems safe to conclude that overdominance plays at best a minor role in heterosis (Charlesworth and Willis 2009; Crow 1998).

Effects of Inbreeding on Rare Harmful Alleles

The harmful effects of inbreeding are seen most dramatically when inbreeding is complete or nearly complete. Although nearly complete inbreeding can be approached in many plant species by self-fertilization and in most other species by many generations of brother–sister mating, autozygosity of whole chromosomes can easily be accomplished in *Drosophila* through the use of special inversion-bearing chromosomes called *balancer chromosomes* that eliminate virtually all recombination. The results of a typical experiment

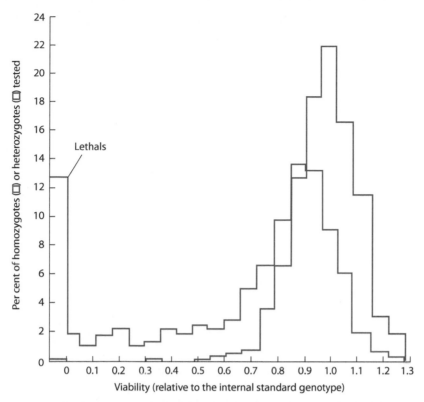

Figure 3.4 Viability distributions of homozygous chromosomes (red outline) and heterozygous chromosomes (blue outline) of second chromosomes extracted from a natural population of *Drosophila pseudoobscura*. The numbers tested were 1063 homozygous genotypes and 1034 heterozygous genotypes. (Data from Dobzhansky and Spassky (1963).)

in which whole chromosomes from natural populations are rendered IBD are shown in Figure 3.4.

The crosses are carried out in such a way as either to make a single natural chromosome homozygous (also IBD) or to make two different natural chromosomes heterozygous (allozygous). The distribution of relative survivorship (viability) of the resulting genotypes is shown. A relative viability of 1.0 corresponds to equality with a standard genotype present in each culture bottle. It is evident that the homozygous genotypes (red outline) are relatively poor in viability. In fact, about 12.5 per cent of the homozygous chromosomes are lethal. Moreover, among the homozygotes that have viabilities within the normal range of heterozygotes (blue outline), virtually all can be shown to have reduced fertility. The harmful inbreeding effects are due mainly to rare recessive alleles that are severely detrimental when homozygous. The inference that each harmful recessive allele is rare is supported by the small proportion of lethal or near-lethal heterozygous chromosomes.

Inbreeding Effects in Human Populations

As in other outcrossing organisms, inbreeding in humans is generally harmful, but the effects are difficult to measure because the degree of inbreeding is often quite small and the effects may also vary from population to population. The effects are again due largely to the increased homozygosity of rare recessive alleles. They are observed most dramatically in the increased frequency of inherited diseases due to harmful recessive alleles among the children of matings between first cousins or other degrees of familial relationship. The frequency of first-cousin matings has fluctuated through time and space. Although first-cousin matings were socially acceptable in mid-nineteenth-century Europe and North America, they began to be shunned in later decades. By the 1880s, a number of U.S. states had passed laws prohibiting cousin marriage, and at this time 31 states have prohibitions or restrictions. First-cousin mating is banned in China, North and South Koreas, the Philippines, and some other countries, but it is quite common across North Africa, the Middle East, Pakistan, and India (Bittles and Black 2010).

For a rare deleterious recessive allele, the frequency of homozygous recessives among the children of first-cousin matings (for which $F = 1/16$) is given by $q^2(1 - 1/16) + q(1/16)$. On the other hand, among the offspring of unrelated parents the frequency of recessive homozygotes is q^2. Therefore, the risk of an affected offspring from a first-cousin mating, relative to that from a mating of nonrelatives, is given by:

$$\text{Relative risk} = \frac{q^2(1 - 1/16) + q(1/16)}{q^2} \approx 0.94 + \frac{0.06}{q} \tag{3.7}$$

For example, when $q = 0.01$, the relative risk is approximately 7, which means that a first-cousin mating has about a sevenfold greater risk of producing a homozygous recessive child than a mating between nonrelatives. This is clearly a large inbreeding effect, and the more rare the frequency of the deleterious recessive allele, the greater the relative risk.

First-cousin mating approximately doubles the risk of birth defects, from about 3 per cent to about 6 per cent (Sheridan et al. 2013). Across a large number of studies comprising more than 2 million offspring, the estimated increase in mortality in the offspring from first-cousin mating is 3.5 per cent. Interpolating in the maize data in Figure 3.3, an inbreeding coefficient of $F = 1/16$ predicts a decrease in yield of 3.75, hence the inbreeding depression observed in human viability is not less than the effects of inbreeding depression in other organisms. Extrapolating linearly from the human data implies that for $F = 0.5$ the increased mortality would be 28 per cent and at $F = 1$ it would be 56 per cent. However, inbreeding coefficients in human populations never get so high.

3.3 Calculation of the Inbreeding Coefficient from Pedigrees

Computation of F from a pedigree is simplified by drawing the pedigree in the form shown in Figure 3.5a, where the lines represent gametes contributed by parents to their offspring. The same pedigree is shown in conventional form in Figure 3.5b. The individuals in gray in (b) are not represented in (a) because they have no ancestors in common and therefore do not contribute to the inbreeding of individual I. The inbreeding coefficient F_I of individual I is the probability that I carries alleles of an arbitrarily chosen gene that are IBD. The first step in calculating F_I is to identify all common ancestors in the pedigree, because an ancestral allele could become IBD in I only if it were inherited through both of I's parents from a common ancestor. In this case there is only one common ancestor, labeled A. The next step in calculating F_I, which is carried out for each common ancestor in turn, is to trace all the paths of gametes that lead backward from one of I's parents to the common ancestor and then down again to the other parent of I. These paths are the paths along which an allele in a common ancestor could become IBD in the individual I.

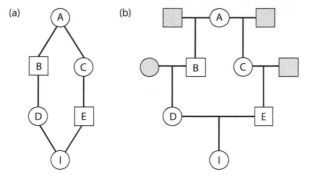

Figure 3.5 (a) Pedigree diagram to facilitate calculation of the inbreeding coefficient. (b) Conventional representation of the same pedigree.

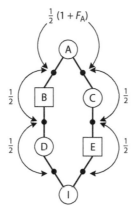

Figure 3.6 Identity loops for the pedigree in Figure 3.5. Each number is the probability of IBD for the alleles indicated.

In Figure 3.5a, there is only one such path: DBACE. The common ancestor has been underlined for bookkeeping purposes, a procedure especially useful in complex pedigrees.

The third step in calculating F_I is to calculate the probability of IBD in I due to each of the paths in turn. For the path DBACE, the reasoning is illustrated in Figure 3.6. Here the black dots represent alleles transmitted along the gametic paths, and the number associated with each loop is the probability of IBD of the alleles indicated. For all individuals except the common ancestor, the probability is 1/2, because with Mendelian segregation the probability that a particular allele present in a parent is transmitted to a specified offspring is 1/2. To understand why $(1/2)(1+F_A)$ is the probability associated with the loop around the common ancestor, denote the alleles in the common ancestor as α_1 and α_2. (The Greek symbols are used to avoid confusion with conventional allele symbols such as A and a because here we are interested only in the ancestry of the alleles.) The pair of gametes contributed by individual A could contain $\alpha_1\alpha_1$, $\alpha_2\alpha_2$, $\alpha_1\alpha_2$, or $\alpha_2\alpha_1$, each with a probability of 1/4 because of Mendelian segregation. In the first two cases, the alleles are clearly IBD; in the second two cases, the alleles are IBD only if α_1 and α_2 are IBD, and α_1 and α_2 are IBD only if individual A is autozygous, which has probability F_A, the inbreeding coefficient of A. Altogether, the required probability for the loop around individual A is $1/4 + 1/4 + (1/4)F_A + (1/4)F_A = 1/2 + (1/2)F_A = (1/2)(1+F_A)$. Because each of the loops in Figure 3.6 is independent of the others, the total probability of autozygosity in individual I due to this path is $1/2 \times 1/2 \times (1/2)(1+F_A) \times 1/2 \times 1/2$, or $(1/2)^5(1+F_A)$. Note that the exponent on the 1/2 is equal to the number of individuals in the path. In general, if a path through a common ancestor A contains i individuals, the probability of autozygosity due to that path is:

$$\left(\frac{1}{2}\right)^i (1+F_A)$$

Thus, the inbreeding coefficient of individual I in Figure 3.5a, assuming $F_A=0$, equals

$$(1/2)^5 = 1/32.$$

In more complex pedigrees, there is more than a single path. The paths are mutually exclusive, because if the alleles are IBD due to being inherited along one path, they cannot at the same time be IBD due to being inherited along a different path. Therefore, the total inbreeding coefficient is the sum of the probability of IBD due to each separate path. The whole procedure for calculating F is summarized in an example involving a first-cousin mating in Figure 3.7. Here there are two common ancestors (A and B) and two paths (one each through A and B). The total inbreeding coefficient of I is the sum of the two mutually exclusive contributions shown in Figure 3.7. If A and B are not themselves inbred, then $F_A = F_B = 0$, and $F_I = (1/2)^5 + (1/2)^5 = 1/16$. This is the probability that an arbitrarily chosen pair of alleles in I are IBD. Alternatively, F_I can be interpreted as the average proportion of all pairs of alleles in I that are IBD.

In general, for any gene except those in the X and Y chromosomes, the formula for calculating the inbreeding coefficient F_I of an individual I is:

$$F_I = \sum_{all\ paths} \left(\frac{1}{2}\right)^i (1+F_A) \tag{3.8}$$

where the summation is over all of the possible paths through all common ancestors, i is the number of individuals in each path, and A is the common ancestor in each path.

The procedure for calculating F for an X-linked gene is similar except for two special provisions. First, ignore all paths with two males in a row because a male cannot transmit

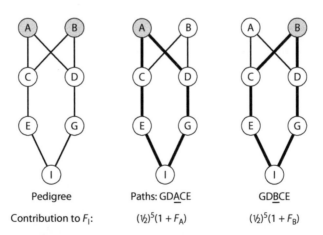

Figure 3.7 Individual I is the offspring of a first-cousin mating. Two paths (heavy lines) through common ancestors A and B are shown at the right.

his X chromosome to his son. Second, if a path has a common ancestor that is male, the loop around the ancestor has probability 1 instead of $(1/2)(1+F_A)$ because a male must transmit his X chromosome to his daughters.

3.4 Regular Systems of Mating

Some domesticated animals and cultivated plants are propagated by a regular system of mating, such as repeated self-fertilization, sib mating, or backcrossing to a standard strain. It is then of interest to know how the inbreeding coefficient increases with time. The reasoning involved for the special case of self-fertilization hinges on the fact that the heterozygosity decreases by half in each generation. Letting H_t/H_0 denote the frequency of heterozygous genotypes in generation t relative to that in generation 0, then:

$$\frac{H_t}{H_0} = \frac{1}{2}\left(\frac{H_{t-1}}{H_0}\right)$$

Because Equation 3.2 implies that $H_t/H_0 = 1 - F_t$ we can also write:

$$1 - F_t = \frac{1}{2}(1 - F_{t-1}) \tag{3.9}$$

It follows that for repeated self-fertilization:

$$1 - F_t = \left(\frac{1}{2}\right)^t (1 - F_0) \tag{3.10}$$

How the inbreeding coefficient increases under various regular systems of mating is shown in Figure 3.8. In most cases, F increases gradually with time. One important general principle is that, no matter what the value of F in a population at any time, and no matter how this value of F was attained, the inbreeding coefficient immediately returns to 0 (no inbreeding) with just one generation of outcrossing to an unrelated strain.

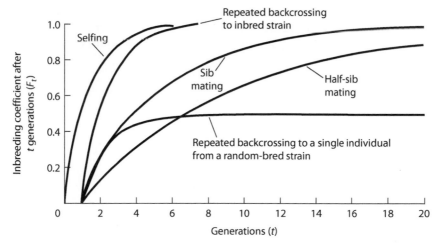

Figure 3.8 Theoretically predicted increase in the inbreeding coefficient for various regular systems of mating.

Partial Selfing

The principle that a single generation of random mating completely undoes any and all previously accumulated inbreeding helps to understand what happens when a regular system of mating is interrupted by an occasional outcross. Partial selfing, in which only a fraction of the population undergoes self-fertilization in each generation, affords an illustration. For full selfing, Equation 3.9 implies that $F_t = (1/2)(1 + F_{t-1})$. If in each generation only a fraction S of the individuals in a population undergo selfing, then the appropriate equation is $F_t = (1 - S) \times 0 + S(1/2)(1 + F_{t-1}) = (S/2)(1 + F_{t-1})$. A neat trick is to multiply both sides by -1 and add $S/(2 - S)$, which leads to:

$$\frac{S}{2-S} - F_t = \frac{S}{2-S} - \left(\frac{S}{2}\right)(1 + F_{t-1}) = \left(\frac{S}{2}\right)\left(\frac{2-2+S}{2-S} - F_{t-1}\right) = \left(\frac{S}{2}\right)\left(\frac{S}{2-S} - F_{t-1}\right)$$

This equation implies that:

$$\frac{S}{2-S} - F_t = \left(\frac{S}{2}\right)\left(\frac{S}{2-S} - F_{t-1}\right) = \left(\frac{S}{2}\right)^2\left(\frac{S}{2-S} - F_{t-2}\right) = \cdots = \left(\frac{S}{2}\right)^t\left(\frac{S}{2-S} - F_0\right)$$

And as t goes to infinity (as you'll see in a few minutes, "infinity" in this case is not many generations), the right-hand side goes to 0 and therefore F goes to a steady-state value denoted with a circumflex:

$$\hat{F} = \frac{S}{2-S} \tag{3.11}$$

The approach to steady state is extremely rapid as shown in Figure 3.9. Even when the probability of self-fertilization is very high, the steady state is attained in fewer than ten generations. The reason for the rapid increase is that in each generation a proportion of the population undergoes outcrossing, which reduces the inbreeding coefficient in those lineages to 0 until self-fertilization takes place once again. In a population of organisms undergoing partial selfing, the probability that a particular lineage undergoes $t = 0, 1, 2, \ldots$ generations of selfing before one generation of outcrossing takes place equals $S^t(1 - S)$. This distribution is a negative binomial distribution, and the average number of generations of self-fertilization before an outcrossing occurs equals $S/(1 - S)$. Even for $S = 0.95$ the average is ten generations of selfing in any one plant before all of this lineage's inbreeding is undone by an outcross. The curves in Figure 3.9 attain their steady state so

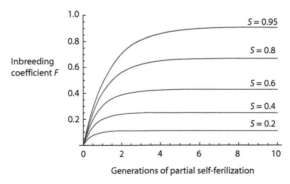

Figure 3.9 Approach to steady-state of the inbreeding coefficient in a population of organisms capable of self-fertilization when a fraction S of the individuals undergo self-fertilization in each generation.

quickly because the number of generations of consecutive self-fertilization is typically small.

Inbreeding reduces the frequency of heterozygous genotypes, and hence it also reduces the frequency of doubly heterozygous genotypes such as $A_1 B_1/A_2 B_2$ and $A_1 B_2/A_2 B_1$. These are the genotypes in which recombination takes place and reduces the extent of linkage disequilibrium. In effect, inbreeding makes the frequency of recombination between genes a function of the inbreeding coefficient, specifically:

$$c^* = c(1-F) \tag{3.12}$$

where c^* is the effective frequency of recombination in an inbred population, F is the inbreeding coefficient, and c is the frequency of recombination in the absence of inbreeding. Note that inbreeding even affects linkage disequilibrium (LD) between genes in different chromosomes ($c = 0.5$).

On the other hand, even a small amount of outcrossing can result in a significant reduction in LD because the progeny of a cross between two homozygous lines are heterozygous for all SNPs at which the parental lines differ. In such a hybrid, every recombination event breaks up some block of SNPs that had been strongly correlated. The diminutive flowering plant *Arabidopsis thaliana*, a member of the mustard family widely used as a model organism in plant biology, serves as an example. In natural populations of *A. thaliana*, about 99 per cent of the plants result from self-fertilization in the previous generation and only about 1 per cent from outcrossing. Although the inbreeding is very intense, the amount of outcrossing is sufficient to reduce the average size of a **haplotype block**, a term referring to a group of linked SNPs that show significant LD as measured by r^2—the square of the correlation coefficient defined in Equation 2.14. Across the genome of *A. thaliana*, the length of the average haplotype block is approximately 10 kb (Kim et al. 2007). This is not much different from the length of haplotype block observed in some human populations (Shifman et al. 2003).

Two factors contribute to the relatively small haplotype blocks observed in *A. thaliana* in spite of its high level of self-fertilization. One is the rapid decrease in the expected r^2 as a function of 4Nc (Equation 2.17). Even though inbreeding decreases the effective value of c by a factor of $1-F$ (Equation 3.12), the global population size of *A. thaliana* is extremely large, hence $4Nc^*$ is likely large. The second factor is that, while high levels of self-fertilization effectively eliminate most highly deleterious mutant alleles from the population, mildly deleterious alleles may persist and become fixed (Bustamante et al. 2002). The heterozygous offspring of outcrosses are therefore likely to be more fit than their self-fertilized counterparts owing to heterosis, and their increased contribution to the gene pool further decreases the extent of linkage disequilibrium.

Repeated Sib Mating

For sexual organisms that cannot undergo self-fertilization, repeated sib mating is one way to steadily increase the inbreeding coefficient. A pedigree of repeated sib mating is shown in Figure 3.10, where the pair of dots in each circle represents the alleles of a gene in that individual, and the theory is straightforward. In Figure 3.10, the symbol F_t is as defined earlier. It is the probability that the alleles present in a single individual (those in black) are IBD. The symbol G_t represents a different but related concept. G_t is the probability that an allele chosen at random from one individual is IBD with an allele chosen at random from a different individual (in Figure 3.10 the chosen alleles are indicated in black). G_t was originally defined by Malécot (1948) and called the *coefficient de parenté*, which

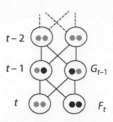

Figure 3.10 Pedigree of repeated mating between siblings. F_t and G_t are respectively the inbreeding coefficient and the kinship coefficient in generation t.

translates to **coefficient of kinship**, but is sometimes referred to as the **coefficient of consanguinity**. Simply put, the coefficient of kinship between two individuals is equal to the inbreeding coefficient of a hypothetical offspring they might have. We can therefore write that $F_t = G_{t-1}$. The probability that a randomly chosen pair of alleles in different individuals in generation t came from the same individual in generation $t-1$ is $1/2$, in which case the probability of IBD equals $(1/2)(1+F_t)$; and the probability that the alleles came from different individuals in generation $t-1$ also equals $1/2$, in which case the probability of IBD equals G_{t-1}.

Putting all this together,

$$F_t = G_{t-1}$$
$$G_t = (1/4)(1+F_{t-1}) + (1/2)G_{t-1}$$

$$(3.13)$$

These equations can be solved simultaneously using methods from linear algebra, but in this case an alternative approach is to substitute F_{t+1} for G_t and F_t for G_{t-1} into the second expression in Equation 3.13, which yields:

$$F_{t+1} = (1/2)F_t + (1/4)(1+F_{t-1})$$

Further simplification results by expressing F_t as $(H_0 - H_t)/H_0$ as in Equation 3.1, where H_t is the proportion of heterozygous genotypes in generation t and H_0 that in generation 0. This substitution along with a little manipulation results in:

$$\frac{H_{t+1}}{H_0} = \frac{H_t}{2H_0} + \frac{H_{t-1}}{4H_0}$$

$$(3.14)$$

Equation 3.14 can be solved recursively. Starting with a population in HWE, $H_0/H_0 = 1$, it follows that $H_1/H_0 = 1$ (or, equivalently, $2/2$), and then, continuing, we have $H_2/H_0 = 3/4$, $H_3/H_0 = 5/8$, $H_4/H_0 = 8/16$, $H_5/H_0 = 13/32$, $H_6/H_0 = 21/64$, and so forth. The numerator of H_t/H_0 increases according to the *Fibonacci* sequence 1, 2, 3, 5, 8, 13, 21, ... , in which each successive number is the sum of the two previous numbers, and the denominator of H_t/H_0 equals 2^t. The exact solution to Equation 3.14 turns out to be:

$$\frac{H_t}{H_0} = \frac{\left(5+3\sqrt{5}\right)}{10}\left(\frac{1+\sqrt{5}}{4}\right)^t + \frac{\left(5-3\sqrt{5}\right)}{10}\left(\frac{1-\sqrt{5}}{4}\right)^t$$

$$(3.15)$$

For practical purposes, the following approximation will suffice:

$$\frac{H_t}{H_0} = 1.171 \times (0.809)^t - 0.171 \times (-0.309)^t$$

$$(3.16)$$

of which successive terms starting with $t=0$ equals approximately 1, 2/2, 3/4, 5/8, 8/16, 13/32, and so forth as required. Note that as t increases, the first term in Equation 3.16 dominates, and this implies that the heterozygosity eventually declines at a steady rate of about 20 per cent per generation. In practice, this steady decline in heterozygosity occurs for $t \geq 7$. Equation 3.16 also implies that, with repeated sib mating, the expected proportion of heterozygous genotypes becomes less than 1 per cent of that in the initial population when $t = 23$.

Recombinant Inbred Lines

Inbred lines have many uses. We've already seen one practical application in the use of inbred lines as parents to produce hybrid offspring exhibiting heterosis for improved agricultural performance and consistency. In research, an important use of inbreds is in the production of **recombinant inbred lines (RILs)**—strains whose genomes are mosaics of segments of the genomes from a set of inbred lines used to establish the RILs (Crow 2007). The underlying principle of why RILs are useful can be understood by considering just two parental inbred lines. The parental lines are crossed to produce a hybrid or F_1 generation, and these individuals are mated randomly among themselves for one or more generations to allow recombination to occur, at which point a new set of inbred lines (the RILs) is established by self-fertilization or sib mating. The genomes of the RILs are different from one another, and each is a mosaic of segments from the two original lines. The genetic basis of phenotypic differences between the original inbred lines can be tracked in the RILs by determining which segments of the parental genomes are consistently associated with differing phenotypes.

Most RILs are created from more than two inbred lines. An example is shown in Figure 3.11, which depicts the creation of a set of RILs known as the *Drosophila* Synthetic Population Resource (King et al. 2012). In this instance, the original inbred lines consisted of eight completely sequenced founders (Figure 3.11), these were crossed in a "round-robin" design (i.e., line $1 \times$ line 2, $2 \times 3, \ldots, 7 \times 8$, 8×1) and the hybrid offspring were combined to create a large, random-mating population (Figure 3.11b). The population was maintained for fifty generations during which recombination progressively split the parental genomes into segments (Figure 3.11c), after which twenty generations of sib mating was used to extract about 1600 RILs (Figure 3.11d), each of which was genotyped to be able to trace each part of the genome back to one of the original inbred founders. Twenty generations is sufficient to reduce the heterozygosity to about 2 per cent of that in the initial population. (With self-fertilization in plants, this would require only six generations.)

Trade-offs in producing RILs are several. First is the number of parental lines: you want enough to include ample phenotypic diversity, but not so many that the representation of each founder in the RILs is so infrequent as to lose statistical power. Then there are the generations of random mating: you want enough recombination to scramble the founder genomes, but no so much that many segments are too small to trace to their founders. Finally there is an issue of how many RILs to produce: you want enough lines to gain statistical power, but not so many as to be redundant.

As a research resource, RILs have many advantages. Chief among them is that, once the founder lines have been sequenced and the RILs genotyped, no further genotyping is necessary on this set of RILs. The same set of lines can be used over and over again in different studies. Another advantage is the near homozygosity of the RILs so that measurements can be replicated as often as necessary to minimize environmental variation and measurement error. The third advantage is that the RILs have their genetic differences spread throughout the genome. This is particularly useful for studies of complex traits that are determined by the interactions of multiple genetics as well as

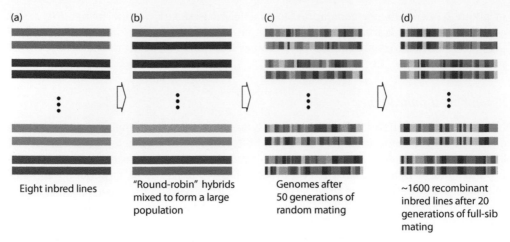

(a)   (b)   (c)   (d)

Eight inbred lines

"Round-robin" hybrids mixed to form a large population

Genomes after 50 generations of random mating

~1600 recombinant inbred lines after 20 generations of full-sib mating

Figure 3.11 Production of recombinant inbred lines constituting the *Drosophila* Synthetic Population Resource. (a) Genomes of original inbred lines; (b) those of F_1 offspring that are mixed to produce a large, random-mating population; (c) recombinant segments in the genome after fifty generations; (d) mosaic structure of the nearly-homozygous RILs after twenty generations of full-sib mating.

environmental factors. In the case of the *Drosophila* Synthetic Population Resource in Figure 3.11, the RILs have a statistical power of 84 per cent to detect a genetic factor that accounts for 5 per cent of the genetic effects on a trait, and the mosaic genomes allow the position of the factor to be resolved to within about 3.75 megabases (which in *Drosophila* corresponds to about 1.5 per cent recombination).

Originally developed in mice in the 1960s (Crow 2007), RILs have been used extensively in mice and *Drosophila* and are especially prominent in research in plant genetics. Even a cursory look at PubMed yields a long list: *Arabidopsis*, bean, cotton, lentil, lotus, maize, pepper, rice, soybean, sunflower, tomato, wheat, and so forth. The widespread use of RILs testifies to their convenience and utility.

3.5 Remote Inbreeding in Finite Populations

Actual populations are limited in size to some finite number of individuals. To be able to deal with finite population size, it is convenient to extend the concept of allele IBD to include the probability of IBD to the case of two randomly chosen alleles from any pair of individuals. By IBD in this context we mean that two randomly chosen alleles are identical in DNA sequence by virtue of being derived by DNA replication from a single ancestral allele (except for a few mutations that may have taken place). This concept generalizes IBD to include any two alleles chosen at random from a population.

One important implication of a finite population persisting through many generations is that, eventually, every member of the population becomes related in some degree or another to every other member of the population. Everyone is related because, the population being limited in size, any two individuals must share at least one recent or remote common ancestor. This is easy to see. In sexual animals, an individual has two parents, four grandparents, eight great-grandparents, and so forth, totaling 2^n ancestors in the nth previous generation. When $n = 20$, 2^n is more than a million, which implies

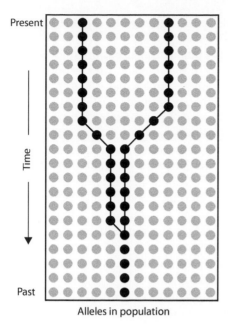

Present

Time

Past

Alleles in population

Figure 3.12 Coalescence of the ancestral histories of two alleles (black) onto a common ancestral allele.

that in a finite population of fewer than a million individuals, many of these ancestors must be the same individual. It also implies that any two individuals in the current generation must have many ancestors in common. It follows that mating pairs must be related in some degree or another, even if the mating pairs are formed "at random." This constitutes a type of inbreeding due to the sharing of remote ancestors, which we shall call *remote inbreeding* to distinguish it from the close inbreeding between immediate relatives discussed in Section 3.3.

The reason why alleles in different individuals can be IBD in a finite population is illustrated in Figure 3.12. Each dot represents an allele in a finite population at a given point in time. The black dots are alleles chosen at random from two different individuals in the present generation, and the solid lines trace their ancestry back in time. The chosen alleles are IBD because their ancestral lineages converge onto a common ancestral allele at some time in the past. The merging of the ancestral histories of two alleles onto a common ancestral allele is known as **coalescence** or a **coalescent event**. Coalescence is a fundamental concept in population genetics and genomics (Wakeley 2009), and we will have much more to say about it in Chapter 4. We introduce the concept here to emphasize how it underlies allele IBD in finite populations.

Identity by Descent in Finite Populations

Because mating pairs in a finite population are still formed at random within the pool of potential partners available, remote inbreeding in a finite population does not cause a departure from Hardy–Weinberg frequencies even though there is increased IBD. The main consequence of remote inbreeding is that, although HWE is maintained in the finite population through all generations, the probability of IBD of the alleles in two randomly chosen individuals steadily increases. In large populations, the increase in IBD occurs relatively slowly, but in extreme cases of chronically small populations over many generations, the IBD eventually increases to such a level that much of the original genetic variation is lost.

Decreased genetic variation due to chronically small population size is often found in island populations, notably the Channel Island fox, *Urocyon littoralis*, off the Southern Coast of California, which includes a population on remote San Nicolas island in which every individual is genetically nearly identical to every other individual (Robinson et al. 2016). Drastically reduced genetic variation is also observed in both major subspecies of the cheetah, *Acinonynx jubatus*. For the East African subspecies (*A. j. raineyi*), estimates of polymorphism and heterozygosity are 0.04 and 0.01, respectively; while for the South African subspecies (*A. j. jubatus*), the corresponding estimates are 0.02 and 0.0004. Most unusual is the finding of skin-graft acceptance between unrelated cheetahs from the South African subspecies. Graft acceptance means that the cheetah population is essentially homozygous for the major histocompatibility locus, which is abundantly polymorphic in other mammals. The cheetah, which was worldwide in its range at one time but presently numbers less than 20,000 animals, evidently underwent at least one very severe constriction in population number at some time in the geologically recent past, probably no later than 10–12 thousand years ago (O'Brien et al. 2017).

The reason for the increase in IBD due to remote inbreeding is shown in Figure 3.13. The alleles are labeled α_1, α_2, α_3, and so forth in order to mask their identities as either A or a, because here we are interested only in IBD. We denote the probability that any two randomly chosen alleles are IBD in generation t as F_t. In much of the population genetics literature the probability of IBD due to remote inbreeding is called the **fixation index** and symbolized as F_{ST} or G_{ST} to distinguish it from the conventional inbreeding coefficient due to close inbreeding. The subscript ST is suppressed in Figure 3.13 and in the discussion that follows only to avoid a proliferation of subscripts.

In Figure 3.13, we assume that a population of size N diploid individuals in generation $t-1$ generates an infinite pool of gametes, from which $2N$ are chosen to be represented in the next generation. Consider any two randomly chosen gametes, which may not necessarily be present in the same zygote. There are only two possibilities for the chosen pair of alleles with regard to their identity in the previous generation. They could either derive from exactly the same allele (α_i) in breeding adults of the previous generation; this event has a probability of $1/(2N)$. Or they could derive from two different alleles (α_i and α_j) in breeding adults of the previous generation; this outcome has a probability of $1 - 1/(2N)$. In the first case ($\alpha_i\alpha_i$), the chosen pair of alleles is certainly IBD. In the second case ($\alpha_i\alpha_j$), they are IBD in the present generation only if they were IBD in the previous generation, and this has the probability F_{t-1}. Putting these possibilities together, the relation between F_t and F_{t-1} is:

$$F_t = \frac{1}{2N} + \left(1 - \frac{1}{2N}\right)F_{t-1} \tag{3.17}$$

To both sides, first multiply by -1 and then add $+1$. This yields:

$$1 - F_t = (1 - F_{t-1})\left(1 - \frac{1}{2N}\right) \tag{3.18}$$

and again by the method of successive substitutions we obtain:

$$1 - F_t = (1 - F_0)\left(1 - \frac{1}{2N}\right)^t \approx (1 - F_0)\,e^{-t/(2N)} \tag{3.19}$$

where the approximation on the right assumes that N is large enough that $[1/(2N)]^2$ and higher powers can be ignored. Equation 3.19 shows that F_t goes to 1 (complete IBD) at

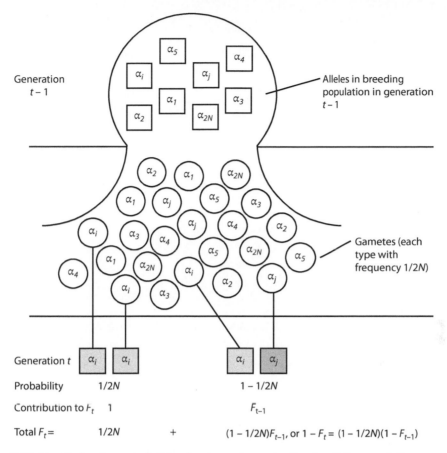

Figure 3.13 Identity by descent of alleles due to random sampling in a finite population.

a rate that depends on the reciprocal of the population size. Eventually, all of the alleles in the finite population become IBD. How can this happen? It happens because the allele frequencies of A and a change randomly from generation to generation, and eventually either a single A allele present in the original population, or a single a allele present in the original population, becomes fixed. When $F_0 = 0$ and N is reasonably large, Equation 3.19 implies that $F_t \approx t/(2N)$.

Decreased Heterozygosity in Admixed Populations

Averaged across two or more random-mating subpopulations, the frequency of heterozygous genotypes is less than would be expected based on the average allele frequencies. This principle is illustrated quantitatively for two subpopulations in Figure 3.14. In the symbols for allele frequencies, the first subscript refers to the allele (1 for A and 2 for a), and the second subscript refers to the subpopulation (1 or 2). The dot in a subscript indicates an average across the subpopulations. Both subpopulations are in HWE, however when the populations are mixed to produce an **admixed population**, the average genotype frequencies no longer conform to HWE. In particular, the admixed population has a deficiency of heterozygous genotypes compared to that expected in a hypothetical total population in which the subpopulations fuse and undergo random mating.

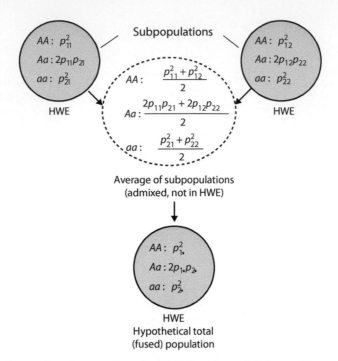

Figure 3.14 A mixture of random-mating subpopulations has a deficiency of heterozygotes relative to HWE.

The deficiency of heterozygous genotypes in the admixed population is a function of F_{ST} and also of the variance in allele frequency among subpopulations. If in Equation 3.1 we substitute H_T for H_0 and $\overline{H}_S$ for H, then we can write:

$$F_{ST} = \frac{H_T - \overline{H}_S}{H_T} \tag{3.20}$$

in which H_T is the frequency of heterozygous genotypes in the hypothetical admixed population in Figure 3.14, and $\overline{H}_S$ is the frequency of heterozygous genotypes in the admixed population. ($\overline{H}_S$ corresponds to the average frequency of heterozygous genotypes in the individual subpopulations.)

Expressing $\left(H_T - \overline{H}_S\right)/H_T$ as a function of the variance in allele frequency yields the relation between F_{ST} and the variance in allele frequency. To find this relation we first need an expression for the variance in allele frequency among the subpopulations in Figure 3.14, which by definition is:

$$\sigma^2 = (1/2)\left[\left(p_{11} - p_{1\bullet}\right)^2 + \left(p_{12} - p_{1\bullet}\right)^2\right] = (1/4)\left(p_{11} - p_{12}\right)^2 \tag{3.21}$$

(Deriving the right-hand side of Equation 3.21 from the left-hand side uses the definition $p_{1\bullet} = \left(p_{11} + p_{21}\right)/2$ and some straightforward but tedious algebra.) Equation 3.21 gives us the variance in the allele frequency of A, however the variance in allele frequency of a is identical and can be written as:

$$\sigma^2 = (1/2)\left[\left(p_{21} - p_{2\bullet}\right)^2 + \left(p_{22} - p_{2\bullet}\right)^2\right] = (1/4)\left(p_{22} - p_{21}\right)^2 \tag{3.22}$$

The equality between Equations 3.21 and 3.22 follows from the relations $p_{11} = 1 - p_{21}$ and $p_{12} = 1 - p_{22}$.

Now we can connect the decrease in heterozygosity to σ^2:

$$\begin{aligned}H_T - \overline{H}_S &= 2p_{1\bullet}p_{2\bullet} - (1/2)\left(2p_{11}p_{21} + 2p_{12}p_{22}\right) \\ &= (1/2)\left(p_{11} - p_{12}\right)\left(p_{22} - p_{21}\right) = 2\sigma^2\end{aligned} \tag{3.23}$$

And therefore:

$$\frac{H_T - \overline{H}_S}{H_T} = \frac{\sigma^2}{p_{1\bullet}p_{2\bullet}} \tag{3.24}$$

Comparing Equation 3.20 with Equation 3.24 leads to:

$$F_{ST} = \frac{\sigma^2}{p_{1\bullet}p_{2\bullet}} \tag{3.25}$$

which says that F_{ST} is simply another way of expressing the variance in allele frequency normalized by the product of the average allele frequencies. Because the variance must always be nonnegative, so also must F_{ST} be nonnegative.

The flip side of a decrease in heterozygosity is an increase in homozygosity. To quantify this effect, let $\overline{R}_S$ be the average frequency of homozygous aa genotypes in the admixed population in Figure 3.14 and R_T be the frequency of homozygous recessives in the hypothetical total population. From Figure 2.14 we have:

$$\begin{aligned}\overline{R}_S - R_T &= (1/2)\left(p_{21}^2 + p_{22}^2\right) - p_{2\bullet}^2 \\ &= (1/4)\left(p_{21} - p_{22}\right)^2 \\ &= \sigma^2 = p_{1\bullet}p_{2\bullet}F_{ST}\end{aligned} \tag{3.26}$$

This equation is one form of **Wahlund's** principle introduced in Chapter 2. This formulation states that the average frequency of homozygous recessive genotypes among a group of subpopulations is always greater than the frequency of homozygous recessive genotypes that would be expected with random mating, and the excess is numerically equal to the variance in the recessive allele frequency. Wahlund's principle is also known as **isolate breaking** because breaking up small-population isolates by mating outside the group reduces the expected frequency of homozygous recessives.

To take a specific example, imagine a population of gray squirrels that by chance has acquired a frequency of recessive albinism equal to 16 per cent. In a nearby forest is another population in which the albino allele is absent, so the allele frequency in this population is 0. Overall, the average frequency of albinos in the two populations is $(0.16 + 0)/2 = 8$ per cent. Were the two populations to fuse and undergo random mating, the allele frequency of the albino allele in the fused population would be $(0.4 + 0)/2 = 0.2$, and the frequency of the homozygous recessive genotype would equal $(0.2)^2 = 4$ per cent, which is in fact less than the average of the separate subpopulations. Furthermore, the variance in allele frequency equals $\left[(0.4 - 0.2)^2 + (0 - 0.2)^2\right]/2 = 0.04$, which does equal the reduction in frequency of the homozygous recessive.

We can also write Equation 3.26 in the form:

$$\overline{R}_S = p_{2\bullet}^2 + p_{1\bullet}p_{2\bullet}F_{ST} \tag{3.27}$$

Note the similarity between this and Equation 3.4, which shows again the inbreeding-like effects of population subdivision. The equation analogous to 3.27 corresponding to the AA genotypes is:

$$\overline{D}_S = p_{1\bullet}^2 + p_{1\bullet}p_{2\bullet}F_{ST} \tag{3.28}$$

which may be compared with Equation 3.3.

Equation 3.25 is not just a trick of algebraic manipulation. It is a fundamental relation in population genetics that connects the fixation index among a set of subpopulations with the variance in allele frequencies among them. The fixation index can also be interpreted in terms of the probability of IBD, and therefore the average genotype frequencies among the subpopulations are given by the counterparts of Equation 3.5:

$$
\begin{aligned}
AA &: p_{1\bullet}^2 + p_{1\bullet}p_{2\bullet}F_{ST} \\
Aa &: 2p_{1\bullet}p_{2\bullet}(1 - F_{ST}) \\
aa &: p_{2\bullet}^2 + p_{1\bullet}p_{2\bullet}F_{ST}
\end{aligned}
\tag{3.29}
$$

These expressions have always struck me as embodying one of the deep paradoxes of population genetics. They say that there is inbreeding in the aggregate of subpopulations, even though each subpopulation itself is undergoing random mating and is in HWE. The reason for the paradox is that the population as a whole is *not* undergoing random mating. In a subdivided population there is remote inbreeding because matings occur only within subpopulations, and because each subpopulation is finite in size, the level of inbreeding as measured by F_{ST} gradually builds up according to Equation 3.19 as $F_{ST} \approx t/(2N)$. This kind of population structure is called a **hierarchical population structure** because it is composed in a hierarchy of subpopulations within larger aggregates.

Hierarchical Population Structure

Hierarchically structured populations may have multiple levels of population structure between subpopulations and the hypothetical total population.

An example is shown in Figure 3.15, which depicts multiple subpopulations (S) clustered within regions (R) clustered within districts (D) all within a larger total population (T). In actual examples, there may be different numbers of subpopulations per district, different numbers of districts per region, and additional levels in the hierarchy. The subpopulations are the actual evolving entities, whereas the higher levels are all hypothetical random-mating units combining subpopulations within districts, districts within regions, and regions within the total. If we were to explicitly define the allele frequencies as in Figure 3.14, we would need symbols like p_{ijkl} to denote the frequency of the ith allele in the jth

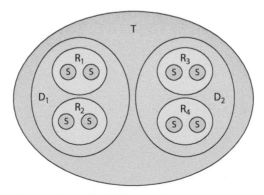

Figure 3.15 A hierarchically structured population composed of subpopulations (S) within regions (R) within districts (D) within a larger total (T).

subpopulation in region k of district l, but this level of detail is unnecessary because we can combine Equations 3.24 and 3.25 to obtain:

$$\frac{H_T - \overline{H}_S}{H_T} = F_{ST} \tag{3.30}$$

and use an analogous equation for each level in the hierarchy. In particular:

$$\frac{\overline{H}_R - \overline{H}_S}{\overline{H}_R} = F_{SR} \tag{3.31}$$

compares the average heterozygosity among subpopulations with the average heterozygosity of hypothetical random-mating populations with allele frequencies equal to the averages across subpopulations within each district. At the next level:

$$\frac{\overline{H}_D - \overline{H}_R}{\overline{H}_D} = F_{RD} \tag{3.32}$$

compares the average heterozygosity among districts with the average heterozygosity among regions. And finally:

$$\frac{H_T - \overline{H}_D}{H_T} = F_{DT} \tag{3.33}$$

compares the average heterozygosity among districts with that of the hypothetical, random-mating total population.

The quantities in Equations 3.30–3.33 are known as **F-statistics**, and their estimation from actual data is challenging because the sampling variances can be large (Weir and Hill 2002). To illustrate the meaning of the F-statistics in context, we'll use the numerical example in Figure 3.16 with the assumption that the allele frequencies in the subpopulations are known without error so that the sampling variances are not an issue.

The layout of Figure 3.16 is the same as that in Figure 3.15 with subpopulations within districts within regions within the total. The number within each subpopulation is the frequency of the A allele in that subpopulation, assuming only two alleles. The allele frequencies among subpopulations are averaged across the levels of the hierarchy to yield the average allele frequencies in districts, regions, and the total. We need to calculate the average frequency of heterozygotes at each level in the hierarchy, namely, $\overline{H}_S$, $\overline{H}_R$, $\overline{H}_D$, and H_T. These are:

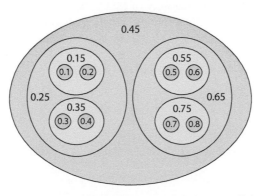

Figure 3.16 Allele frequencies among subpopulations, and average allele frequencies in districts, regions, and the total in a hierarchically structured population. Layout as in Figure 3.15.

$$\overline{H}_S = [2 \times 0.1 \times (1 - 0.1) + 2 \times 0.2 \times (1 - 0.2) + \cdots + 2 \times 0.8 \times (1 - 0.8)] / 8 = 0.390$$
$$\overline{H}_R = [2 \times 0.15 \times (1 - 0.15) + 2 \times 0.35 \times (1 - 0.35) + \cdots + 2 \times 0.75 \times (1 - 0.75)] / 4 = 0.395$$
$$\overline{H}_D = [2 \times 0.25 \times (1 - 0.25) + 2 \times 0.65 \times (1 - 0.65)] / 2 = 0.415$$
$$H_T = 2 \times 0.45 \times (1 - 0.45) = 0.495$$

From these we can calculate:

$$\frac{H_T - \overline{H}_S}{H_T} = 0.212 = F_{ST}$$

$$\frac{\overline{H}_R - \overline{H}_S}{\overline{H}_R} = 0.013 = F_{SR}$$

$$\frac{\overline{H}_D - \overline{H}_R}{\overline{H}_D} = 0.048 = F_{RD}$$

$$\frac{H_T - \overline{H}_D}{H_T} = 0.162 = F_{DT}$$

The high value of F_{ST} indicates the presence of substantial genetic differentiation among the subpopulations in Figure 3.16, and the large value of F_{DT} compared with F_{SR} and F_{RD} indicates that most of the genetic differentiation is due to differences among districts within the total population and not to differences at lower levels in the population hierarchy.

The F-statistics are related among themselves in a straightforward manner, as a little manipulation of Equations 3.30–3.33 leads to:

$$1 - F_{ST} = (1 - F_{SR})(1 - F_{RD})(1 - F_{DT}) \tag{3.34}$$

Mating Between Relatives in a Structured Population

In regard to population subdivision, we've so far assumed that mating within each subpopulation is random. The reduced average heterozygosity among subpopulations results from the differences in allele frequency that increase with time owing to random genetic drift in finite populations.

What happens when there is close inbreeding within a subpopulation of the sort discussed earlier, when you can calculate the inbreeding coefficient from the pedigree? For example, what is the inbreeding coefficient of the offspring of a first-cousin mating within one of the subpopulations in a hierarchically structured population? The inbreeding coefficient is calculated as discussed in Section 3.2, and for the offspring of first cousins, $F = 1/16$. This is assigned a special symbol to distinguish it from the other F-statistics, namely F_{IS}. In words, F_{IS} is the probability of IBD of two alleles in an individual resulting from nonrandom mating within a subpopulation. F_{IS} is the only one of the F-statistics that can be negative, and only then if for some reason the correlation between uniting gametes is negative.

In terms of reduction in heterozygosity in an individual due to nonrandom mating:

$$\frac{\overline{H}_S - H_I}{\overline{H}_S} = F_{IS} \tag{3.35}$$

and F_{IS} has the same properties as the other F-statistics, most importantly:

$$1 - F_{IT} = (1 - F_{IS})(1 - F_{SD})(1 - F_{DR})(1 - F_{RT}) \tag{3.36}$$

where F_{IT} is the reduction in heterozygosity in an individual due both to nonrandom mating within a subpopulation and to all levels of population subdivision in the hierarchy.

Problems

3.1 The gel shown here depicts the banding patterns observed among genotypes for a two-allele gene in a population of *Phlox cuspidata*. Estimate the allele frequencies and test for goodness of fit to HWE. *P. cuspidata* can undergo self-fertilization. In light of these data, what would you conclude about whether or not self-fertilization actually takes place? If there is evidence of inbreeding, estimate F.

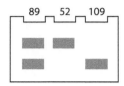

3.1 ANSWER Let p be the frequency of the allele yielding the smaller DNA fragment (the one that migrates nearest the bottom of the gel) and q that of the allele yielding the larger DNA fragment. The sample size is $89 + 52 + 109 = 250$, or 500 alleles sampled. The allele frequencies are estimated as $p = (89 + 2 \times 109)/500 = 0.614$ and $q = (89 + 2 \times 52)/500 = 0.386$. With HWE the expected numbers of the genotypes, left-to-right across the gel, are $2(0.614)(0.386) \times 250 = 94.25$, $(0.614)^2 \times 250 = 37.25$, and $(0.386)^2 \times 250 = 118.50$. The chi-square equals 15.50 with one degree of freedom, for which $P = 8.3 \times 10^{-5}$. The null hypothesis of HWE must clearly be rejected. The main reason for the poor fit is that there are too few observed heterozygous genotypes, relative to the number expected with HWE. A deficiency of heterozygous genotypes might well be expected from inbreeding, and would be consistent with some degree of self-fertilization. To estimate F, set $2pq(1 - F) \times 250 = 89$, yielding $F = 0.2490$. With this value of F, the expected numbers agree perfectly with the observed, so there is no opportunity to do a chi-square test. The reason for the perfect fit is that both degrees of freedom were consumed in estimating p and F.

3.2 An agronomic trait is affected by alleles A_1 and A_2 of a gene such that A_1A_1 and A_1A_2 genotypes have a phenotypic mean of a and A_2A_2 genotypes have a phenotypic mean of $-a$. Hence A_2 is a recessive allele that is unfavorable for the trait. Show that, if the allele frequencies remain constant, the mean phenotypic value of the trait in the population decreases linearly in F.

3.2 ANSWER Let D, H, and R represent the genotype frequencies of A_1A_1, A_1A_2, and A_2A_2, respectively. Then the mean of the trait m is given by $m = (D + H)a - Ra = a(D + H - R)$. Substituting from Equations 3.2–3.4 yields $m = a(p^2 + pqF + 2pq - 2pqF - q^2 - pqF) = a[(p - q) + 2pq(1 - F)]$, which is manifestly a linear function of F.

3.3 The following data show the average number of pups surviving to winter in first-born litters, relative to the inbreeding coefficient of the pups. The results are from observations of 24 breeding pairs in a wild population of grey wolves (*Canis lupus*) in Scandinavia during the period 1983–2002 (Liberg et al. 2005).

Average no. surviving	6.0	5.2	6.3	4.0	3.1	2.8	2.5	3.0
F	0	0.12	0.19	0.21	0.25	0.30	0.36	0.40

Find the slope and intercept of the regression line of the average number of surviving pups on the inbreeding coefficient. Plot the average number of surviving pups as a function of F along with the regression line. (Note: for the line $y = a + bx$, the slope is given by:

$b = Covariance(x, y) / Variance(x)$ and the intercept by $a = E(y) - bE(x)$.)

3.3 ANSWER Let y be the average number of surviving pups and x be the inbreeding coefficient. Then $E(x) = 0.22875$, $E(y) = 4.1125$, $Cov(x, y) = -0.164268$, $Var(x) = 0.0168696$. The slope and intercept of the regression line are -9.73748 and 6.33995, respectively. The plot is as follows:

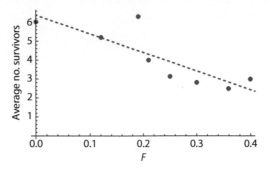

3.4 A recessive allele for albinism has a frequency of $q = 0.0025$ in the United States. What is the expected proportion of affected offspring among unrelated parents ($F = 0$)? What is it among the offspring of first cousins ($F = 1/16$)?

3.4 ANSWER For unrelated parents, $q^2 = (0.0025)^2 = 6.25 \times 10^{-6}$ or about 1 in 160,000 births. For first-cousin parents, $q^2(1 - F) + pqF = 1.62 \times 10^{-4}$ or about 1 in 6169 births. First-cousin mating increases the risk by a factor of about 26.

3.5 What is the inbreeding coefficient of individual I in the following pedigree assuming that $F_A = F_B = 0$? (Note: the "mating" that produced individual I is the self-fertilization of individual E.)

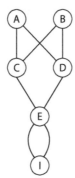

3.5 ANSWER The key to this kind of problem is to realize that all necessary information about the ancestry of individual E is included in the inbreeding coefficient of E, hence the first thing to do is to calculate the inbreeding coefficient of E. This is $F_E = (1/2)^3(1 + F_A) + (1/2)^3(1 + F_B) = 1/4$ since $F_A = F_B = 0$. Then $F_I = (1/2)(1 + F_E) = (1/2)(5/4) = 5/8$. Note that the equation for F_I is in accord with Equation 3.8 because there is only one individual (namely, E) in the loop passing through E.

3.6 In the Mohave Desert, local populations of the diminutive annual plant *Linanthus parryae* ("desert snow") are polymorphic for white versus blue flowers. Blue flowers result from homozygosity for a recessive allele. The geographical distribution of the frequency q of the recessive allele across a region of the Mohave Desert is shown in the accompanying illustration. Each allele frequency is based on an examination of approximately 4000 plants over an area of about 30 square miles (Epling and Dobzhansky 1942). The highest frequencies of the blue-flower allele are largely concentrated at the west and east ends of the region in question. Treat each of the three regions as a single random-mating unit in HWE for the flower-color alleles. Estimate the average allele frequency in each region and in the population as a whole. From these data:

a. Estimate H_S and H_T for the flower-color gene.
b. Estimate F_{ST} for the flower-color gene.

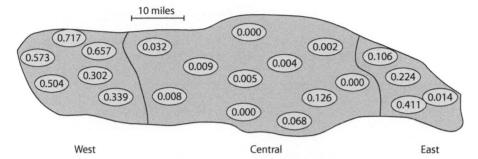

West   Central   East

3.6 ANSWER Let q_W, q_C, and q_E be the average frequency of the blue-color allele in the West, Central, and East regions, respectively, and q_T be the overall average allele frequency, weighting each population equally. Then $q_W = 3.092/6 = 0.5153$, $q_C = 0.254/11 = 0.0231$, $q_E = 0.755/4 = 0.1888$, and $q_T = 4.101/21 = 0.1953$. **a.** The average subpopulation heterozygosity is the weighted average of the regional heterozygosities, where the weight for each region equals the number of subpopulations in the region. In effect, we treat the number of individuals in each region as proportional to the number of subpopulations sampled. Therefore $H_S = [6 \times 2(0.5153)(1 - 0.5153) + 11 \times 2(0.0231)(1 - 0.0231) + 4 \times 2(0.1888)(1 - 0.1888)]/21 = 0.2247$. The total heterozygosity is the heterozygosity that would be expected were the population one large unit in HWE, or $H_T = 2(0.1953)(0.8047) = 0.3143$. **b.** $F_{ST} = (H_T - H_S)/H_T = 0.2851$, which means that the population substructure decreases the average frequency of heterozygous genotypes by almost 30 per cent as compared with a single population in HWE.

3.7 A group of subpopulations of the house mouse has a fixation index of $F_{ST} = 0.12$. A mouse within one subpopulation is the product of a full-sib mating, so that its inbreeding coefficient relative to its own subpopulation equals $F_{IS} = 0.25$. What is F_{IT}, the inbreeding coefficient of this mouse relative to the population as a whole?

3.7 ANSWER Use a simplified version of Equation 3.36 that applies to two population levels, namely $1 - F_{IT} = (1 - F_{IS})(1 - F_{ST})$ with $F_{IS} = 0.25$ and $F_{ST} = 0.12$. In this case $F_{IT} = 1 - (0.75)(0.88) = 0.34$. Note that unless $F_{ST} = 0$, F_{IT} is always greater than F_{IS}. This is because F_{IT} takes into account not only the inbreeding in the local population but also the remote inbreeding due to population structure and the finite size of the subpopulations.

References

Abbott, S. and Fairbanks, D. J. (2016), 'Experiments on Plant Hybrids by Gregor Mendel', *Genetics*, 204 (2), 407–22.

Bittles, A. H. and Black, M. L. (2010), 'Evolution in health and medicine Sackler colloquium: Consanguinity, human evolution, and complex diseases', *Proc Natl Acad Sci U S A*, 107 (Suppl 1), 1779–86.

Bustamante, C. D., et al. (2002), 'The cost of inbreeding in Arabidopsis', *Nature*, 416 (6880), 531–4.

Charlesworth, D. and Willis, J. H. (2009), 'The genetics of inbreeding depression', *Nat Rev Genet*, 10 (11), 783–96.

Crow, J. F. (1998), '90 years ago: the beginning of hybrid maize', *Genetics*, 148 (3), 923–8.

Crow, J. F. (2007), 'Haldane, Bailey, Taylor and recombinant-inbred lines', *Genetics*, 176 (2), 729–32.

Dobzhansky, T. and Spassky, B. (1963), 'Genetics of natural populations. XXXIV. Adaptive norm, genetic load, and genetic elite in Drosophila pseudoobscura', *Genetics*, 48, 1467–85.

Epling, C. and Dobzhansky, T. (1942), 'Genetics of natural populations. VI. Microgeographic races in Linanthus parryae', *Genetics*, 27, 317–32.

Kim, S., et al. (2007), 'Recombination and linkage disequilibrium in Arabidopsis thaliana', *Nat Genet*, 39 (9), 1151–5.

King, E. G., Macdonald, S. J., and Long, A. D. (2012), 'Properties and power of the Drosophila Synthetic Population Resource for the routine dissection of complex traits', *Genetics*, 191 (3), 935–49.

Liberg, O., et al. (2005), 'Severe inbreeding depression in a wild wolf (Canis lupus) population', *Biol Lett*, 1 (1), 17–20.

Malécot, G. (1948), *Les mathématiques de l'hérédité* (Paris: Masson et Cie).

Neal, N. P. (1935), 'The decrease in yielding capacity in advanced generations of hybrid corn', *J Am Soc Agron*, 27, 666–70.

O'Brien, S. J., et al. (2017), 'Conservation genetics of the cheetah: lessons learned and new opportunities', *J Hered*, 108 (6), 671–7.

Paige, K. N. (2010), 'The functional genomics of inbreeding depression: a new approach to an old problem', *BioScience*, 60, 267–77.

Robinson, J. A., et al. (2016), 'Genomic flatlining in the endangered island fox', *Curr Biol*, 26 (9), 1183–9.

Sheridan, E., et al. (2013), 'Risk factors for congenital anomaly in a multiethnic birth cohort: an analysis of the Born in Bradford study', *Lancet*, 382 (9901), 1350–9.

Shifman, S., et al. (2003), 'Linkage disequilibrium patterns of the human genome across populations', *Hum Mol Genet*, 12 (7), 771–6.

Wakeley, J. (2009), *Coalescent Theory* (Greenwood Village, CO: Roberts).

Weir, B. S. and Hill, W. G. (2002), 'Estimating F-statistics', *Annu Rev Genet*, 36, 721–50.

Wright, S. (1922), 'Coefficients of inbreeding and relationship', *Am Nat*, 56, 330–8.

CHAPTER 4

Mutation, Gene Conversion, and Migration

The models we've considered so far ignore most of the complexities of actual populations. The alleles of genes are changed by mutation and gene conversion. Allele frequencies among subpopulations are altered by migration. Random fluctuations in allele frequency can occur solely by chance because populations are not infinitely large. Natural selection underlies changes in allele frequency resulting from differences in survival or reproduction among organisms. All of these processes contribute to evolution because, in the widest sense, **evolution** can be defined as cumulative change in the genetic composition of a population. Some authors prefer a narrower definition that includes the stipulation that the genetic changes must be adaptive. One problem with a too-narrow definition is that there are many examples of genetic changes in populations, as well as genetic differences between species, whose adaptive significance is uncertain. In this chapter, we consider the basic evolutionary processes of mutation, gene conversion, and migration.

4.1 Mutation

New genetic variation is created by changes in the genetic material; hence, mutation is the ultimate source of genetic variation. The term **mutation** is used here in a widest sense to mean all genetic changes, including nucleotide substitutions, insertions and deletions, changes in the genomic location of transposable genetic elements, and chromosome rearrangements. Single nucleotide substitutions are the most common type of mutation, but their rate of occurrence differs among organisms. Among the highest rates are those in RNA viruses and retroviruses such as HIV, which are of the order of 10^{-3}–10^{-5} per nucleotide pair per round of replication. These rates are so high because the reverse transcriptase activity needed for viral replication lacks a proofreading activity that in cellular organisms detects and corrects such errors. Organisms with DNA genomes have rates of nucleotide substitution of the order of 10^{-9}–10^{-11} per nucleotide pair per round of replication. In organisms with large genomes and long generation times, the germ cells undergo a large number of divisions before gamete formation and many new mutations can occur per genome per generation. In bacteria and yeast, for example, the number of new mutations per genome per generation is about 0.003, in nematode worms and *Drosophila* the number is 0.1–1.0, and in mice and humans it is 10–30. A human diploid zygote contains an average of about 60 new mutations, with an average of 15 present in the egg and 45 in the sperm. The number present in the egg is about 15 irrespective of the mother's age, but the number present in the sperm increases from 30 in males aged

A Primer of Population Genetics and Genomics. Fourth Edition. Daniel L. Hartl, Oxford University Press (2020). © Daniel L. Hartl.
DOI: 10.1093/oso/9780198862291.003.0004

20 years to 70 at age 40 (Kong et al. 2012). The age dependence in males reflects the ongoing germ-cell divisions that take place throughout a male's lifetime.

Although precious few nucleotide-substitution mutations are beneficial, the creative role of mutation in evolution is exemplified by a study of 195 new genes that originated in *Drosophila* 3–35 million years ago. About 30 per cent of these recent genes are essential for survival, which is about the same as the proportion of ancient genes that are essential (Chen et al. 2010).

Irreversible Mutation

In this section and the next we consider irreversible and reversible mutation. The term **irreversible mutation** means that mutation occurs from wildtype to mutant with no possibility of reverse mutation; **reversible mutation** means that mutation can occur from wildtype to mutant and the other way around. These are classical models that were proposed when genes were recognized as entities that could undergo change from one allele to another, but long before genes were known to consist of stretches of DNA. Hence "mutation" in the context of these models refers to mutation anywhere in a gene, and because a gene consists of thousands of nucleotide sites (and in multicellular eukaryotes sometimes hundreds of thousands), the mutation rates are far greater than those of individual nucleotides.

The approach for analyzing changes in allele frequency due to mutation is to invoke the Hardy–Weinberg model but relax the assumption of no mutation. Consider, therefore, a gene with two alleles A and a, and suppose that the mutation rate per generation from A to a is μ. This process is called **forward mutation** if A is the prevalent wildtype allele. A mutation rate of μ per generation means that, in the transition from any one generation to the next, a fraction μ of A alleles undergoes mutation to become a alleles, whereas a fraction $1 - \mu$ of A alleles escapes mutation and remain A. Therefore, if p is the allele frequency of A in any generation, then the allele frequency in the subsequent generation will be $p' = p(1 - \mu)$. Hence:

$$p_t = p_{t-1}(1 - \mu) = p_{t-2}(1 - \mu)^2 = \cdots = p_0(1 - \mu)^t \tag{4.1}$$

where t is time in generations and p_0 is the allele frequency in the original population.

Suppose also that $p_0 \approx 1$ (i.e., the initial population is nearly fixed for A) and that t is not too large relative to $1/\mu$. Then $p_t \approx p_0 - t\mu$, and so the allele frequency q_t of the mutant a allele is given to a good approximation by:

$$q_t = q_0 + t_\mu \tag{4.2}$$

This equation implies that the frequency of the mutant a allele increases linearly with time and that the slope of the line equals μ. Because μ is small, the linear increase in q_t is difficult to detect experimentally except in very large populations of the magnitude achievable in experimental populations of bacteria. An example is shown in Figure 4.1. Note the abrupt increase in mutation rate (indicated by the increase in slope) shortly after the addition of caffeine, a bacterial mutagen.

Because spontaneous forward mutation rates are typically rather small (of the order of 10^{-4} to 10^{-6} mutations per allele per generation), the tendency for allele frequencies to change as a result of recurrent mutation (**mutation pressure**) is very small over the course of a few generations. On the other hand, the cumulative effects of mutation over long periods of time can become appreciable, as shown in Figure 4.2 where the solid curve has been calculated from Equation 4.1 for $\mu = 10^{-5}$. Because A undergoes recurrent

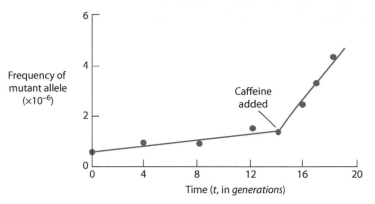

Figure 4.1 Linear increase in the frequency of mutants resistant to bacteriophage T5 in a culture of *Escherichia coli* due to recurrent mutation. The mutation rate is estimated as the slope, in this case 7.2×10^{-8} per generation initially, and 6.6×10^{-7} per generation after the addition of caffeine. (Based on data in Novick (1955).)

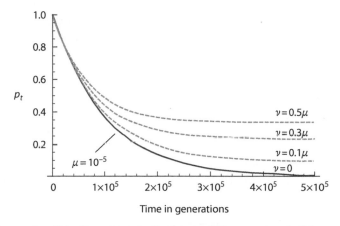

Figure 4.2 Changes in allele frequency under irreversible mutation (solid curve) or reversible mutation (dashed curves).

mutation to *a*, but *a* never reverse mutates to *A*, the allele frequency of *A* eventually goes to zero, but very slowly. For realistic values of μ, it requires $t = 0.693/\mu$ generations to decrease the value of *p* by half, which in Figure 4.2 means 69,300 generations.

Reversible Mutation

Something different happens when mutation is reversible. In this case the population eventually reaches a stable equilibrium at which both *A* and *a* are present at frequencies that remain constant through time. Let us examine the same model as in the previous section, but allow reverse mutation from *a* to *A* at the rate *v* per generation. In any generation, an *A* allele could have only two origins relative to the previous generation. It could have been an *A* allele in the previous generation that escaped mutation to *a*, or it could have been an *a* allele in the previous generation that underwent reverse mutation to *A*. These possibilities are captured by the equation:

$$p_t = p_{t-1}(1-\mu) + (1-p_{t-1})v \tag{4.3}$$

A little algebraic manipulation puts this in the form:

$$p_t - \frac{v}{\mu + v} = \left(p_{t-1} - \frac{v}{\mu + v}\right)(1 - \mu - v) \tag{4.4}$$

which, by repeated substitution as in Equation 4.1, leads to:

$$p_t - \frac{v}{\mu + v} = \left(p_0 - \frac{v}{\mu + v}\right)(1 - \mu - v)^t \tag{4.5}$$

What happens to the allele frequencies in the long run is shown by the dashed lines in Figure 4.2. The frequency of A no longer goes to zero but to an equilibrium value given by:

$$\hat{p} = \frac{v}{\mu + v} \tag{4.6}$$

In the dashed curves, $v = 0.1\mu$, 0.3μ, and 0.5μ, corresponding to the equilibrium frequencies $\hat{p} = 1/11, 3/13,$ and $5/15$, respectively. The equilibrium is called a **stable equilibrium** because the allele frequencies converge to the equilibrium given in Equation 4.6 irrespective of the starting frequencies. Once again, for realistic values of the mutation rates, it takes a long time to reach the vicinity of the equilibrium.

Gene Duplication and Functional Divergence

In this section, we adopt a broader view of mutation to include large-scale changes in the genome that result in the duplication (or deletion) of one or more existing genes. Duplications and deletions are relatively common. On average, a eukaryotic gene is duplicated at the rate of at least 1 per cent per million years, which is of the same order of magnitude as the rate of nucleotide substitution at synonymous nucleotide sites (Lynch and Conery 2003). Across the genomes of diverse prokaryotes and eukaryotes, the percentage of genes that originated by duplication is typically in the range 20–40 per cent (Zhang 2003). In the long run, the rate of deletion must roughly match the rate of duplication unless the genome is increasing or decreasing in size. Many duplications or deletions are doomed from the start and persist for at most a few generations because they are deleterious for survival or reproduction. For this reason, most copy-number variations (CNVs) in the human genome are restricted to a single individual or a small group of individuals. But some gene duplications are beneficial and persist for many generations. Among many examples, we can single out a polymorphism for the number of salivary amylase genes in human populations (Santos et al. 2012). Salivary amylase aids in the digestion of starch, and a greater copy-number of the gene is found in populations with diets historically rich in high-starch foods such as rice and corn. A gene duplication also contributes to drug resistance in the malaria parasite. Pyrimethamine is an antimalarial drug that inhibits an enzyme essential for parasite survival. A duplication of the gene located upstream in the metabolic pathway of the essential enzyme allows more substrate to flow through the pathway, thereby reducing drug efficacy (Nair et al. 2008).

Some gene duplications persist long enough that their duplicates diverge in function, either through changes in amino acid sequence of the encoded protein or through changes in the expression of the gene at different times or in different cell types. Some mechanisms by which gene duplications may diverge in function and show long-term persistence are discussed later in this section. First, we consider one mathematical model of gene duplication and deletion. This is one of many ways to model the process (Lynch and

Conery 2003; Rogers et al. 2009; Zhao et al. 2015), and while less realistic than some alternatives, our approach has the virtue of simplicity.

Suppose that gene duplication takes place at a constant rate b per generation and that deletion occurs at a rate d per generation. You may think of b as the birth rate of new genes and d as the deletion rate of existing genes, where both b and d are very much smaller than 1. A difference equation for the number of gene copies n_t at time t can be written as:

$$n_t = b + n_{t-1}(1-d) \tag{4.7}$$

Equation 4.7 assumes that, whereas the input of new copies is constant, the loss of existing copies is proportional to copy number. This is an assumption of convenience justified in part by the notion that deletion of excess copies may have a fitness advantage over the addition of new copies.

Equation 4.7 implies that:

$$n_t = \frac{b}{d} - \frac{(b-d)(1-d)^t}{d} \tag{4.8}$$

When $t = 0$, the number of gene copies equals 1; and as t goes to infinity, n_t goes to b/d, which is the expected maximum copy number maintained at equilibrium. The time required for the number of gene copies to increase from 1 to half of the maximum equals $-\log[2]/\log[1-d]$, which equals approximately $0.69/d$ because $d \ll 1$.

Some examples are shown in Figure 4.3, where the parameters have been chosen to illustrate different rates of increase in copy number and maxima of copy number. The timescale has been chosen to roughly match that of realistic gene duplication-deletion processes. The black curves have maxima at 5 or 10 copies and half-maxima at 100 million years, and the gray curves have maxima at 15 or 20 copies and half-maxima at 50 million years.

Sometimes extra copies of a gene alone may increase fitness, as in the example of salivary amylase in which the extra copies increase the energy extracted from starchy foods. On an evolutionary timescale, mutations may result in some copies of a gene acquiring new functions. At equilibrium in the model expressed in Equation 4.8, the number of copies of a gene of age t is proportional to $(1-d)^t$, and because d is so small, some copies are able to accumulate mutations that change their time of expression, their catalytic capabilities, or both.

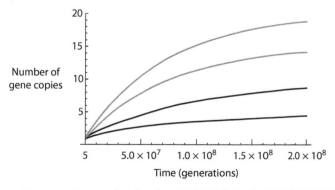

Figure 4.3 Increase in gene copy number for the model in Equation 4.7 for differing rates of birth (b) and deletion (d). The values of (b, d) for the curves, from top to bottom, are $(27.6 \times 10^{-8}, 1.38 \times 10^{-8})$, $(20.7 \times 10^{-8}, 1.38 \times 10^{-8})$, $(10 \times 10^{-8}, 1 \times 10^{-8})$, and $(5 \times 10^{-8}, 1 \times 10^{-8})$.

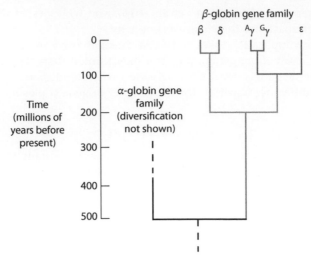

Figure 4.4 Evolution of the alpha and beta hemoglobin gene families by means of gene duplication and functional divergence.

A classic example of duplication and functional divergence is found in the hemoglobin genes, part of whose ancestral history is shown in Figure 4.4. Adult hemoglobin is a tetramer composed of two alpha (α) chains and two beta (β) chains; however, alternative versions of both the α and β chains are found in the embryo and fetus. The α and β genes originated from a common ancestral gene that duplicated about 500 million years ago (Figure 4.4), and the gene in both branches underwent further duplication and functional divergence to form the α-globin gene family (not shown in Figure 4.4) and the β-globin gene family. (A **gene family** is a set of genes whose similarity in DNA in sequence implies that they arose by duplication and divergence.) In the β-globin branch of the gene tree, another duplication took place about 200 million years ago that gave rise to the adult β-globin called β and δ as well as the embryonic globin ε and the fetal globins $^A\gamma$ and $^G\gamma$ (so called because of an amino acid difference that distinguishes them). The $\beta - \delta$ duplication occurred about 40 million years ago and the $^A\gamma$–$^G\gamma$ duplication about 30 million years ago. The gene tree in Figure 4.4 pertains to the globins in humans; other vertebrates have different numbers of copies and times of expression during development.

What fate awaits a duplicate gene? It may be deleted, as assumed in the model in Equation 4.7, or it may undergo mutations that abolish its function, such as promoter mutations that prevent its expression or internal deletions or multiple stop codons in the reading frame that knock out its activity. Mutational inactivation is known as **nonfunctionalization**, and large genomes including the human genome contain large numbers of **pseudogenes** resulting from nonfunctionalization.

Figure 4.5 illustrates two alternatives to nonfunctionalization. The diagram at the top depicts a gene prior to duplication, and the stars in the promoter region indicate sequences that bind different transcription factors resulting in the gene being expressed early as well as late in development. In the process of subfunctionalization, one duplicate copy has a mutation in the "early" binding site and the other a mutation in the "late" binding site. In other words, **subfunctionalization** means that each copy of the duplicate gene retains only a subset of the functionalities of the original. Both copies are essential; however, their coding sequences are now free to diverge to specialize their function to either early

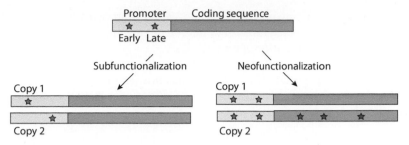

Figure 4.5 Subfunctionalization and neofunctionalization are two mechanisms for functional divergence of gene copies.

or late expression. The probability of gene preservation by subfunctionalization has been analyzed by Lynch and Force (2000).

In **neofunctionalization** (Figure 4.5), one copy of the duplication acquires mutations that effectuate a new function, such as promoter mutations that cause the gene to be expressed in different tissues or times in development, or coding sequence mutations that alter the biochemical properties of the protein. In the evolution of the β-globin gene family in Figure 4.4, the gene copies have undergone neofunctionalization in both senses: they are expressed at different times in development and they have different affinities for oxygen appropriate to their time of expression.

Equilibrium Heterozygosity with Mutation

So far, we've examined mutation of individual genes and mutation resulting in CNV. In this section we examine one model of mutation at the molecular level. We start from where we left off in Chapter 3 when we examined F_t, the probability of identity by descent (IBD) in generation t due to random genetic drift in a finite population of N diploid individuals that produces a theoretically infinite pool of gametes of which $2N$ are chosen at random to produce each successive generation. The assumptions of a constant population size of N diploid organisms and random sampling of $2N$ gametes to form the genotypes of the next generation constitutes the **Wright–Fisher model** of random genetic drift.

In the Wright–Fisher model, a fraction $1/(2N)$ of the randomly chosen alleles in any generation are replicas of literally the same allele in generation $t-1$, hence their probability of identity by decent equals 1; and a fraction $1 - 1/(2N)$ are replicas of different alleles in generation $t-1$, and their probability of identity by decent equals F_{t-1}. These two possibilities taken together result in the equation:

$$F_t = \frac{1}{2N} + \left(1 - \frac{1}{2N}\right)F_{t-1} \tag{4.9}$$

which is Equation 3.17 in Chapter 3. Equation 4.9 assumes no mutation, and because in each generation F_t is augmented by the amount $1/(2N)$, it follows that F_t eventually goes to 1.

When mutation is allowed, then in each generation mutation creates some number of new alleles. Under what is known as the **infinite-alleles model** (or, more accurately, the infinite *number* of alleles model), each mutation creates an allele that is unique to the population (Kimura and Crow 1964). The model was originally devised to calculate the expected number of alleles maintained in a finite population at steady state when the stochastic loss of preexisting alleles is balanced by the input of new alleles arising from mutation. The model may have reached its high-water mark with Ewens' sampling formula for calculating the expected number of neutral alleles and their expected frequencies in a sample from a large population (Ewens 1972).

In the infinite-alleles model, any new mutant allele is regarded as being not identical by descent (IBD) with the parental allele. Taking this consideration into account, Equation 4.9 is valid only if neither of the chosen alleles underwent mutation in the previous generation, and therefore we should write:

$$F_t = \left[\frac{1}{2N} + \left(1 - \frac{1}{2N} \right) F_{t-1} \right] (1 - \mu)^2 \tag{4.10}$$

In this case, F_t does not go to 1 but to an equilibrium value $\hat{F}$ found by solving Equation 4.10 with $F_t = F_{t-1} = \hat{F}$. Ignoring terms in μ^2 because $\mu^2 << \mu$, the result sought is, to an excellent approximation:

$$\hat{F} = \frac{1}{1 + 4N\mu} \tag{4.11}$$

Furthermore, because each new mutation creates a unique allele, $\hat{F}$ is also the equilibrium frequency of homozygous genotypes. Hence the equilibrium frequency of heterozygous genotypes, $\hat{H}$, equals $1 - \hat{F}$, or:

$$\hat{H} = 1 - \hat{F} = \frac{4N\mu}{1 + 4N\mu} = \frac{\theta}{1 + \theta} \tag{4.12}$$

where $\theta = 4N\mu$ is the nucleotide nutation rate scaled by population size, a parameter that arises in many contexts in population genetics theory.

As one application of Equation 4.12, consider **allozyme polymorphisms**, which are nonsynonymous mutations in which the amino acid replacement results in a change in the ionic charge of a protein and therefore alters its electrophoretic mobility. In *Drosophila*, allozyme polymorphisms have an average heterozygosity of about 0.14 (Nevo 1978), yielding an estimate of $\langle \theta \rangle = 0.163$; however, the 95 per cent confidence interval around this estimate is very large. In most higher organisms, allozyme heterozygosities are in the range of 6–19 per cent, yielding $\langle \theta \rangle$ estimates in the range 0.06–0.23 (Lewontin 1974). Taking $\langle \theta \rangle = 0.163$ and $N = 10^6$ for *Drosophila* (Akashi 1997) results in an estimate of $\langle \mu \rangle = 4 \times 10^{-8}$ for allozyme alleles. This estimate is about two orders of magnitude smaller than the rate estimated directly from the appearance of new allozyme alleles (Voelker et al. 1980), and it suggests that the great majority of electrophoretic enzyme variants must be harmful and eliminated so rapidly by selection that they do not persist for long as allozyme polymorphisms. This issue has important implications for the maintenance of polymorphism in natural populations, which is discussed at greater length in later chapters.

4.2 The Coalescent

With the advent of high-throughput DNA sequencing, a model of nucleotide mutation that regards a gene as a sequence of nucleotide sites rose to prominence. In this model, known as the **infinite-sites model** (or, better, the infinite *number* of sites model), a gene is made up of a very large number of nucleotide sites, each with a mutation rate so small that any nucleotide substitution occurs at a different site. These stipulations imply that every new mutation is a forward mutation at a different nucleotide site, and reverse mutation at a site cannot occur. One implication of this model is that, if two alleles share a mutation at a particular nucleotide site, they must derive from a common ancestral allele. The infinite-sites model is therefore well suited for applications of coalescence theory, which is the subject of this section.

The infinite-sites model was originally devised to calculate the expected frequency of heterozygous nucleotide sites in a finite population, and it assumed free recombination between nucleotide sites (Kimura 1969). A later incarnation allowing no recombination between sites was used to estimate a number of important parameters such as the scaled nucleotide mutation rate $\theta = 4N\mu$ (Watterson 1975). When recombination between the nucleotide sites is disallowed, the infinite-sites model is equivalent in many respects to the infinite-alleles model. The assumption of no recombination can be and often is relaxed in the analysis of sequence data (Rasmussen et al. 2014; Wakeley 2009).

Coalescence in the Wright–Fisher model

The elegant concept of coalescence was already introduced in Chapter 3 in connection with remote inbreeding in finite populations. In tracing the ancestral histories alleles backward through time, pairs of lineages occasionally merge into a common ancestral allele as illustrated in Figure 3.12 in Chapter 3. A merger of two allele ancestries is a **coalescence**, and coalescence theory deals with the probabilities and timing of such coalescent events (Hudson 1991; Kingman 1982, 2000; Wakeley 2009). The relative timing of coalescent events is illustrated in Figure 4.6 for a sample of $n = 6$ alleles whose sequence identities are masked by the Greek symbols. As one proceeds backward in time, the lineages come together and undergo coalescence with the alleles' most recent common ancestor. The coalescence times are indicated by the dashed lines. The symbol T_i represents the random coalescence time at which the lineages of two of i alleles in the population coalesce into $i-1$ lineages. Ultimately, all of the gene genealogies coalesce into one common ancestor of all of the alleles in the sample of n of the current population's alleles.

The total time back to the common ancestor is given by the sum of the coalescence times T_i. The analysis of coalescence times has become an important part of theoretical population genetics because powerful inferences can be made. Here we provide an introduction to the theory as it pertains to the Wright–Fisher model.

Consider i alleles present at some time in a population of constant size N. What is the probability that a coalescence took place in the immediately preceding generation? This

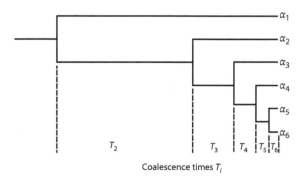

Coalescence times T_i

Figure 4.6 Gene genealogy of six alleles in a sample at the present time showing the coalescence times T_i of the allele lineages as they are traced backward in time.

must equal 1 minus the probability that there was no coalescence in the immediately preceding generation, a probability that is quite straightforward to calculate. Equation 4.9 implies that the probability that two alleles in the present generation have distinct ancestors in the previous generation is $1 - 1/(2N) = (2N - 1)/(2N)$. It follows that the probability of a third allele having an ancestor distinct from the first two is $(2N - 2)/(2N) = 1 - 2/(2N)$, because once the distinct ancestors of the first two alleles are chosen, there are only $2N - 2$ distinct ancestors left to choose from. Similarly, the probability that a fourth allele has an ancestor distinct from the first three is $(2N - 3)/(2N) = 1 - 3/(2N)$, and so forth. Since these events are independent, the overall probability that i distinct alleles present in any generation have i distinct ancestors in the previous generation is:

$$\left(1 - \frac{1}{2N}\right)\left(1 - \frac{2}{2N}\right)\cdots\left(1 - \frac{i-1}{2N}\right) \approx 1 - \frac{1}{2N}(1 + 2 + \cdots + i - 1) = 1 - \frac{i(i-1)}{4N} \quad (4.13)$$

where the approximation assumes that $1/N^2$ is small compared with $1/N$. The simplification on the far right comes from the fact that the sum of the first $i - 1$ integers equals $i(i-1)/2$.

Since the right-hand side of Equation 4.13 is the probability of the *absence* of a coalescence, the probability of the *presence* of a coalescence, C, is equal to:

$$C = 1 - \left(1 - \frac{i(i-1)}{4N}\right) = \frac{i(i-1)}{4N} \quad (4.14)$$

Therefore, for i alleles, the probability of no coalescence for the first $t - 1$ generations followed by coalescence in the tth generation is $(1 - C)^{t-1}C$, and the mean of this geometric distribution of coalescence times is $\bar{T}_i$ where:

$$\bar{T}_i = \sum_{t=1}^{\infty} t(1 - C)^t C = \frac{1}{C} = \frac{4N}{i(i-1)} \quad (4.15)$$

One implication of Equation 4.15 is that the average coalescence times become longer toward the base of the genealogy of the alleles, as indicated by the length of the horizontal lines in Figure 4.6. For $n = 6$, as in Figure 4.6, the average time for the first coalescence is $4N/30$ generations, and the average time for the final coalescence is $4N/2$ generations.

The expected total length of all the branches in the genealogy in a sample of size n is equal to:

$$E[T_{total}] = \sum_{i=2}^{n} i\bar{T}_i = \sum_{i=2}^{n} i\frac{4N}{i(i-1)} = 4N\left(1 + \frac{1}{2} + \frac{1}{3} + \cdots + \frac{1}{n-1}\right) \quad (4.16)$$

$E[T_{total}]$ is different from the expected time from the present time back to the **most recent common ancestor (MRCA)** of the alleles in a sample. For a sample of size n, the expected time back to the MRCA equals:

$$E[T_{MRCA}] = \sum_{i=2}^{n} \bar{T}_i = \sum_{i=2}^{n} \frac{4N}{i(i-1)} = 4N\sum_{i=2}^{n}\left(\frac{1}{i-1} - \frac{1}{i}\right)$$

$$= 4N\left(1 - \frac{1}{2} + \frac{1}{2} - \frac{1}{3} + \cdots - \frac{1}{n-1} + \frac{1}{n-1} - \frac{1}{n}\right) \quad (4.17)$$

$$= 4N\left(1 - \frac{1}{n}\right)$$

Expressions for the variance of $E[T_{Total}]$ and $E[T_{MCRA}]$ can be found in Wakeley (2009). Figure 4.7 shows the expected values of T_{Total} (blue) and T_{MCRA} (red) as a function of

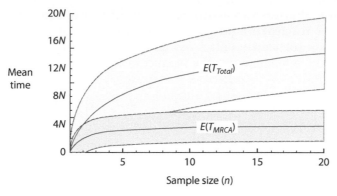

Figure 4.7 Values of the expected total branch length (blue) and the expected time to the most recent common ancestor (red) in a sample of size n. The shaded areas demarcate the means ± one standard deviation.

sample size, where time on the y-axis is scaled in units of N. The solid lines are the means and the dashed lines are the means ± their standard deviations. The main point is that while the mean T_{Total} is larger than T_{MCRA} and continues to increase slowly with sample size, T_{MCRA} rather rapidly approaches its asymptotic value of $4N$.

An important assumption of coalescence theory is that the lineages must coalesce in pairs. The model does not allow for simultaneous coalescence of three or more lineages. On the other hand, in the Wright–Fisher model simultaneous coalescence of three or more lineages can occur, because a parental allele in any generation can leave multiple descendant alleles in the next generation. The probability of such multiple coalescences decreases as the population size increases, hence to apply coalescence theory we must assume that the population size N is large, in practice $N \geq 1000$ (Wakeley 2009).

We now turn to important implications of Equation 4.16 that serve to emphasize the power of coalescence theory in the analysis of sequence data.

Nucleotide Polymorphism

In the infinite-sites mutation model, each new mutation in the branches of the gene tree results in a distinct allele in the sample. This is a plausible assumption for DNA sequences, where a mutation at a particular nucleotide site yields a single-nucleotide polymorphism in the sample. If the mutations occur uniformly in time, then in a sample of n alleles with L aligned nucleotide sites from each, the expected number of segregating sites, $E(S)$, must equal the mutation rate across the L sites, μ, times the total length of all the branches in the tree ($E[T_{total}]$ in Equation 4.16). The expected number of segregating sites, $E(S)$, is therefore given by:

$$E(S) = \mu \sum_{i=2}^{n} i\bar{T}_i = 4N\mu \left(1 + \frac{1}{2} + \frac{1}{3} + \cdots + \frac{1}{n-1}\right) = a_n\theta \tag{4.18}$$

where $\theta = 4N\mu$ and $a_n = 1 + \frac{1}{2} + \frac{1}{3} + \cdots + \frac{1}{n-1}$.

Note that μ is the probability of mutation in any of the L nucleotide sites per generation, hence μ is the per-nucleotide mutation rate multiplied by L.

Equation 4.18 implies that an estimate of θ can be formulated as:

$$\theta_W = S/a_n \tag{4.19}$$

which is known as **Watterson's estimator** (Watterson 1975) and corresponds to what is often called the **nucleotide polymorphism**. One implicit assumption is that the nucleotide sites are completely linked (no recombination). Watterson (1975) has also shown that, with no recombination, the variance of θ_W can be estimated as:

$$Var\left(\frac{S}{a_n}\right) = Var(\theta_W) = \frac{\theta_W}{a_n} + \frac{a_2\theta_W^2}{a_n^2} \tag{4.20}$$

where a_2 is given by $a_2 = 1 + \frac{1}{2^2} + \frac{1}{3^2} + \cdots + \frac{1}{(n-1)^2}$.

Nucleotide Diversity

The sequences of alleles can also be compared in pairs to derive an expression for the average number of pairwise differences among the L aligned nucleotide sites, typically denoted as π, as a function of θ. Assuming complete linkage, Tajima (1983) showed that the expected number of pairwise differences per nucleotide site, $E(\pi)$, satisfies:

$$E(\pi) = \theta \tag{4.21}$$

An alternative estimator of θ is therefore:

$$\theta_\pi = \pi \tag{4.22}$$

which in this context is often called the **nucleotide diversity**. The variance of θ_π can be estimated as:

$$Var(\pi) = Var(\theta_\pi) = b_n\theta_\pi + b_2\theta_\pi^2 \tag{4.23}$$

(Tajima 1983), where $b_n = \frac{n+1}{3(n-1)}$ and $b_2 = \frac{2(n^2+n+3)}{9n(n-1)}$.

In principle, if all mutations are selectively neutral and the population has been of constant size long enough to reach mutation-drift equilibrium, then the estimate θ_W based on number of polymorphic sites should equal the estimate θ_π based on the average number of pairwise differences among sites, within the limits of sampling error. A test for a significant difference between θ_W and θ_π is the basis of one type of "test of neutrality," where the quotes serve as a caveat that in reality the comparison simultaneously tests all of the simplifying assumptions related to selective neutrality, constant population size, and mutation-drift equilibrium. Various other "tests of neutrality" comparing estimated of θ based on different attributes of sequence data are also in use. These are discussed in more detail in Chapter 7.

Estimating θ and π from Sequence Data

Figure 4.8 summarizes data that can be used to illustrate application of the key concepts of nucleotide polymorphism and nucleotide diversity. The data in comprise 500 bp (base pairs) of the coding sequence from five naturally occurring alleles of the *Rh3* (rhodopsin 3) gene of *Drosophila simulans*, extracted from a much larger dataset (Ayala et al. 1993). Several types of nucleotide sites may be distinguished.

- A **segregating site** is one that is polymorphic in the sample. In this case only the $S = 16$ segregating sites are listed. They are numbered consecutively, but in fact they are scattered throughout the whole sequence, separated by distances of 2–104 bp, with an average spacing of 22 bp. The sample also contains 484 sites that are monomorphic, each of which is a nonsegregating site.

		Polymorphic site number (also 484 monomorphic sites)															
		1	2	3	4	5	6	7	8	9	10	11	12	13	14	15	16
	α_1	T	C	T	A	C	C	T	C	C	T	C	G	G	T	T	A
	α_2	T	C	C	T	A	C	C	T	C	C	T	G	G	T	T	T
Allele	α_3	C	T	C	C	C	C	C	T	C	T	T	T	G	C	T	A
	α_4	C	T	C	C	C	C	C	T	T	C	T	G	A	C	T	T
	α_5	C	T	C	C	C	T	C	T	T	T	T	G	G	C	C	A
Pairwise mismatches		6	6	4	7	4	4	4	4	6	6	4	4	4	6	4	6
Sample configurations		(3,2)	(3,2)	(4,1)	(3,1,1)	(4,1)	(4,1)	(4,1)	(4,1)	(3,2)	(3,2)	(4,1)	(4,1)	(4,1)	(3,2)	(4,1)	(3,2)

Figure 4.8 DNA polymorphisms among alleles a_1–a_5 of the *Rh3* (rhodopsin 3) gene of *Drosophila simulans*. Each sequence consists of 500 bp of coding sequence. Only polymorphic sites are shown. (Data from Ayala et al. (1993).)

- A **pairwise difference** between any two sequences is a nucleotide site at which the sequences differ. Comparing the sequences in all possible pairs and averaging the number of differences yields the proportion or pairwise differences in the sample. The *Rh3* example comprises five sequences that can be paired in ten ways. More generally, among n sequences there are $n(n-1)/2$ possible pairwise comparisons. The number of pairwise differences at each polymorphic site is listed across the bottom of the table. For example, site 1 has 2 T's and 3 C's, and hence $2 \times 3 = 6$ mismatches in pairwise comparisons. The proportion of pairwise mismatches, denoted π, is obtained by summing the mismatches across the sample and dividing by the total number of pairwise comparisons. In this case $\pi = (4 \times 9 + 6 \times 6 + 7 \times 1)/10 = 7.9$.
- The **sample configuration** or **site frequency spectrum** of a site is the set of numbers giving, in decreasing order, the count of each different kind of element present at a particular site in a sample. Site 1 in the *Rh3* data has the configuration (3, 2, 0, 0), but the 0s are normally omitted and this is written as (3, 2). The symbol (3, 2) means that the sample site includes 3 with the majority nucleotide (in this case C) and 2 with the minority nucleotide (in this case T). Site 2 also has the sample configuration (3, 2), although in this case the identities of the major and minor nucleotides are reversed. This means that the sample configurations are indifferent to the identity of the nucleotides at a site, but depend only on the relative numbers. When ties occur both numbers are listed. For example, site 4 has the configuration (3, 1, 1), where each 1 represents a **singleton** occurring only once at the site. In this case the singletons happen to be A and T, but the sample configuration would be (3, 1, 1) regardless. Note that, strictly speaking, site 4 violates the assumption of the infinite-sites model that each nucleotide in a sequence can mutate no more than once, but in the real world such events occasionally occur. The 484 monomorphic sites all have the sample configuration (5), and this would normally be written (5, 0) to emphasize that the sites are monomorphic.

Applying Equations 4.19 and 4.20 to the *Rh3* data in Figure 4.8, we have $n = 5$, $S = 16$, $a_n = 2.0833$, and $a_2 = 1.4236$. Hence $\theta_W = 7.68$ and $\text{Var}(\theta_W) = 23.03$, which yields a standard error equal to 4.80.

Likewise, applying Equations 4.22 and 4.23 to the pairwise differences in the *Rh3* data, we have $\theta_\pi = 7.9$, $b_n = 0.5000$, and $b_2 = 0.3667$, hence $\theta_\pi = 7.9$ and $\text{Var}(\theta_\pi) = 26.8337$, which yields a standard error equal to 5.18.

The *Rh3* data shows very good agreement between the estimate of θ based on nucleotide polymorphism S, from which $\theta_W = 7.68 \pm 4.80$, or based on pairwise mismatches, from which $\theta_\pi = 7.9 \pm 5.18$. But such good agreement is expected only when the assumptions of the model are satisfied. These include the stipulation that the sample is taken from a population that is constant in size and that has attained a steady state between mutation and random genetic drift. Another important assumption is that the polymorphisms are selectively neutral. To seek evidence for natural selection, statistical tests of several key properties of coalescent gene trees have been devised (Chapter 7).

As noted earlier, the values of θ_W and θ_π in Equations 4.19 and 4.22 are the estimates of the true θ across the entire set of L nucleotides. It is straightforward to convert these estimates into values per nucleotide site, which facilitates comparison among data sets for which L is likely to differ. For convenience, let's denote the estimates of θ per nucleotide site as $\hat{\theta}_W$ and $\hat{\theta}_\pi$. Then $\hat{\theta}_W = \theta_W/L$ and $\hat{\theta}_\pi = \theta_\pi/L$, where θ_W and θ_π are as given in Equations 4.19 and 4.22. Likewise the per-site variance estimates are $Var(\hat{\theta}_W) = Var(\theta_W)/L^2$ and $Var(\hat{\theta}_\pi) = Var(\theta_\pi)/L^2$, where $Var(\theta_W)$ and $Var(\theta_\pi)$ are as given in Equations 4.20 and 4.23. For the *Rh3* data in Figure 4.8, $L = 500$ and hence $\hat{\theta}_W = 0.01536$ and $\hat{\theta}_\pi = 0.01580$. The corresponding variances are $Var(\theta_W) = 9.213 \times 10^{-5}$ and $Var(\theta_\pi) = 1.073 \times 10^{-4}$.

The Moran Model

Coalescence theory assumes that the lineages must coalesce in pairs, and that the Wright–Fisher model approximates this ideal only when the population size is large. In an alternative model the lineages are constrained to coalesce in pairs. This is known as the **Moran model** (Moran 1958, 1962), and we introduce it in this chapter mainly to show how it differs from the Wright–Fisher model.

The Moran model assumes a haploid population of constant size N that reproduces continuously in time. At each instant, two individuals are chosen. Sampling is with replacement, and therefore the same individual may be chosen twice. The first individual chosen reproduces and the second dies. If the individuals are the same, nothing changes. If the first individual carries allele A and the second allele a, the number of A alleles increases by 1; and if the first individual carries a and the second A, the number of A alleles decreases by 1.

In terms of allele frequency, if p is the frequency of A and $1 - p$ that of a, the sampling process in the Moran model implies that, in the next time step:

$$p \to p + \frac{1}{N} \text{with probability } p(1-p)$$

$$p \to p \text{ with probability } p^2 + (1-p)^2 \tag{4.24}$$

$$p \to p - \frac{1}{N} \text{with probability } (1-p)p$$

Theoreticians often prefer the Moran model because some results that can be obtained in the Wright–Fisher model only as approximations may be derived exactly in the Moran model. There are, however, some important differences (Wakeley 2009). The first difference is in how time is measured. The natural measure of time in the Wright–Fisher model is one generation because, at the end of each generation, the entire population reproduces and then dies. In the Moran model, reproduction occurs in instantaneous increments (δt). One measure of time that makes sense is to consider one generation in the Moran as consisting of N such increments because in $N\delta t$ steps N individuals will have been replaced.

The second difference is the distribution of offspring number. In the Wright–Fisher model for large N, the number of offspring per generation has a Poisson distribution with mean and variance equal to 1. In the Moran model for large N, the number of offspring per generation (N time steps) has a geometric distribution with mean 1 and a variance of 2. The difference in the variance accelerates the rate of random genetic drift in the Moran model by a factor of 2. To see this for yourself, consider that in a haploid Wright–Fisher model, the probability of randomly choosing two different alleles is given by the following expression:

$$H_t = 2p(1-p)(1-F_t) = 2p(1-p)\left(1-\frac{1}{N}\right)(1-F_{t-1}) = H_{t-1}\left(1-\frac{1}{N}\right) \tag{4.25}$$

where the relation between $1-F_t$ and $1-F_{t-1}$ is by analogy with Equation 4.9. H_t is called the **heterozygosity** and its decrease is one measure of the rate of random genetic drift in a finite population. (In a haploid model the H_t is *virtual heterozygosity* because there are no heterozygous genotypes.)

Starting with an allele frequency of A equal to p in the Moran model, after one time step:

$$H_{t+\delta t} = 2\left(p+\frac{1}{N}\right)\left(1-p-\frac{1}{N}\right)\left[p(1-p)\right] + 2p(1-p)\left[p^2+(1-p)^2\right]$$
$$+ 2\left(p-\frac{1}{N}\right)\left(1-p+\frac{1}{N}\right)\left[(1-p)p\right]$$

where the factors in square brackets are the probabilities of the transitions in Equation 4.24. This expression simplifies to:

$$H_{t+\delta t} = 2p(1-p)\left(1-\frac{2}{N^2}\right) = H_t\left(1-\frac{2}{N^2}\right)$$

It follows that:

$$H_{t+2\delta t} = H_{t+\delta t}\left(1-\frac{2}{N^2}\right) = H_t\left(1-\frac{2}{N^2}\right)^2$$
$$H_{t+3\delta t} = H_{t+2\delta t}\left(1-\frac{2}{N^2}\right) = H_t\left(1-\frac{2}{N^2}\right)^3$$

and so forth. Because there are N time steps per generation in the Moran model, we can write:

$$H_{t+1} = H_{t+N\delta t} = H_t\left(1-\frac{2}{N^2}\right)^N \approx H_t\left(1-\frac{2N}{N^2}\right) = H_t\left(1-\frac{2}{N}\right) \tag{4.26}$$

where the approximation ignores terms of order $1/N^3$ and smaller. Comparing Equation 4.26 with Equation 4.25 verifies that heterozygosity in the Moran model decreases twice as fast as in the Wright–Fisher model.

As already noted, the increased rate of random genetic drift in the two models is due to their differing distributions of offspring number—a Poisson distribution in the Wright–Fisher model and a geometric distribution in the Moran model. With a Poisson distribution of offspring number, the probability that an individual leaves exactly k offspring is given by $\Pr\{k \text{ offspring}\} = \lambda^k e^{-\lambda}/k!$, where λ is the mean and equals 1 in the Wright–Fisher model. The probability of leaving exactly k offspring after one generation in the Moran model is somewhat more complex. Moran (1962) gives a generating function

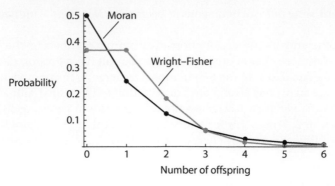

Figure 4.9 Distribution of offspring numbers in the Wright–Fisher model (gray) and in the Moran model (black). In both cases the mean equals 1. A population size of $N = 1000$ is assumed in the Moran model.

from which the probabilities may be deduced. For a mean offspring number of 1, these probabilities are:

$$\Pr\{0 \text{ offspring}\} = \frac{N-1}{2N-1} \quad \Pr\{k \text{ offspring}\} = \frac{N^2(N-1)^{k-1}}{(2N-1)^{k+1}} \text{ for } k = 1,2,3,\ldots$$

In Figure 4.9 these distribution are compared for the case $N = 1000$. About half of the individuals in the Moran model produce no offspring versus 37 per cent in the Wright–Fisher model. This difference helps explain the more rapid decrease in heterozygosity in the Moran model. The Poisson distribution associated with the Wright–Fisher model is also intuitively appealing, the modified geometric distribution less so. But as Moran has noted, "the model sacrifices some empirical applicability in order to obtain a random process easier to deal with analytically" (Moran 1962). On the other hand, despite the intuitive appeal of the Wright–Fisher model, it applies literally to very few organisms.

Effective Population Number

To this point in dealing with random genetic drift and the coalescent we have assumed that the populations being considered are ideal in the sense that the population number remains constant from one generation to the next and that each individual has the same distribution of offspring number as any other. Populations in the real world are rarely ideal in these respects. One approach for understanding how deviations from constant population size or other assumptions may affect some specified quantity is to deduce an **effective population number**, symbolized N_e, that an ideal population would need to have for the specified quantity to match that in an actual population. The effective population number is usually smaller, and sometimes much smaller, than the actual population number (i.e., the census number or head count). In this section we consider some examples. It is worth emphasizing at the outset that there are several types of effective population number that differ according to the specified quantity that is of interest. We'll begin by considering a population that violates the assumption of constant population size.

CHANGING POPULATION SIZE

Imagine a subpopulation that is ideal in all respects except that its number changes from generation to generation, and compare this with an ideal population of constant size N. In an ideal Wright–Fisher population of diploid organisms, the heterozygosity is given by:

$$H_t = 2pq\,(1 - F_t) = 2pq\,(1 - F_{t-1})\left(1 - \frac{1}{2N}\right) = H_{t-1}\left(1 - \frac{1}{2N}\right) \tag{4.27}$$

where p and q are the allele frequencies of A and a and the relation between $1 - F_t$ and $1 - F_{t-1}$ follows from Equation 4.9. Equation 4.27 implies that:

$$H_t = H_0\left(1 - \frac{1}{2N}\right)^t \approx H_0\left(1 - \frac{t}{2N}\right) \tag{4.28}$$

where the approximation ignores terms of $1/N^2$ or smaller.

Suppose now that an actual population has a population size of N_0 in generation 0, N_1 in generation 1, N_2 in generation 2, and so forth. Aside from the changes in population number, the population is ideal in all other respects. Equation 4.27 implies that:

$$H_t = H_{t-1}\left(1 - \frac{1}{2N_{t-1}}\right) = H_{t-2}\left(1 - \frac{1}{2N_{t-1}}\right)\left(1 - \frac{1}{2N_{t-2}}\right)$$

$$= \cdots = H_0\left(1 - \frac{1}{2N_{t-1}}\right)\left(1 - \frac{1}{2N_{t-2}}\right)\cdots\left(1 - \frac{1}{2N_0}\right) \tag{4.29}$$

$$\approx H_0\left(1 - \sum_{i=0}^{t-1}\frac{1}{2N_i}\right)$$

where the approximation ignores terms of magnitude $1/(N_i N_j)$ and smaller. Replacing N in Equation 4.28 with N_e and equating H_t in Equations 4.28 and 4.29 leads to:

$$N_e = \frac{1}{\left(\frac{1}{t}\right)\sum_{i=0}^{t-1}\frac{1}{N_i}} \tag{4.30}$$

Equation 4.30 says that the effective population number of a population that varies in population size is calculated as the reciprocal of the average of the reciprocals. This is a special sort of average called the **harmonic mean**, which tends to be dominated by the smallest terms. Suppose, for example, that $N_0 = 1000$, $N_1 = 10$, and $N_2 = 1000$ in a population that underwent a severe temporary reduction in population size (a **population bottleneck**) in generation 1. Then $1/N_e = (1/3)(1/1000 + 1/10 + 1/1000) = 0.034$. The average effective number over the three-generation period is only $N_e = 29.4$, whereas the average actual number is $(1/3)(1000 + 10 + 1000) = 670$. A severe population bottleneck often occurs in nature when a small group of emigrants from an established subpopulation founds a new subpopulation. The random genetic drift accompanying such a founder event is known as a **founder effect**.

The effective population number in Equation 4.30 is the **inbreeding effective size** (or *inbreeding effective population number*) because the quantity of interest is the decrease in heterozygosity, which corresponds to an increase in identity-by-descent of alleles (inbreeding) in the population (Equation 3.18). In most equations in population genetics in which the symbol N_e appears, it means the inbreeding effective size. Calculation of N_e in several special cases is described in the following sections.

UNEQUAL SEX RATIO

An inequality in the sex ratio creates a peculiar sort of "bottleneck" because half of the alleles in any generation must come from each sex no matter how few individuals of the minority sex there are. We are again interested in the inbreeding effective number. Note that in Equation 4.9 that the first term, $1/(2N)$, is the probability that two randomly chosen alleles in generation t are replicas of the same allele in generation $t-1$. In a population with N_m males, 1/4 of the randomly chosen pairs of alleles in generation t come from a male in generation $t-1$, $1/N_m$ is the probability that they are from the same male, and 1/2 is the probability that they are replicas of the same allele in the male; altogether the probability that a randomly chosen pair of alleles in generation t are replicas of the same allele in a male in generation $t-1$ is $1/(8N_m)$. Likewise, the probability that the alleles are replicas of the same allele in a female in generation $t-1$ is $1/(8N_f)$. Since these are mutually exclusive events, the probability corresponding to $1/(2N)$ is $1/(8N_m) + 1/(8N_f)$. Replacing N with N_e and setting these probabilities equal to each other leads to:

$$N_e = \frac{4N_mN_f}{N_m + N_f} \tag{4.31}$$

In light of Equation 4.30, it is unsurprising that N_e equals two times the harmonic mean of N_m and N_f.

To take an example from wildlife management, suppose that deer hunting is permitted to a level at which the number of surviving males is one-tenth of the number of surviving females. Then $N_m = 0.1 \times N_f$ and Equation 4.31 implies that N_e is reduced to only one-third of the census number: $N_e = 0.33 (N_m + N_f)$.

UNIFORM POPULATION DISPERSION

For a population spread out uniformly in two dimensions, the inbreeding effective number depends on (1) the number of breeding individuals per unit area, usually denoted by the symbol δ, and (2) the amount of dispersion between an individual's own birthplace and that of its offspring. If dispersion follows a normal, bell-shaped curve in both dimensions with standard deviation σ, then 39 per cent of all individuals have their offspring within a circle of radius σ centered at their own birthplace, 87 per cent have their offspring within a circle of radius 2σ, and 99 per cent have their offspring within a circle of radius 3σ. In terms of δ and σ, the inbreeding effective number of the population (often called **neighborhood size** in this context) is given by:

$$N_e = 4\pi\delta\sigma^2 \tag{4.32}$$

(Wright 1946), where π is the conventional $\pi = 3.14159$.

Equation 4.32 can be applied to data on the abundant prairie deer mouse *Peromyscus maniculatus*. In a large area in Southern Michigan, Dice and Howard (1951) estimated the density of breeding individuals δ to be in the range of 5–7 per hectare (1 hectare equals 10,000 square meters or about 2.5 acres). By following the movement of marked animals from birth to breeding sites, they estimated σ as 114 meters, yielding $\sigma^2 = 1.3$ hectares. With these parameters, Equation 4.32 yields an inbreeding effective size in the range 81.7–114.3, which is perhaps surprisingly small for such an abundant animal.

SUBDIVIDED POPULATIONS

Many natural populations are subdivided into local interbreeding groups called **demes** among which individual organisms can migrate. To be concrete, imagine a population consisting of D demes, each containing N diploid organisms. Migration among demes is measured by a quantity m, which equals the probability that a randomly chosen allele in any deme originates from one of the remaining $D-1$ demes in the previous generation. In effect, each deme is equally likely to contribute migrants to any other deme.

In this situation, random genetic drift takes place within demes and among demes, and their timescales are different. Within demes, random drift occurs relatively rapidly and heterozygosity decreases at the rate $1/(2N)$ per generation as expressed in Equation 4.27. Among demes, random drift is limited by the amount of migration and occurs relatively slowly. If the number of demes is relatively large, then heterozygosity decreases in the population as a whole at the rate $1/(2N_e)$ per generation, where:

$$N_e = ND \left(1 + \frac{1}{4Nm}\right) \qquad (4.33)$$

(Wakeley 1999, 2000). In Equation 4.33, the factor ND comes from the within-deme phase of the random genetic drift, and the factor $1 + 1/(4Nm)$ comes from the among-deme phase. This is one of the unusual cases in which the effective size of a population is always larger than the census size. When Nm is large, then N_e is not much larger than ND—for example, if $Nm = 0.1$, then $N_e = 3.5 \times ND$; but when Nm is small, then N_e can be very much larger than ND—for example, if $Nm = 0.001$, then $N_e = 251 \times ND$. The main implication is that when many demes are connected by low rates of migration, alleles spread among demes very slowly, and it may take a long time for two alleles in different demes to coalesce into a common ancestral allele in the same deme in the remote past.

VARIANCE EFFECTIVE NUMBER

As we've seen in the case of a population subdivided into demes, random genetic drift can have different timescales depending on the context. One of the shortest timescales relates to the variance in allele frequency from generation to generation. In a Wright–Fisher model consisting of N diploid individuals, at the end of each generation the individuals produce an enormous pool of gametes, among which $2N$ are chosen in pairs to produce the N genotypes in the next generation. This sampling process is binomial, and if the frequency of an allele A is p in the parental generation, then among a large number of replicate samples, the variance in p among in offspring generation is given by $p(1-p)/(2N)$. This variance requires that the population be ideal in the sense that the distribution of offspring number for large N is Poisson with mean 1. Most natural populations fall short of this ideal in that the actual variance in allele frequency is larger than that expected in an ideal population.

Suppose an actual population has a variance in allele frequency from one generation to the next of σ^2. Then, writing $p(1-p)/(2N_e)$ as the expected variance with effective size N_e and setting this equal to σ^2 results in:

$$N_e = \frac{p(1-p)}{2\sigma^2} \qquad (4.34)$$

This N_e is called the **variance effective size**, and it may or may not match the inbreeding effective size. The variance effective size usually changes on a much faster timescale than the inbreeding effective size, and it is often much smaller than the inbreeding effective size. For example, samples of the malaria parasite over a number of years from a clinic in Senegal yielded an average variance effective size of smaller than 100 (Chang et al. 2012), whereas the inbreeding effective size estimated from the same samples averaged greater than 10^6 (Daniels et al. 2013). The discrepancy implies that recent intervention to eliminate malaria in this region has been effective in reducing the population number, as reflected in the variance effective size; however, the reduction has been too recent to greatly affect the inbreeding effective size.

COALESCENT EFFECTIVE NUMBER

We're already seen how the variance in the distribution of offspring number affects the rate of random genetic drift. In particular, the decrease in heterozygosity in a haploid Wright–Fisher model (Equation 4.25) is slower than that in a haploid Moran model (Equation 4.25) by a factor of 2, which corresponds to the difference in the variance in offspring number per generation in the two models. A more rapid decrease in heterozygosity implies shorter coalescence times, and hence a shorter time to the MRCA in sample of size n alleles. We can therefore generalize Equation 4.17 to read:

$$E[T_{MRCA}] = 4N_e \left(1 - \frac{1}{n}\right) \tag{4.35}$$

in which $N_e = N/\sigma^2$ where in this case σ^2 represents the variance in offspring number per generation. For large N in the Wright–Fisher model, the offspring number per generation has a Poisson distribution with mean 1 and variance $\sigma^2 = 1$, and therefore $N_e = N$. For the Moran model, σ^2 per generation equals 2, and hence $N_e = N/2$. A natural population with a mean offspring number of 2 per mating pair and a variance in offspring number of σ^2 has an effective number of $N_e = 4N/(2 + \sigma^2)$ (Crow and Kimura 1970).

The effective population number in Equation 4.35 is distinct from both the inbreeding effective number and the variance effective number. Because it depends on the expected coalescence time, it is known as the **coalescent effective size**. The coalescent effective size need not be equal to either the inbreeding effective size or the variance effective size. In some cases a coalescence effective size does not exist—in particular, when population size undergoes large fluctuations with about the same rate as coalescences, or when a population is subdivided into multiple demes and migration occurs at about the same rate as coalescences (Sjödin et al. 2005; Wakeley and Sargsyan 2009).

4.3 Gene Conversion

The repair of double-stranded breaks in DNA is one of the key mechanisms by which genomes maintain their integrity through time. In meiosis, repair of double-stranded breaks can result in **crossing-over**, in which alleles in homologous chromosomes far removed from the site of repair undergo **recombination**, an apparently precise breakage-and-reunion event. But at the molecule level the breakage-and-reunion event is not as precise as first appears. In the repair of the double-stranded break, single-stranded regions from homologous DNA molecules come together to produce a **heteroduplex molecule** with strands differing in parental origin. This heteroduplex will contain mismatches at every site at which the parental molecules differ in sequence. A DNA-repair system called **mismatch repair** excises one of the mismatched bases and replaces it with the correctly paired partner base (A with T, or G with C). The mismatch repair ultimately results in one of the parental strands undergoing a change from one nucleotide pair to another, which is a kind of mutation. Changes in nucleotide sequence mediated by mismatch repair are known as **gene conversion**.

Biased Gene Conversion

Gene conversion is illustrated in Fig 4.10, where the highlighted G–T mismatch is a potential site of gene conversion. An estimated 90 per cent of gene conversion events occur in the absence of recombination (indicated in Figure 4.10a by the same blue color

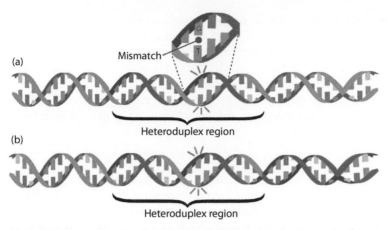

Figure 4.10 Heteroduplex regions formed during double-strand break repair. Note the G–T mismatch in the magnified portion. The heteroduplexes are shown in (a) a nonrecombinant molecule and (b) a recombinant molecule.

to the left and right of the heteroduplex) and only 10 per cent with recombination (indicated by the flanking blue and red duplexes in Figure 4.10b) (Jeffreys and May 2004). The heteroduplex tracts in noncrossover gene conversion (hundreds of basepairs) tend to be shorter than those in crossover gene conversion (thousands of basepairs), nevertheless across the human genome the number of gene-conversion events per gamete per generation is about 20 (Williams et al. 2015), which is comparable to the number of new mutations (Kong et al. 2012). This comparison makes it clear that gene conversion is a mechanism whose effects on genetic variation should not be ignored.

If gene conversion were indifferent to the identity of the mismatched bases, and the mismatches were repaired randomly, then the process might not matter much. But in some organisms, especially in warm-blooded vertebrates, mismatch repair is by no means random. In particular, when a heteroduplex contains a mismatched G–T (as in Figure 4.10) or G–A, the mismatch tends to be resolved in favor of G–C, and when the mismatch is C–T or C–A, it tends to be resolved in favor of C–G. Mismatches therefore tend to be repaired in favor of GC, which constitutes **biased gene conversion**.

A Model of Biased Gene Conversion

A–T and T–A base pairs are often called *weak* because they have two hydrogen bonds, whereas G–C and C–G are called *strong* because they have three hydrogen bonds. Spontaneous nucleotide-substitution mutations in many organisms are biased in favor of weak base pairs; that is, mutation is AT-biased. With a consistent AT bias in mutation, regions of non-protein-coding DNA in genomes would progressively become more AT rich through time, unless some other process counteracts the increase in weak base pairs. (The GC content of protein-coding regions is constrained by the nature of the genetic code.) Some organisms do indeed have highly AT-rich genomes. For example, noncoding DNA in the genome of the malaria parasite *Plasmodium falciparum* is approximately 90 per cent AT (Hamilton et al. 2017).

One process that can offset the AT bias of mutation is a GC bias of gene conversion. Eventually equilibrium is reached when gain of AT base pairs by mutation is balanced by loss of AT base pairs through biased gene conversion. To make this argument more quantitative, suppose that μ is the mutation rate of strong to weak base pairs per generation and that $a\mu$ is the mutation rate of weak to strong base pairs. Then a is a measure of mutation bias, and $a < 1$ means that mutation is biased toward AT. In the human genome, a equals approximately 1/2 (Kong et al. 2012). Suppose further that gene conversion occurs at a rate of $c\mu$ per generation, where c is a measure of gene-conversion bias. When $c = 0$ gene conversion is unbiased, and when $c > 0$ gene conversion is GC-biased. These stipulations imply that the change in the proportion of GC base pairs in one generation is given by:

$$g' = g\left(1 - \mu\right) + (1 - g)\,a\mu + 2g\left(1 - g\right)c\mu \qquad (4.36)$$

where g is the proportion of G–C base pairs in the DNA. The last term corresponds to gene conversion, and it includes the factor $2g\left(1 - g\right)$ because sites subject to gene conversion must be polymorphic in the population for weak versus strong base pairs. Setting $g' = g$ results in a quadratic equation for the equilibrium GC content $\hat{g}$, however the relevant root is

$$\hat{g} = \frac{-\left(1 + a - 2c\right) + \sqrt{\left(1 + a - 2c\right)^2 + 8ac}}{4c} \qquad (4.37)$$

Figure 4.11 is a plot of equilibrium GC content when $a = 1/2$ with c ranging from 0.05 to 1.05. Across this range of GC-biased gene conversion, the equilibrium GC content ranges from 35 per cent to 65 per cent.

The model gains another wrinkle if you assume that the bias in gene conversion can vary across the genome, especially if the bias is correlated with the proportion of GC base pairs in the region surrounding the mismatch. In this case, random variations in GC content gradually become amplified by biased gene conversion, which ultimately leads to inhomogeneities in GC content across regions larger than would be expected by chance.

Such inhomogeneities in base composition are well known in the genomes of humans and other warm-blooded vertebrates (Bernardi 2000; Costantini et al. 2006; Eyre-Walker and Hurst 2001). Approximately 60 per cent of the human genome consists of stretches of DNA (domains) greater than 10 kb that are compositionally homogeneous in GC content but differ in GC content from neighboring domains (Elhaik et al. 2010). About half of these are longer than 100 kb, and one-third are longer than 300 kb. The domains longer

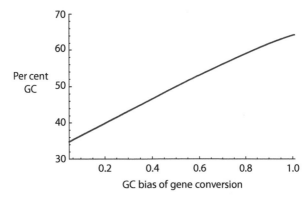

Figure 4.11 Equilibrium GC content of a long DNA molecule when AT-biased mutation is offset by GC-biased gene conversion. In this example $a = 1/2$, which implies a twofold excess of strong-to-weak mutations.

than 300 kb are termed **isochores** (Bernardi 2000; Eyre-Walker and Hurst 2001). Among all the domains, the per cent GC shows approximately the same range of values as shown in Figure 4.11 (Costantini et al. 2006; Elhaik et al. 2010), which implies that the GC bias in gene conversion may differ substantially from one domain to the next across the human genome.

4.4 Migration

The term **migration** refers to the movement of individuals among subpopulations. It is a sort of genetic glue that holds subpopulations together and sets a limit to how much genetic divergence can occur. To understand the homogenizing effects of migration, it is useful to study migration in simple models of population structure.

Models of Migration

The individuals in many populations are distributed homogeneously in space, but much more often populations are subdivided into subpopulations or demes within which most individuals find their mates but among which migration can occur. Figure 4.12 illustrates the most common models of population structure. Part (a) depicts **one-way migration**, in which a large founder population provides migrants for one or more small satellite populations, but migration in the reverse direction does not take place. Part (b) illustrates the **island model**, a population structure in which any subpopulation is equally likely to send or receive migrants from any other subpopulation. Parts (c) and (d) each a depict **stepping-stone model**, in which migration is restricted to adjacent subpopulations spread out along one direction like a seashore or riverbank or distributed in two dimensions like patches of favorable habitat on a plain.

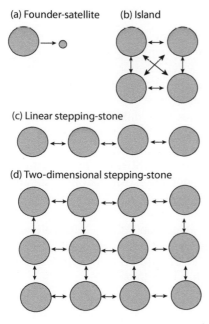

Figure 4.12 Models of migration.

One-Way Migration

To model the effects of one-way migration (Figure 4.12a), let p_t be the frequency of an allele in a satellite population of interest and p_f be the allele frequency in the founder population and source of migrants. Then:

$$p_t = p_{t-1}(1-m) + p_f m \tag{4.38}$$

where m is the probability that an allele in the satellite population comes from a migrant in the previous generation. Equation 4.38 is similar in form to Equation 4.3 for reversible mutation, and the solution in terms of the initial allele frequency in the satellite population, p_0, is

$$p_t - p_f = (p_{t-1} - p_f)(1-m) = (p_{t-2} - p_f)(1-m)^2 = \cdots = (p_0 - p_f)(1-m)^t$$

or, equivalently:

$$p_t = p_f + (p_0 - p_f)(1-m)^t \tag{4.39}$$

Not surprisingly, the allele frequency in the satellite population becomes progressively more similar to that in the founder population. On the other hand, when m is small, the convergence can take a long time. For example, when $m = 0.01$ it requires 69 generations to reduce the difference in allele frequency by half. In human populations, 69 generations is more than 1700 years.

Because migration is only one way or, more realistically, when it is disproportionally from founder to satellite populations, mutations that arise in a satellite population are unlikely to make it back to the founder population. One might therefore expect genetic variation to increase from central to peripheral populations. This is often the case, although the peak of diversity is not necessarily at the very edge of the species range but just behind it (Kark et al. 2008).

The Island Model of Migration

In the island model of migration (Figure 4.12b), a large population is split into many subpopulations dispersed geographically like islands in an archipelago. Examples of an island population structure might include fish in freshwater lakes or snails in dispersed garden plots. Each subpopulation is assumed to be so large that random changes in allele frequency can be neglected.

Consider two alleles A and a with allele frequencies that differ among a set of subpopulations. Migration is assumed to occur in such a way that the allele frequencies among migrants equal the average values among subpopulations, designated $\bar{p}$ and $\bar{q}$. The amount of migration is again symbolized by the parameter m, which in this case equals the probability that a randomly chosen allele in any subpopulation comes from a migrant in the previous generation. For any randomly chosen allele in any subpopulation in generation t, the allele could have come from the same subpopulation in generation $t-1$ (with probability $1-m$), in which case it is an A allele with probability p_{t-1}. Alternatively, the allele could have come from a migrant in generation $t-1$ (with probability m), in which case it is an A allele with probability $\bar{p}$. Since all assumptions of the HWE are in force except that of no migration, $\bar{p}$ stays the same in all generations.

Putting this this all together:

$$p_t = p_{t-1}(1-m) + \bar{p}m \tag{4.40}$$

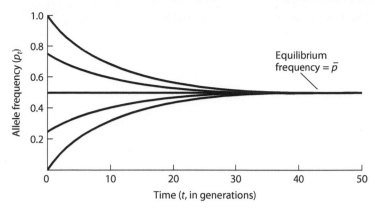

Figure 4.13 Change of allele frequency in each of five subpopulations in the island model of migration.

Equation 4.40 is identical to Equation 4.38 with $\bar{p}$ replacing p_f, and therefore:

$$p_t = \bar{p} + (p_0 - \bar{p})(1-m)^t \tag{4.41}$$

where p_0 is the initial frequency of A in the subpopulation of interest. Suppose, for example, that there were only two subpopulations, with initial allele frequencies of A of 0.2 and 0.8 and with $m = 0.10$. This value of m means that 10 per cent of the alleles in either subpopulation in any generation are from migrants in which the allele frequency of A equals $\bar{p} = (0.2+0.8)/2 = 0.5$. Suppose we wish to deduce the allele frequency of A in the two populations after ten generations. For the population with initial allele frequency 0.2, we substitute $p_0 = 0.2$, $\bar{p} = 0.5$, and $m = 0.10$ into Equation 4.39 to obtain $p_{10} = 0.5 + (0.2-0.5)(1-0.10)^{10}$, or $p_{10} = 0.395$. The allele frequency in the other population is obtained similarly as $p_{10} = 0.5 + (0.8-0.5)(1-0.10)^{10} = 0.605$.

A graphical example using Equation 4.39 is shown in Figure 4.13, where again $m = 0.10$, but this time with five subpopulations having initial frequencies 1, 0.75, 0.50, 0.25, and 0. Note how rapidly the allele frequencies converge to an equilibrium value equal to the average among the initial subpopulations, in this case 0.5.

The island model of migration has the property that the effective population size is always greater than the actual population size, and very much greater if m is small (see Equation 4.33).

How Migration Limits Genetic Divergence

In this section we return to the concept of identity by descent (IBD) to examine the offsetting effects of migration and random change in allele frequency due to finite population size. As in Chapter 3, we use F_t to denote the fixation index F_{ST} in generation t. F_t is the probability that any two randomly chosen alleles from the same subpopulation are IBD. Among a set of subpopulations that do not exchange migrants, F_t gradually increases in each subpopulation. This means that the allele frequencies change among the subpopulations—in other words, the subpopulations undergo random genetic divergence. In the absence of migration, the genetic divergence continues until ultimately each subpopulation becomes fixed for either the A allele or the a allele due to the accumulated effects of remote inbreeding.

It is remarkable how little migration is required to prevent significant genetic divergence due to remote inbreeding within the subpopulations. The effect can be seen quantitatively by considering the effects of migration at a rate m according to the island model. Equation 4.9 is still valid for the increase in F_t in one generation; however, it holds only so long as neither of the randomly chosen alleles comes from a migrant. Taking migration into account therefore yields:

$$F_t = \left[\frac{1}{2N} + \left(1 - \frac{1}{2N} \right) F_{t-1} \right] (1-m)^2 \tag{4.42}$$

Equation 4.42 is identical to Equation 4.10 except that the mutation parameter μ is replaced with the migration parameter m. It follows that, for reasonably small values of m, the equilibrium fixation index is approximately:

$$\hat{F} = \hat{F}_{ST} = \frac{1}{1 + 4Nm} \tag{4.43}$$

Not surprisingly, this looks exactly like Equation 4.11 except that Nm replaces $N\mu$. This emphasizes the theoretical similarity between the effects of mutation and migration. The practical difference is that rates of migration between subpopulations are typically very much greater than rates of mutation, so the implications of Equations 4.11 and 4.43 are very different.

Equation 4.43 implies that $\hat{F}_{ST}$ decreases as the number of migrants increases, but the decrease is extremely rapid, as shown in Figure 4.14. The dashed horizontal lines demarcate regions advised by Wright (1978) for qualitative interpretation of the fixation index in terms of degree of genetic divergence among the subpopulations. A value of Nm as small as 2 already brings $\hat{F}_{ST}$ into the zone of "moderate divergence." Because N is the size of each subpopulation, and m is the probability that an individual in a subpopulation is a migrant, Nm corresponds to the absolute number of migrants into each subpopulation in each generation. This means that, independent of subpopulation size, two or more migrants per generation severely restricts genetic divergence.

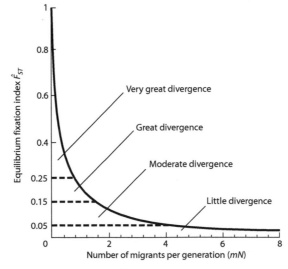

Figure 4.14 The equilibrium fixation index ($\hat{F}_{ST}$) plotted against the number of migrants per generation, assuming the island model of migration.

Table 4.1 Estimates of Nm and $\hat{F}_{ST}$

Species	Type of organism	Estimated Nm	Estimated $\hat{F}_{ST}$
Stephanomeria exigua	Annual plant	1.4	0.152
Mytilus edulis	Mollusc	42.0	0.006
Drosophila willistoni	Insect	9.9	0.025
Drosophila pseudoobscura	Insect	1.0	0.200
Chanos chanos	Fish	4.2	0.056
Hyla regilla	Frog	1.4	0.152
Plethodon ouachitae	Salamander	2.1	0.106
Plethodon cinereus	Salamander	0.22	0.532
Plethodon dorsalis	Salamander	0.10	0.714
Batrachoseps pacifica ssp. 1	Salamander	0.64	0.281
Batrachoseps pacifica ssp. 2	Salamander	0.20	0.556
Batrachoseps campi	Salamander	0.16	0.610
Lacerta melisellensis	Lizard	1.9	0.116
Peromyscus californicus	Mouse	2.2	0.102
Peromyscus polionotus	Mouse	0.31	0.446
Thomomys bottae	Gopher	0.86	0.225

Source: Data from Slatkin 1985.

How large is Nm in natural populations? One method of estimating genetic migration relies on the finding that, in theoretical models, the logarithm of Nm decreases approximately as a linear function of the average frequency of "private" alleles that are unique to individual samples from the subpopulations (Slatkin 1985). Estimates based on this method and the resulting equilibrium values of F_{ST} are summarized in Table 4.1. There is obviously considerable variation among organisms, but many of the values of Nm are approximately 2 or smaller, which gives considerable scope for genetic divergence resulting from random genetic drift.

The Fixation Index F_{ST} in Relation to Coalescence

Equation 4.43 applies when the number of demes in the island model is large. In this section we derive an expression for F_{ST} when the number of demes is specified, and along the way we'll also see how F_{ST} relates to coalescence times. The development here largely follows that of Slatkin (1991); for further results see Cox and Durrett (2002), Duforet-Frebourg and Slatkin (2016), and Wakeley (2009).

In Chapter 3 we defined the fixation index as the ratio:

$$F_{ST} = \frac{H_T - \bar{H}_S}{H_T} \tag{4.44}$$

where $\bar{H}_S$ is the average heterozygosity among subpopulations (demes) in a subdivided population and H_T is the heterozygosity in a hypothetical total population formed by pooling all of the demes into one large random-mating unit. On the other hand, at each level of population structure, $1 - H$ is proportional to the probability of IBD, and therefore we can write:

$$F_{ST} = \frac{(1 - \Pr\{IBD_T\}) - (1 - \Pr\{IBD_S\})}{1 - \Pr\{IBD_T\}} = \frac{\Pr\{IBD_S\} - \Pr\{IBD_T\}}{1 - \Pr\{IBD_T\}} \tag{4.45}$$

where $\Pr\{IBD_S\}$ is the average probability of IBD of two alleles drawn from the same subpopulation and $\Pr\{IBD_T\}$ is the probability of IBD of two alleles drawn from the pooled total population.

In tracing the ancestry of two alleles backward through time, two alleles that are IBD remain IBD only if, in each generation, neither allele has undergone mutation. The probability that neither allele mutates in one generation equals $(1-\mu)^2 \approx 1-2\mu$ where μ is the mutation rate per generation and μ^2 is negligible. The probability that neither allele mutates all the way back to their coalescence equals $\Pr\{IBD\} = (1-2\mu)^t \approx 1-2\mu t$ where t is the coalescence time. Substituting this into Equation 4.45 leads to:

$$F_{ST} = \frac{\bar{t}_s - \bar{t}}{\bar{t}} \tag{4.46}$$

where $\bar{t}_s$ is the average time to coalescence of two alleles drawn from the same deme and $\bar{t}$ is the average time to coalescence of two alleles drawn at random from the total population.

Suppose a subdivided population consisting of D demes, each of size N, where migration occurs according to the island model with migration rate m. The average coalescence time in the total population can be written as:

$$\bar{t} = \left(\frac{1}{D}\right)\bar{t}_s + \left(\frac{D-1}{D}\right)\bar{t}_d \tag{4.47}$$

where $\bar{t}_d$ is the average time to coalescence of two alleles drawn from different demes. The reasoning behind Equation 4.47 is that, when two alleles are drawn at random from the total population, the probability is $1/D$ that they were present in the same deme and $(D-1)/D$ that they were present in different demes.

The next important observation is that the average coalescence time for two alleles in the same deme is:

$$\bar{t}_s = 2ND \tag{4.48}$$

This relation comes from Equation 4.35 with N replaced by ND (the total population size) and n replaced by 2 (because we are sampling only two alleles).

For alleles drawn from different demes, the average coalescence time is:

$$\bar{t}_d = \frac{D-1}{2m} + 2ND \tag{4.49}$$

in which the first term is the average waiting time for two alleles in different demes to trace their ancestry back to the same deme, and the second term is the average time to coalescence once the alleles are traced back to the same deme.

Substituting from Equations 4.48 and 4.49 into Equation 4.47 yields the average coalescence time in the total population:

$$\bar{t} = 2ND + \frac{(D-1)^2}{2mD} \tag{4.50}$$

and this, together with $\bar{t}_s$ from Equation 4.48 being substituted into Equation 4.46, gets us the result we were driving at, namely:

$$F_{ST} = \frac{1}{1 + 4Nm[D/(D-1)]^2} \tag{4.51}$$

When D is large, $[D/(D-1)]^2 \approx 1$ and $F_{ST} = 1/(1+4Nm)$, which reproduces Equation 4.43 using coalescence theory.

Stepping-Stone Models

The discussion in the preceding section provides a good foundation for understanding the consequences of stepping-stone models, in which each deme can send or receive migrants only from adjacent demes. To keep the migration parameter commensurate with that in the island model, the migration rate between demes must be adjusted so that the net migration in and out of any deme in each generation equals m. In the linear stepping-stone model (Figure 4.12c), this means that the migration rate between adjacent demes equals $m/2$. This stipulation creates a problem at the ends of the array of demes, because the demes at the end exchange migrants with only one neighbor, not two, and therefore have effectively a different migration rate than the other demes (Maruyama 1970). A convenient solution is to imagine that the demes are arranged in a ring, so that each deme has two neighbors irrespective of its position.

Suppose a linear stepping stone model with D demes, each of size N, arranged in a ring, and for convenience suppose D is an even number. Using Equation 4.46 we can write $F_{ST}(i)$ for alleles sampled from demes i steps apart as:

$$F_{ST}(i) = \frac{\bar{t}_i - \bar{t}_s}{\bar{t}_i + \bar{t}_s} \quad \text{for} \quad i = 1, 2, \ldots, D/2 \tag{4.52}$$

because $\bar{t} = \bar{t}_i + \bar{t}_s$ when the comparison is between only two demes.

The value of $\bar{t}_s = 2ND$ again (Equation 4.48), and the analog of Equation 4.49 is:

$$\bar{t}_i = \frac{(D-i)i}{2m} + 2ND \tag{4.53}$$

in which the first term is the average time in generations for the ancestry of alleles in two demes i steps apart to trace back to a single deme and the second part is the average coalescence time once the alleles are in the same deme.

Substituting $\bar{t}_s = 2ND$ and the right-hand side of Equation 4.53 into Equation 4.52 yields the analog of Equation 4.51, namely:

$$F_{ST}(i) = \frac{1}{1 + 8NmD/[(D-i)i]} \tag{4.54}$$

(Slatkin 1991). Furthermore, when D is large and i is much smaller than D, then:

$$F_{ST}(i) = \frac{1}{1 + 8Nm/i} \tag{4.55}$$

Figure 4.15 shows $F_{ST}(i)$ as a function of distance (i) for various values of Nm. The steady-state value of F_{ST} in the island model for each value of Nm is the point at which the curve intersects the dashed line, which corresponds to $i = 2$. The take-home message from Figure 4.15 is that population structure can result in steady-state values of F_{ST} that are substantially greater than those predicted by the island model. The increased values of F_{ST} due to population structure are often referred to as **isolation by distance**.

In the analysis of real data, it is convenient to note that isolation by distance in a one-dimensional stepping-stone model is expected to result in the linear relation:

$$\frac{F_{ST}(i)}{1 - F_{ST}(i)} = \frac{i}{8Nm} \tag{4.56}$$

which has intercept 0 and slope $1/(8Nm)$.

An application of Equation 4.56 shown in Figure 4.16. The data pertain to populations of the blue manakin bird (*Chiroxiphia caudata*) that are spread out along the slopes of mountains facing the Atlantic Ocean along the southeast coast of Brazil (Francisco et al.

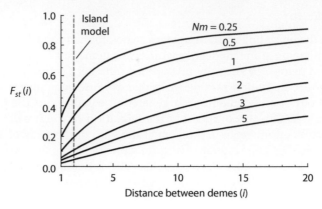

Figure 4.15 The fixation index F_{ST} plotted against distance between demes in the linear stepping-stone model for various values of Nm.

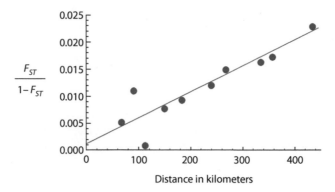

Figure 4.16 Isolation by distance in the blue manakin (*Chiroxiphia caudata*) along the Atlantic coast of Brazil. (Data from Francisco et al. (2007).)

2007). The estimated Nm for neighboring populations from these data is smaller than 2 (Francisco et al. 2007).

The analysis of two-dimensional stepping-stone models (Figure 4.12d) is complex (Charlesworth and Charlesworth 2012; Maruyama 1971; Slatkin 1991). In this case the migration rate between adjacent demes is $m/4$, but again this creates an end-deme problem that can be avoided by joining the long edge of the array to form a tube and then joining the edges of the tune to form a doughnut (technically a torus). As might be expected, F_{ST} in the two-dimensional stepping-stone model is somewhere between the island model and the one-dimensional stepping-stone model, but there seems to be no simple expression for F_{ST} (Slatkin 1991). Roughly speaking, if the original array of demes is approximately square, then the result is more like the island model, and if the original array of demes is a long, thin rectangle, then the result is similar to that in the linear stepping-stone model (Charlesworth and Charlesworth 2012).

Problems

4.1 A mutant allele of the gene for red eyes in *Drosophila mauritiana* has an insertion of the transposable element *mariner*. This transposable element undergoes spontaneous excision and loss at a frequency of approximately 1 per cent per generation, restoring the wildtype red-eye allele. In a population in which the mutant is fixed (homozygous), how many generations would be required for the expected frequency of flies homozygous for the wildtype allele to exceed 5 per cent? Assume that the population is large, that mating is random, and that deletion of the transposable element does not affect survival or reproduction.

4.1 ANSWER Excision of the transposable element is an irreversible mutation process, hence Equation 4.1 with $p_0 = 1$ for the initial frequency of the mutant and $\mu = 0.01$. The required answer is the smallest value of t for which $q_t^2 = 0.05$ because of random mating and HWE. Hence $q_t = \sqrt{(0.05)} = 0.2236$ or $p_t = 0.7764$, which means that $(1 - \mu)^t = 0.7764$ or $t = \ln(0.7764)/\ln(0.99) = 25.182$. Therefore $t = 26$ generations is the required answer. A quick check for verification shows that $q_{25}^2 = 0.0494$ whereas $q_{26}^2 = 0.0529$.

4.2 Stocker (1949) studied a case in the bacterium *Salmonella typhimurium* in which the mutation rates were sufficiently large that the equation for mutational equilibrium could be tested. The gene in question controls a protein component of the cellular flagella. There are two alleles, which we can call A and a. The mutation rate from A to a was estimated as $\mu = 8.6 \times 10^{-4}$ per generation, and that of a to A as $v = 4.7 \times 10^{-3}$ per generation. (These mutation rates are orders of magnitude larger than typically observed for other genes. The reason is that the change from A to a and back again does not involve mutation in the conventional sense, but results from intrachromosomal recombination.) In cultures initially established with a frequency of A at $p_0 = 0$, Stocker found that the frequency increased to $p = 0.16$ after 30 generations and to $p = 0.85$ after 700 generations. In cultures initiated with $p_0 = 1$, the frequency decreased to 0.88 after 388 generations and to $p = 0.86$ after 700 generations.

a. Are these values in accord with those calculated from Equation 4.5 using the estimated mutation rates?

b. What is the predicted equilibrium frequency of A?

4.2 ANSWER a. The recombinational switch is formally equivalent to reversible mutation. Apply Equation 4.5 with $v/(\mu + v) = 0.8453$ and $1 - \mu - v = 0.9944$. For initial $p_0 = 0$, $p_{30} = 0.1302$ (versus observed 0.16) and $p_{700} = 0.8283$ (versus observed 0.85). For initial $p_0 = 1$, $p_{388} = 0.8631$ (versus observed 0.88) and $p_{700} = 0.8484$ (versus observed 0.86). The fit to the observed values is very good. **b.** The predicted equilibrium frequency of A is given by Equation 4.6 as $v/(\mu + v) = 0.8453$.

4.3 The per-nucleotide mutation rate in humans has been estimated as about 1.65×10^{-8} per generation, and the long-term effective population size is about 10^4. For nucleotides whose mutation is likely to have negligible effects on survival or reproduction, what is the expected average heterozygosity? How many heterozygous single-nucleotide polymorphisms does this estimate represent, given a genome size of 10^9 base pairs?

4.3 ANSWER Use Equation 4.11 with $\theta = 4N\mu = 0.00066$, which because $\theta << 1$ yields $H = 0.00066$ also. This number corresponds to $0.00066 \times 10^9 \approx 660,000$ heterozygous sites in the average genome, or about one heterozygous nucleotide site per 1500 nucleotide pairs.

4.4 In the Wright–Fisher model, calculate the average time to coalescence, the expected sum of the lengths of all branches in the gene genealogy, and the expected time to the MRCA for samples of size 2, 4, 8, 16, and 32 in an ideal diploid population of size N.

4.4 ANSWER The average times to coalescence (Equation 4.15) for samples of size 2, 4, 8, 16, and 32 are given by $2N$, $0.333N$, $0.0714N$, $0.017N$, and $0.004N$, respectively. Note that these decrease dramatically as the sample size increases. The expected sums of the lengths of all branches in the gene genealogy (Equation 4.16) are given by $4N$, $7.33N$, $10.37N$, $13.27N$, and $16.11N$. These increase with sample size, but not dramatically. The expected times to the most recent common ancestor (Equation 4.17) are given by $2N$, $3N$, $3.5N$, $3.75N$, and $3.88N$, which change only modestly with sample size.

4.5 A highly isolated colony of the moth *Panaxia dominula* near Oxford, England, was intensively studied by Ford and collaborators over the period 1928–1968. This species has one generation per year, and estimates of population size were carried out yearly beginning in 1941. For the years 1950 to 1961, inclusive, estimates of population size were as follows:

1950:	4,100	1951:	2,250	1952:	6,000	1953:	8,000
1954:	11,000	1955:	2,000	1956:	11,000	1957:	16,000
1958:	15,000	1959:	7,000	1960:	2,500	1961:	1,400

Assuming that the actual size of the population in any year equals the effective size in that year, estimate the average effective number over the entire 12-year period.

4.5 ANSWER Use the harmonic mean $1/N_e = (1/t) \Sigma (1/N_i)$ as in Equation 4.30. The estimated $N_e = 1/0.0002540 = 3936.8$.

4.6 A dairy farmer has a herd consisting of 200 cows and 2 bulls. What is the effective size of the population?

4.6 ANSWER Apply Equation 4.31 with $N_m = 2$ and $N_f = 200$. In this case $N_e = 7.92$, or only 3.9 per cent of the actual population size.

4.7 The average GC content (proportion of G–C base pairs) differs among mammalian genomes from a low of 0.4389 in the opossum *Monodelphis domesticus* to a high of 0.5789 in the platypus *Ornithorhynchus anatinus*. The human genome has an average GC content of 0.461, which is near the average of all mammals and also near the inferred average GC content of the ancestral mammalian genome. Although there are large regional differences in GC content across the genome owing to regional differences in gene conversion, use the averages given to deduce the overall GC bias of gene conversion assuming an AT bias in mutation with $a = 1/2$. (Hint: use Equation 4.36.)

4.7 ANSWER In Equation 4.6, set $g' = g$ and solve for c, which yields the equation $c = [g - (1 - g)a] / [2g(1 - g)]$. Then set $a = 1/2$ and g equal to the average GC content of each of the species in turn. The results are $c = 0.3215$ for the opossum, 0.7555 for the platypus, and 0.3853 for humans.

4.8 Let σ_t^2 be the variance in allele frequency among an ensemble of subpopulations. In the island model of migration with migration rate m, show that $\sigma_t^2 = (1 - m)^2 \sigma_{t-1}^2$.

4.8 ANSWER Equation 4.40 implies that $p_t - \bar{p} = (p_{t-1} - \bar{p})(1 - m)$. Squaring both sides and taking the expectation leads to $E(p_t - \bar{p})^2 = (1 - m)^2 E(p_{t-1} - \bar{p})^2$, and therefore, by the definition of the variance, $\sigma_t^2 = (1 - m)^2 \sigma_{t-1}^2$.

References

Akashi, H. (1997), 'Synonymous codon usage in Drosophila melanogaster: natural selection and translational accuracy', *Genetics*, 136 (3), 927–35.

Ayala, D. J., Chang, B. S. W., and Hartl, D. L. (1993), 'Molecular evolution of the Rh3 gene in Drosophila', *Genetica*, 92 (1), 23–32.

Bernardi, G. (2000), 'Isochores and the evolutionary genomics of vertebrates', *Gene*, 241 (1), 3–17.

Chang, H. -H., et al. (2012), 'Genomic sequencing of P. falciparum malaria parasites from Senegal reveals the demographic history of the population', *Mol Biol Evol*, 29 (11), 3427–39.

Charlesworth, B. and Charlesworth, D. (2012), *Elements of Evolutionary Genetics* (Greenwood Village, CO: Roberts and Company).

Chen, S., Zhang, Y. E., and Long, M. (2010), 'New genes in Drosophila quickly become essential', *Science*, 330 (6011), 1682–5.

Costantini, M., et al. (2006), 'An isochore map of human chromosomes', *Genome Res*, 16 (4), 536–41.

Cox, J. T. and Durrett, R. (2002), 'The stepping stone model: new formulas expose old myths', *Ann Appl Prob*, 12 (4), 1348–77.

Crow, J. F. and Kimura, M. (1970), *Introduction to Population Genetics* (New York: Harper & Row).

Daniels, R., et al. (2013), 'Genetic surveillance detects both clonal and epidemic transmission of malaria following enhanced intervention in Senegal', *PLoS One*, 8 (4), e60780.

Dice, L. R. and Howard, W. B. (1951), 'Distance of dispersal by prairie deer mice from birthplace to breeding sites', *Cont Lab Vert Bio Univ Mich*, 50, 1–15.

Duforet-Frebourg, N. and Slatkin, M. (2016), 'Isolation-by-distance-and-time in a stepping-stone model', *Theor Popul Biol*, 108, 24–35.

Elhaik, E., et al. (2010), 'Identifying compositionally homogeneous and nonhomogeneous domains within the human genome using a novel segmentation algorithm', *Nucleic Acids Res*, 38 (15), e158.

Ewens, W. J. (1972), 'The sampling theory of selectively neutral alleles', *Theor Popul Biol*, 3 (1), 87–112.

Eyre-Walker, A. and Hurst, L. D. (2001), 'The evolution of isochores', *Nat Rev Genet*, 2 (7), 549–55.

Francisco, M. R., et al. (2007), 'Genetic structure in a tropical lek-breeding bird, the blue manakin (Chiroxiphia caudata) in the Brazilian Atlantic Forest', *Mol Ecol*, 16 (23), 4908–18.

Hamilton, W. L., et al. (2017), 'Extreme mutation bias and high AT content in Plasmodium falciparum', *Nucleic Acids Res*, 45 (4), 1889–901.

Hudson, R. R. (1991), 'Gene genealogies and the coalescent process', *Oxf Surv Evol Biol*, 7, 1–44.

Jeffreys, A. J. and May, A. C. (2004), 'Intense and highly localized gene conversion activity in human meiotic crossover hot spots', *Nat Genet*, 36 (2), 151–6.

Kark, S., et al. (2008), 'How does genetic diversity change towards the range periphery? An empirical and theoretical test', *Evol Ecology Res*, 10 (3), 391–414.

Kimura, M. (1969), 'The number of heterozygous nucleotide sites maintained in a finite population due to steady flux of mutations', *Genetics*, 61 (4), 893–903.

Kimura, M. and Crow, J. F. (1964), 'The number of alleles that can be maintained in a finite population', *Genetics*, 49 (4), 725–38.

Kingman, J. F. C. (1982), 'On the genealogy of large populations', *J Applied Probab*, 19A, 27–43.

Kingman, J. F. C. (2000), 'Origins of the coalescent 1974–1982', *Genetics*, 156 (4), 1461–3.

Kong, A., et al. (2012), 'Rate of de novo mutations and the importance of father's age to disease risk', *Nature*, 488 (7412), 471–5.

Lewontin, R. C. (1974), *The Genetic Basis of Evolutionary Change* (New York: Columbia University Press).

Lynch, M. and Conery, J. S. (2003), 'The evolutionary demography of duplicate genes', *J Struct Funct Genomics*, 3 (1–4), 35–44.

Lynch, M. and Force, A. (2000), 'The probability of duplicate gene preservation by subfunctionalization', *Genetics*, 154 (1), 459–73.

Maruyama, T. (1970), 'Analysis of population structure. I. One-dimensional stepping-stone models of finite length and other geographically structures populations', *Ann Hum Genet*, 34 (2), 201–19.

Maruyama, T. (1971), 'Analysis of population structure. II. Two-dimensional stepping-stone models of finite length and other geographically structured populations', *Ann Hum Genet*, 35 (2), 179–96.

Moran, P. A. P. (1958), 'Random processes in genetics', *Proc Camb Phil Soc*, 54, 60–71.

Moran, P. A. P. (1962), *Statistical Processes of Evolutionary Theory* (Oxford: Clarendon Press).

Nair, S., et al. (2008), 'Adaptive copy number evolution in malaria parasites', *PLoS Genet*, 4 (10), e1000243.

Nevo, E. (1978), 'Genetic variation in natural populations: patterns and theory', *Theoret Popul Biol*, 13 (1), 121–77.

Novick, A. (1955), 'Mutagens and antimutagens', *Brookhaven Symp Biol*, 8, 201–15.

Rasmussen, M. D., et al. (2014), 'Genome-wide inference of ancestral recombination graphs', *PLoS Genet*, 10 (5), e1004342.

Rogers, R. L., Bedford, T., and Hartl, D. L. (2009), 'Formation and longevity of chimeric and duplicate genes in Drosophila melanogaster', *Genetics*, 181 (1), 313–22.

Santos, J. L., et al. (2012), 'Copy number polymorphism of the salivary amylase gene: implications in human nutrition research', *J Nutrigenet Nutrigenomics*, 5 (3), 117–31.

Sjödin, P., et al. (2005), 'On the meaning and existence of an effective population size', *Genetics*, 169 (2), 1061–70.

Slatkin, M. (1985), 'Rare alleles as indicators of gene flow', *Evolution*, 39 (1), 53–65.

Slatkin, M. (1991), 'Inbreeding coefficients and coalescence times', *Genet Res Camb*, 58 (5–6), 167–75.

Stocker, B. A. D. (1949), 'Measurements of rate of mutation of flagellar antigenic phase in Salmonella typhimurium', *J Hyg (Lond)*, 47 (4), 398–412.

Tajima, F. (1983), 'Evolutionary relationships of DNA sequences in finite populations', *Genetics*, 105 (5), 437–60.

Voelker, R. A., Schaffner, H. E., and Mukai, T. (1980), 'Spontaneous allozyme mutations in Drosophila melanogaster: rate of occurrence and nature of the mutants', *Genetics*, 94 (4), 961–8.

Wakeley, J. (1999), 'Nonequilibrium migration in human history', *Genetics*, 153 (4), 1863–71.

Wakeley, J. (2000), 'The effects of subdivision on the genetic divergence of populations and species', *Evolution*, 54 (4), 1092–101.

Wakeley, J. (2009), *Coalescent Theory* (Greenwood Village, CO: Roberts).

Wakeley, J. and Sargsyan, O. (2009), 'Extensions of the coalescent effective population size', *Genetics*, 181 (1), 341–5.

Watterson, G. A. (1975), 'On the number of segregating sites in genetical models without recombination', *Theor Popul Biol*, 7 (2), 256–76.

Williams, A. L., et al. (2015), 'Non-crossover gene conversions show strong GC bias and unexpected clustering in humans', *Elife*, 4.

Wright, S. (1946), 'Isolation by distance under diverse systems of mating', *Genetics*, 31, 39–59.

Wright, S. (1978), *Evolution and the Genetics of Populations. Vol. 4. Variability Within and Among Natural Populations* (Chicago, IL: University of Chicago Press).

Zhang, J. (2003), 'Evolution by gene duplication: an update', *Trends Ecol Evol*, 18 (12), 292–8.

Zhao, J., et al. (2015), 'A generalized birth and death process for modeling the fates of gene duplication', *BMC Evol Biol*, 15, 275.

CHAPTER 5

Natural Selection in Large Populations

As Charles Darwin correctly proposed more than 150 years ago, natural selection is the driving force of evolution (Darwin 1859). By **natural selection** we mean inherited differences in the ability of organisms to survive and reproduce, so that through time the genotypes that are superior in survival and reproduction increase in frequency in the population. Natural selection is the principal process that results in greater adaptation of organisms to their environment, because what we really mean by **adaptation** is the acquisition of traits that enhance survival and reproduction in a given environment. In any species, genetic variation produced by mutation is organized, maintained, eliminated, or dispersed among subpopulations according to a complex balance of migration, random changes in allele frequency, and natural selection acting on many traits.

5.1 Selection in Haploids

Selection in its simplest form occurs in haploid organisms such as bacteria that reproduce asexually by binary fission. In Chapter 1 we considered models of population growth in such organisms. In this section we extend these models to cases in which the population of organisms contains two genotypes that differ in their likelihood of survival and reproduction. The ability of an organism to survive and reproduce, relative to other organisms in the same population, constitutes its **fitness**. The definition of fitness is subtly different depending on whether population growth is continuous or discrete, and we will consider these cases separately to make the distinctions precise and establish some basic terminology. As in Chapter 4, we will be more concerned with changes in allele frequency than with the absolute numbers of organisms; however, the models of population growth in Chapter 1 afford a convenient place to start.

Continuous-Time Model of Haploid Selection

Equation 1.5 states that, for a population undergoing exponential growth, the change in population number at an instant in time t is given by $dN(t)/dt = rN(t)$ where r is the intrinsic rate of increase. Consider a population of haploid, asexual organisms with either of two genotypes A_1 and A_2, present in the numbers $N_1(t)$ and $N_2(t)$ at time t, and suppose that they differ in their intrinsic rates of increase, m_1 and m_2. (We use m instead of r in this context to be consistent with much of the literature in population genetics.) The growth

A Primer of Population Genetics and Genomics. Fourth Edition. Daniel L. Hartl, Oxford University Press (2020). © Daniel L. Hartl.
DOI: 10.1093/oso/9780198862291.003.0005

rates m_1 and m_2 are measures of the relative fitness of the genotypes A_1 and A_2; m_1 and m_2 define the **Malthusian fitness** of each genotype.

Suppose now that $p(t)$ is the frequency of A_1 in the population at time t, so that $p(t) = N_1(t)/[N_1(t) + N_2(t)]$. Then $q(t) = 1 - p(t)$ is the frequency of A_2 at time t. The change in $p(t)$ at an instant in time t is given by:

$$\frac{dp(t)}{dt} = p(t)q(t)(m_1 - m_2) \tag{5.1}$$

Equation 5.1 is an elementary differential equation with the solution:

$$\ln\left(\frac{p_t}{q_t}\right) = \ln\left(\frac{p_0}{q_0}\right) + (m_1 - m_2)t \tag{5.2}$$

where p_t and q_t are used as shorthand for $p(t)$ and $q(t)$, respectively.

Equation 5.2 says that, with selection under exponential growth, a plot of $\ln(p/q)$ against time is expected to be linear with a slope equal to the difference in Malthusian parameters $m_1 - m_2$, which measures the difference in the relative fitnesses of the genotypes. The difference $m_1 - m_2$ is often denoted s and called the **selection coefficient**. If $s > 0$ then allele A_1 is favored and $\ln(p/q)$ increases with time; if $s < 0$ then allele A_2 is favored and $\ln(p/q)$ decreases; and if $s = 0$ the alleles are **selectively neutral** and $\ln(p/q)$ remains constant.

An example of such a plot is shown in Figure 5.1, which shows the result of competition between pairs of genotypes of *Escherichia coli* that differ in one amino acid in a single enzyme but are otherwise genetically identical (Hartl and Dykhuizen 1981). The gray dots are the data from a pair of strains differing in fitness by a selection coefficient

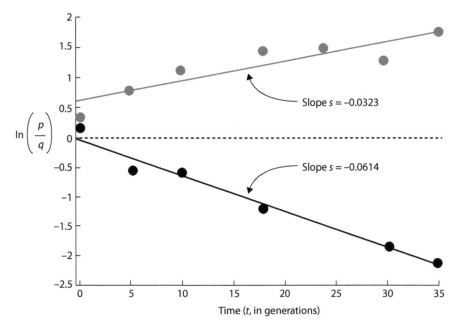

Figure 5.1 Selection in chemostats between strains of *E. coli* that are genetically identical except for alleles encoding different naturally occurring mutants of 6-phosphogluconate dehydrogenase. The selection coefficient is estimated as the slope of the straight line. The allele designations are arbitrary. In one pair of strains (gray dots), the allele designated A_1 is favored; in the other pair (black dots), the allele designated A_2 is favored. (Data from Hartl and Dykhuizen (1981).)

$s = 3.23$ per cent per generation, with selection favoring the strain with the allele designated A_1. The black dots are data for another pair of strains, in this case with an amino acid replacement at a different site in the same enzyme. Here the selection coefficient is 6.14 per cent per generation, but in this case the selection favors the strain with the allele designated A_2.

Discrete-Generation Model of Haploid Selection

In this section we derive an expression corresponding to Equation 5.1 for a model of population growth with discrete, nonoverlapping generations. Our jumping-off point is again the population growth model in Chapter 1, in which Equation 1.3 asserts that the population size N_t in generation t is related to that in generation $t-1$ by the relation $N_t = N_{t-1}(1+\rho)$ where ρ is the per capita increase in population size on one generation. To simplify some of the formulas that follow, we will write this as $N' = Nw$ where N is the population size in some generation, N' the population size a generation later, and w the growth rate. (In population genetics, $1+\rho$ is usually denoted as w.)

Proceeding as we did earlier, imagine two genotypes A_1 and A_2 with respective numbers N_1 and N_2 in some generation with growth rates w_1 and w_2. The growth rates w_1 and w_2 are the relative fitnesses of A_1 and A_2. In the context of discrete, nonoverlapping generations, w_1 and w_2 define the **Darwinian fitness** of each genotype.

Letting p equal the allele frequency of A_1 implies that $p = N_1/(N_1 + N_2)$, and then the allele frequency of A_2 is $q = 1-p$. Discrete population growth implies that the allele frequency in the next generation, p', is given by:

$$p' = \frac{N_1'}{N_1' + N_2'} = \frac{N_1 w_1}{N_1 w_1 + N_2 w_2} = \frac{pw_1}{pw_1 + qw_2} \tag{5.3}$$

and the change in p in one generation is given by:

$$\Delta p = p' - p = \frac{pw_1}{pw_1 + qw_2} - p = \frac{pw_1 - p^2 w_1 - pqw_2}{pw_1 + qw_2} = \frac{pq(w_1 - w_2)}{\overline{w}} \tag{5.4}$$

where $\overline{w} = pw_1 + qw_2$ is the **average fitness** in the population.

Equation 5.4 is the discrete-generation analog of Equation 5.1 and it differs in the presence of the denominator $\overline{w}$, which is necessary as a normalizing factor to ensure that the allele frequencies sum to 1 in each generation. A key point about Equation 5.4 is that both the numerator and denominator can be divided by either w_1 or w_2 without changing the magnitude. This feature means that only the relative fitnesses matter, and so we could write the relative fitnesses as $w_1/w_1 = 1$ and $w_1/w_2 = 1 - s$ without changing anything. The parameter s is the selection coefficient in the discrete-generation case.

If s is small, then $\overline{w} \approx 1$ and the denominator of Equation 5.4 disappears. The change Δp is also small enough that $\Delta p \approx dp/dt$. Under these conditions Equation 5.4 is so close to Equation 5.1 that the equations can be considered equivalent.

It is often much simpler to think of differences in fitness as due to differential survival rather than differential reproduction. Figure 5.2 is a graphical representation of the logic behind Equation 5.3 for a specific numerical example when the genotypes differ only in their probability of survival. Each little box represents an organism, and the alleles are represented by the colored dots. The progeny undergo a process of maturation in which some of the offspring fail to survive. In Figure 5.2 we assume that the survival probabilities are 9/10 for A_1 and 6/10 for A_2. Selection based on differential survival is usually called **viability selection**.

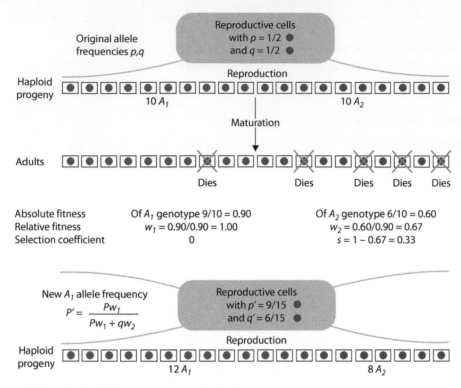

Figure 5.2 Model of viability selection in a haploid organism.

5.2 Selection in Diploids

The major concepts of diploid selection are identical to those for haploid selection, but models that track the numbers of each genotype continuously through time are somewhat complex because the genotypes in diploids do not "breed true." The allele combination in each genotype is broken up in each generation by Mendelian segregation and reassembled in the following generation when the gametes come together in fertilization. Models with nonoverlapping generations are therefore conceptually simpler.

Directional Selection

The complications resulting from Mendelian segregation in heterozygous genotypes are illustrated in Figure 5.3, which has the same overall layout as Figure 5.2. Each little box represents a diploid organism, and each organism contains two alleles of the gene. Selection is assumed to occur on the diploid genotypes, not on the gametes, and so with random mating a large pool of gametes with allele frequencies p and q produces a pool of zygote genotypes that are in Hardy–Weinberg equilibrium for the same allele frequencies. Viability selection occurs between fertilization and maturation.

There are three fitnesses corresponding to the genotypes A_1A_1, A_1A_2, and A_2A_2, but taking A_1A_1 as the reference standard, the relative fitnesses can be written as 1 for A_1A_1, w_{12} for A_1A_2, and w_{22} for A_2A_2. In the example in Figure 5.3, the relative fitnesses with A_1A_1 as the standard are $w_{12} = 8/10$ and $w_{22} = 6/10$. By analogy with haploid selection, we can also define a selection coefficient against A_1A_2 and against A_2A_2. The selection

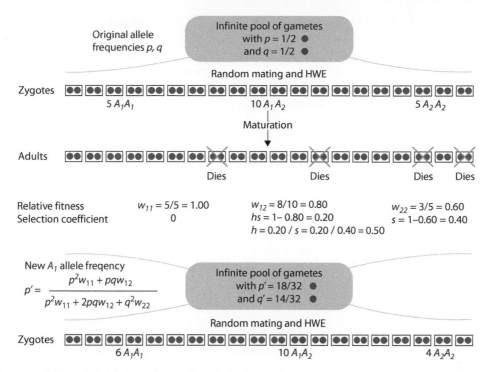

Figure 5.3 Model of viability selection in a diploid organism.

coefficient against homozygous A_2A_2 is denoted s, and $s = 1 - w_{22}$, which in this case equals 0.40. The selection coefficient against the heterozygous genotype is often expressed as a multiple h of s, and in this example $hs = 1 - w_{12} = 0.20$. Hence $h = hs/s = 0.20/0.40 = 0.50$. The parameter h is called the **degree of dominance** of A_2, and it is a convenient parameter in diploid models because:

- $h = 0$ means that A_2 is recessive to A_1.
- $h = 1/2$ means that the heterozygous fitness is the arithmetic average of the homozygous fitnesses; in this case, the effects of the alleles are said to be **additive effects**. Additive effects are exemplified in Figure 5.3. This case is also called **genic selection**.
- $h = 1$ means that A_2 is dominant to A_1.

As we shall see later, it is also possible that $h < 0$ or $h > 1$, but for the moment we will consider only the range $0 \leq h \leq 1$ and also assume that $s > 0$. These stipulations mean that A_1 is favored irrespective of its allele frequency, and therefore A_1 is expected to eventually become fixed. Selection consistently favoring one allele is known as **directional selection**.

In Figure 5.3, the genotype frequencies after the viability selection are 5 A_1A_1, 8 A_1A_2, and 3 A_2A_2, which corresponds to allele frequencies of $p = 18/32 = 0.56$ of A_1 and $q = 14/32 = 0.44$ of A_2. As before, we assume that these remain unchanged in the gametes and zygotes of the next generation (no differential mating success or fertility selection). It may be verified by numerical substitution that the allele frequency in the next generation prior to selection is given by:

$$p' = \frac{p^2 w_{11} + pq w_{12}}{p^2 w_{11} + 2pq w_{12} + q^2 w_{22}} \tag{5.5}$$

where the pqw_{12} in the numerator is not multiplied by 2 because only half of the gametes from an A_1A_2 genotype carry A_1. Because each term in both the numerator and denominator in Equation 5.5 have fitness as a factor, any of the genotypes can be regarded as the reference standard and its fitness used as a divisor for both numerator and denominator. In effect, this division sets the relative fitness of the reference genotype equal to 1. In Figure 5.3, the reference genotype is A_1A_1.

Note that the denominator of Equation 5.5 is the average fitness in the population, usually symbolized $\overline{w}$:

$$\overline{w} = p^2 w_{11} + 2pqw_{12} + q^2 w_{22} \tag{5.6}$$

Equation 5.5 does not usually have an analytical solution, and for this reason it is usually more useful to calculate the difference in allele frequency between two successive generations, namely $\Delta p = p' - p$. This can be found by subtracting p from both sides of Equation 5.5, multiplying p on the right-hand side by $\overline{w}/\overline{w}$ to obtain a common denominator, and simplifying. The result is:

$$\Delta p = \frac{pq\left[p\left(w_{11} - w_{12}\right) + q\left(w_{12} - w_{22}\right)\right]}{\overline{w}} \tag{5.7}$$

To illustrate the use of these equations we will use data on change in the frequency of the allele Cy (*Curly* wings) in a laboratory population of *Drosophila melanogaster* (Teissier 1942). These are plotted in Figure 5.4 as the frequency of $Cy/+$ adults in the population. Since Cy/Cy homozygotes die as embryos, the allele frequency of Cy equals the frequency of $Cy/+$ adults divided by 2.

In order to apply the equations directly, we set A_1 to correspond to the Cy allele and A_2 to the $+$ allele. In this case A_2A_2 is the reference genotype for calculating the relative fitnesses, and therefore we set $w_{22} = 1$. Because A_1A_1 (i.e., Cy/Cy) is lethal, $w_{11} = 0$. The best-fitting curve for these data is obtained from Equation 5.5 with $w_{12} = 0.50$, thus the effects of Cy on fitness are additive. The expected genotype frequencies can be calculated recursively from Equation 5.5. The initial population was established with a

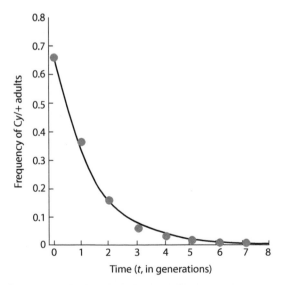

Figure 5.4 Change in frequency of adult *Drosophila melanogaster* heterozygous for the dominant mutation Cy (*Curly* wings) in an experimental population. (Data from Teissier (1942).)

frequency of $Cy/+$ adults of 2/3, and so $p = 1/3$ and $q = 2/3$. In the next generation we expect:

$$p' = \frac{(1/3)^2(0) + (1/3)(2/3)(0.5)}{(1/3)^2(0) + 2(1/3)(2/3)(0.5) + (2/3)^2(1)} = 1/6$$

This predicts the frequency of $Cy/+$ adults in the next generation to be $2p' = 1/3 = 0.33$, which is reasonably close to the observed value of 0.368.

Time Required for Changes in Allele Frequency

If selection is weak enough that $\overline{w} \approx 1$, then Δp is small enough that it can be treated as the derivative dp/dt where time is measured in generations. Equation 5.7 then becomes a differential equation that can be integrated to deduce the time required for a given change in allele frequency. The equations and integrals are most easily presented in terms of the selection coefficient and the degree of dominance. With these as the parameters and $\overline{w} \approx 1$, Equation 5.7 becomes:

$$\frac{dp_t}{dt} = p_t q_t s \left[p_t h + q_t (1 - h) \right] \tag{5.8}$$

where p_t and q_t are used as shorthand for $p(t)$ and $q(t)$, respectively. The solutions to Equation 5.8 in three cases of special interest are as follows.

1. **A_1 is a favored dominant**. In this case $h = 0$ and $dp/dt = pq^2 s$, which yields

$$\ln\left(\frac{p_t}{q_t}\right) + \left(\frac{1}{q_t}\right) = \left[\ln\left(\frac{p_0}{q_0}\right) + \left(\frac{1}{q_0}\right)\right] + st \tag{5.9}$$

2. **A_1 is favored and the alleles are additive**. Here $h = 1/2$ and $dp/dt = pqs/2$. The integral is

$$\ln\left(\frac{p_t}{q_t}\right) = \ln\left(\frac{p_0}{q_0}\right) + \left(\frac{s}{2}\right)t \tag{5.10}$$

The equation for additive alleles is identical to Equation 5.2 for haploid selection if, in the haploid model, we set $m_1 - m_2 = s/2$ and stipulate that s is small.

3. **A_1 is a favored recessive**. In this case $h = 1$ and $dp/dt = p^2 qs$, which produces:

$$\ln\left(\frac{p_t}{q_t}\right) - \left(\frac{1}{p_t}\right) = \left[\ln\left(\frac{p_0}{q_0}\right) - \left(\frac{1}{p_0}\right)\right] + st \tag{5.11}$$

It is worthwhile to emphasize a particular implication of Equation 5.9 for a rare harmful recessive, in which case A_1 is a favored dominant and $dp/dt = pq^2 s$. When a harmful allele is rare, q is close to 0, and q^2 is therefore extremely small. Consequently, an increase in s from an already large value (e.g., 0.5) to a still larger value (e.g., 1) has a trivial effect on the change in allele frequency because, with q^2 so small, the actual value of s matters little. In other words, the change in allele frequency of a rare harmful recessive is slow whatever the value of the selection coefficient. For this reason, an increase in selection against rare homozygous recessive genotypes has almost no effect in changing the allele frequency. The implication for human population genetics is that the forced sterilization of rare homozygous recessive individuals—a procedure advocated in a number of outdated eugenic programs to "improve" the "genetic quality" of human beings—is genetically unsound as well as being morally and ethically abhorrent.

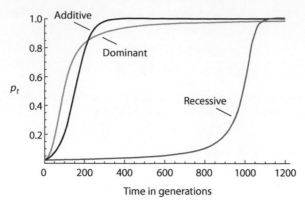

Figure 5.5 Change in frequency of a favored allele with dominant, additive, or recessive effects on fitness. In each case, there is a 5 per cent difference in relative fitness between the homozygous genotypes.

Changes in allele frequency from Equations 5.9–5.11 are shown in Figure 5.5. The change in frequency of a favored dominant allele is slowest when the allele is common, and the change in frequency of a favored recessive allele is slowest when the allele is rare. The explanation is the same in both cases. Rare recessive alleles are present mainly in heterozygous genotypes and thereby are hidden from selection. With a favorable additive allele, the initial increase in frequency is slower than that of a favored dominant, but the additive allele eventually overtakes and goes to fixation faster, because additive selection continues to distinguish between the homozygous and heterozygous genotypes. Note that in all cases the allele frequency of the favored allele increases most rapidly when the allele frequency is in the range 0.2–0.8. What this means is that, with directional selection acting on a trait influenced by many genes, those genes with intermediate allele frequencies contribute disproportionally to phenotypic changes in the population.

Some of the most dramatic examples of evolution in action result from the natural selection for chemical pesticide resistance in natural populations of insects and other agricultural pests. In the 1940s, when chemical pesticides were first applied on a large scale, an estimated 7 per cent of the agricultural crops in the United States were lost to insects. Initial successes in chemical pest management were followed by gradual loss of effectiveness. By 1985, more than 400 pest species had evolved significant resistance to one or more pesticides, and 13 per cent of the agricultural crops in the United States were lost to insects. In many cases, significant pesticide resistance had evolved in 5–50 generations, irrespective of the insect species, geographical region, pesticide, frequency and method of application, and other seemingly important variables (May 1985).

Because many of the pesticide resistance phenotypes result from single mutant genes, Equations 5.9–5.11 help to understand why pesticide resistance evolves so rapidly. The resistance genes are often partially or completely dominant, and so Equations 5.9 and 5.10 are applicable. Prior to application of the pesticide, the allele frequency p_0 of the resistant mutant is usually close to 0. Application of the pesticide increases the allele frequency, sometimes by many orders of magnitude, but significant resistance is noticed in the pest population even before the allele frequency p_t increases above a few per cent. Thus, as rough approximations, we may assume that q_0 and q_t are both close enough to 1 that $\ln(p_0/q_0) \approx \ln(p_0)$ and $\ln(p_t/q_t) \approx \ln(p_t)$. Using these approximations, Equation 5.10 (additive case) implies that $t \approx (2/s)\ln(p_t/p_0)$, and Equation 5.9 (dominant case) implies that $t \approx (1/s) \ln(p_t/p_0)$. In many cases, the ratio p_t/p_0 may range from 1×10^2 to perhaps 1×10^7, and s may typically be 0.5 or greater. Over this wide range of parameter values, the time t is effectively limited to 5–50 generations for the appearance of a significant degree of pesticide resistance. The time actually required will depend on such details as population size and the extent of migration between local populations, and the evolution of resistance due to multiple interacting genes may be expected to take somewhat longer than single-gene resistance.

Selective Sweeps: Hard Sweeps and Soft Sweeps

If directional selection occurs rapidly, as in the examples in Figure 5.5, the process leaves a statistical signature in the chromosomal region of the selected allele. Rapid fixation or near-fixation of a single, strongly beneficial mutation in a population is called a **selective sweep**. During a selective sweep of a beneficial mutation, any neutral alleles that are sufficiently tightly linked in the chromosome will also increase in frequency, a process called **genetic hitchhiking** (Maynard Smith and Haigh 1974). The hitchhiking of linked genes reduces the amount of genetic variation in a chromosomal region flanking the beneficial mutation (Kim and Stephan 2002; Pavlidis and Alachiotis 2017). If selective sweeps occur frequently enough, regions of low recombination will seldom, if ever, have a chance to regain genetic variation by mutation and random genetic drift before the recovery process is interrupted by the next beneficial mutation (Perlitz and Stephan 1997).

The characteristic genomic signature of a selective sweep is illustrated in Figures 5.6a and 5.6b. Prior to the occurrence of a favorable mutation destined to be fixed (Figure 5.6a), each chromosome in the nearby region of the genome (gray bars) will have accumulated neutral or nearly neutral mutations (white stars) because of mutation and random genetic drift. The black star in the chromosomal region at the bottom denotes a newly arising beneficial mutation destined to be fixed.

As the beneficial mutation increases in frequency because of selection, the nearby region of the genome hitchhikes along, and the particular combination of mutations (the **haplotype**) linked to the beneficial mutation will also increase in frequency, displacing other haplotypes in the population. As the selective sweep progresses (Figure 5.6b), the hitchhiking haplotype comes to be disproportionally represented, which means that the level of genetic variation in the region of the selective sweep becomes restricted relative to other regions of the genome. The length of the hitchhiking haplotype depends on the strength of selection and the level of recombination. The stronger the selection and the less recombination, the longer the hitchhiking haplotype will be.

Figure 5.6c illustrates reduced genetic variation in a region of chromosome 2 in *D. melanogaster* associated with a selective sweep of a novel gene designated *Quetzalcoatl* (*Qtzl*), located precisely at the point of minimum genetic variation (Rogers et al. 2010). The y-axis is π, the average number of mismatches across a sliding window of 10 kb in pairwise comparisons among a sample of 39 sequences. (The appropriateness of π as an estimate of genetic variation is discussed in Chapter 4.) The x-axis shows the nucleotide position along chromosome 2 as a function of distance from *Qtzl* (star), and it indicates a significant reduction in genetic variation in the 10-kb window around *Qtzl*. The dashed curve shows the expected values after a selective sweep with a value of $s \approx 0.01$ that took place about 15,000 years ago (Rogers et al. 2010).

A selective sweep of the type illustrated in Figure 5.6 is more properly called a **hard selective sweep** caused by selection of a single, strongly beneficial mutation. Another type of sweep is known as a **soft selective sweep**, in which multiple favorable mutations

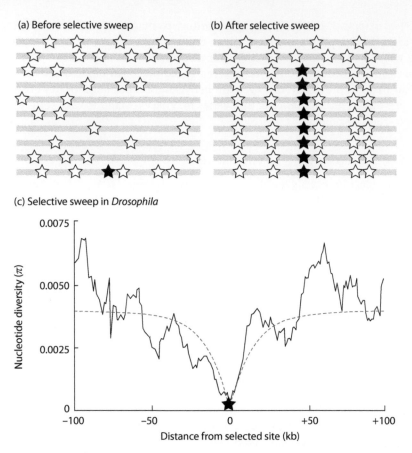

Figure 5.6 Selective sweep. (a) Haplotypes in a chromosomal region prior to a selective sweep; stars represent SNP polymorphisms. (b) Haplotypes after a selective sweep of a favorable allele (black star). Note the reduction in genetic variation due to the high frequency of one particular haplotype. (c) Reduced genetic variation around the gene *Quetzalcoatl* in *Drosophila* is consistent with a selective sweep. (Data in (c) from Rogers et al. (2010).)

increase in the population at the same time. The multiple favorable mutations may arise as the result of a high mutation rate, mutations brought in by migration from other populations, neutral or nearly neutral mutations that become favorable, or adaptive mutations that occur in a deteriorating environment (Pennings and Hermisson 2006; Wilson et al. 2017). The hallmark of a soft selective sweep is that multiple haplotypes can each carry a favorable allele and therefore increase in frequency simultaneously.

Some authors have argued that the conditions favoring soft sweeps over hard sweeps are unlikely to occur in natural populations (Jensen 2014), and furthermore that even favorable alleles are unlikely to be fixed unless their initial frequency is appreciable (Orr and Betancourt 2001). The likelihood that a favorable allele becomes fixed is examined in the following section.

Probability of Survival of a Favorable Mutation

Figure 5.5 may seem to imply that favorable mutations are virtually certain to increase in frequency in a population and eventually become fixed. On the contrary, mutations with such a small selective advantage as those in Figure 5.5 are much more likely to be lost than fixed, even in an infinite population. To understand why, we use an argument originally put forward by R. A. Fisher (1922). For simplicity we consider a discrete-generation haploid population that is infinite in size, and let N_t be the number of individuals in the population at time t carrying a mutant allele that is identical by descent with a new mutation that originally occurred in generation 0, so that $N_0 = 1$. Then:

$$N_t = \sum_{i=1}^{N_{t-1}} \Pr\{\text{Allele } i \text{ leaves } k \text{ offspring alleles}\}$$

where $k = 0, 1, 2, 3, \ldots$ with probability $p_0, p_1, p_2, p_3, \ldots$. Assume further that each mutant allele leaves descendant alleles independently of other mutant alleles and that the distribution of k is identical for every mutant allele in every generation.

Here we introduce the **probability generating function (pgf)** for the distribution of number of offspring. The pgf is typically represented as $\varphi(z)$ where:

$$\varphi(z) = p_0 + p_1 z + p_2 z^2 + p_3 z^3 + \cdots \tag{5.12}$$

The utility of the pgf in this context is that $\varphi(0) = p_0$, $\varphi'(0) = p_1$, $\varphi''(0) = 2!p_2$, $\varphi'''(0) = 3!p_3$, and so forth. In general, the kth derivative evaluated at 0 is $\varphi^{(k)}(0) = k!p_k$. The pgf is therefore a convenient way to compress all of the probabilities into one compact expression from which the individual probabilities can be extracted as needed.

Now let $\varphi_t(z)$ be the pgf of the number of mutant alleles in generation t. The assumption that a single new mutant allele occurs at time 0 means that $N_0 = 1$ and therefore $\varphi_0(z) = z$. Then:

$$\varphi_1(z) = \sum_{k=0}^{\infty} \Pr\{k \text{ mutant alleles in generation } 1\} z^k = \varphi(z) = \varphi(\varphi_0(z))$$

This expression uses the definition of the pgf in Equation 5.12. By the same logic, $\varphi_2(z) = \varphi(\varphi_1(z))$, $\varphi_3(z) = \varphi(\varphi_2(z))$, and in general:

$$\varphi_t(z) = \varphi(\varphi_{t-1}(z)) \tag{5.13}$$

Assuming now that a mutant allele has a nonzero probability of leaving no descendants $(0 < p_0 < 1)$, we can calculate the probability than a new mutant allele eventually goes extinct (i.e., all copies of the mutant are eventually lost). Let the probability of extinction by generation t equal d_t. Then:

$$d_t = \varphi_t(0) = \varphi(\varphi_{t-1}(0)) = \varphi(d_{t-1}) \tag{5.14}$$

where the equality in the middle is justified by Equation 5.13. The probability of eventual extinction d is found by letting t go to infinity in Equation 5.14. The result implies that the probability of eventual extinction of a new mutant allele is given by the smallest positive solution of:

$$d = \varphi(d) \tag{5.15}$$

where $\varphi(z)$ is the pgf of the distribution of the number of offspring alleles per mutant parental allele.

Two special cases warrant consideration:

Poisson distribution of offspring number. As a specific example, suppose that the mutant allele has no selective advantage (i.e., a neutral mutation) and that the distribution of offspring number is Poisson. The pgf of a Poisson distribution with mean λ is $\varphi(z) = e^{\lambda(z-1)}$, and for $\lambda = 1$ this becomes $\varphi(z) = e^{z-1}$. The probability of ultimate extinction is given by the smallest positive solution of $d = e^{d-1}$, which has only one solution, namely, $d = 1$. In other words, any new neutral mutation in an infinite population is destined for eventual extinction. (This is not true in a finite population, as we shall see in Chapter 6.)

Equation 5.15 also implies that most new mutations, even those with a selective advantage, are likely to become extinct. Let's assume the new mutation has a selective advantage of s and that the offspring distribution is again Poisson with mean $1 + s$. The pgf is $\varphi(z) = e^{(1+s)(z-1)}$ and the probability of ultimate loss is the smallest positive solution of:

$$d = \varphi(d) = e^{-(1+s)(d-1)} \tag{5.16}$$

If s is small, then $d \approx 1 - 2s$, which can be seen by substituting $1 - 2s$ into Equation 5.16 and expanding the right-hand side: $e^{(1+s)(-2s)} \approx e^{(-2s)} \approx 1 - 2s$. Since d is the probability of ultimate loss, the probability of ultimate survival of a mutant with selective advantage s is approximately:

$$\Pr\{\text{Survival} \,|\text{small } s \ \& \ \text{Poisson}\} \approx 2s \tag{5.17}$$

For example, a mutant with a 1 per cent selective advantage has only a 2 per cent chance of ultimate survival, and a mutant with a selective advantage of 5 per cent has only a 10 per cent chance of ultimate survival. More generally the right-hand side of Equation 5.17 can be written as $2s/\sigma^2$, where σ^2 is the variance in offspring number. Any increase in the variance in offspring number therefore reduces the probability of survival.

Mutant alleles that are destined to be lost are usually lost very quickly. The probabilities of loss as a function of time can be calculated from Equation 5.14, but for illustration consider a new neutral mutation that has a Poisson distribution of offspring number with a mean of 1. In this case the mutation has a 53 per cent chance of being lost within the first two generations and a 73 per cent chance of being lost within the first five generations.

The likelihood of a mutant surviving is improved if the selective advantage is large, but in that case Equation 5.16 must be solved numerically. Figure 5.7 shows the result of such calculations. Unless the selective advantage s is greater than an average of 1.2 descendant alleles per mutant allele, a new mutation is more likely to be lost than to survive. This represents a selective advantage of 16 per cent of the mutant over the nonmutant, which is already a very large selective advantage.

Geometric distribution of offspring number. Equation 5.17 is Fisher's classic result, but it is optimistic in assuming a Poisson distribution of offspring number. A Poisson distribution is reasonable in light of the Wright–Fisher model. In the spirit of the Moran model, an alternative is to assume a geometric distribution of offspring number. In this case the probability that an allele has k descendants in the next generation is given by $[(1+s)/(2+s)]^k [1/(2+s)]$, where $k = 0, 1, 2, \ldots$ This distribution has a mean of $1 + s$ and a variance of $(1+s)(2+s)$. The pgf is given by $\varphi(z) = 1/[2 + s - z(1+s)]$. In this case, the equation $d = \varphi(d)$ has an explicit solution, namely, $d = 1/(1+s)$, hence the probability of ultimate survival equals $s/(1+s)$. When s is small, the probability of ultimate survival is given by:

$$\Pr\{\text{Survival} \,|\text{small } s \ \& \ \text{geometric}\} \approx s \tag{5.18}$$

This is less than Fisher's value by a factor of 2. As shown in the dashed curve in Figure 5.7, the probability of ultimate survival with a geometric distribution of offspring number is always substantially smaller than expected with a Poisson distribution.

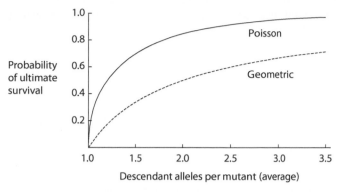

Figure 5.7 The probability that a new mutant allele ultimately survives as a function of the average number of descendant alleles per mutant allele per generation. The average number of descendent alleles per nonmutant allele is assumed to be 1.0. The solid curve assumes a Poisson distribution of offspring number, the dashed curve a geometric distribution.

The take-home message is that, for a selectively favored mutation to survive in an infinite population, it must either:

- Have a large selective advantage—but large selective advantages are rare in evolution except under such extreme conditions as selection for drug resistance in pathogens or selection of insecticide resistance in insect pests or vectors of disease;
- Happen over and over again if the selective advantage is small; or
- Be very lucky.

Kimura (1962) found methods to deduce the probabilities of fixation and loss of an allele in a finite population. We will examine this approach in Chapter 6.

Overdominance and Heterozygote Inferiority

Equation 5.7 implies that certain combinations of relative fitness may result in an equilibrium value of p that is between 0 and 1. An equilibrium value of p is a value for which $\Delta p = 0$. An **equilibrium** in allele frequency means that the allele frequency remains at a constant value generation after generation. An alternative term is **steady state**, which means that any evolutionary force that changes the allele frequency in one direction is balanced by a counteracting force that changes the allele frequency in an equal and opposite direction.

There are several types of equilibrium depending on what happens to allele frequency when the allele frequency does not equal the equilibrium value. Consider first the case when the initial allele frequency is near (but not equal to) the equilibrium. If the allele frequency moves progressively further away from the equilibrium in subsequent generations, the equilibrium is said to be **unstable**. If the allele frequency moves progressively closer to the equilibrium in subsequent generations, the equilibrium is said to be **locally stable**.

A locally stable equilibrium might also be **globally stable**, which means that, whatever the initial allele frequency, it always moves progressively closer to the equilibrium. The term **neutrally stable** applies to cases such as the Hardy–Weinberg equilibrium in which any allele frequency represents an equilibrium because, whatever its value, the allele frequency does not change.

These concepts of stability can be applied to the case of directional selection for the A allele governed by Equation 5.7. Directional selection for A means that $w_{11} \geq w_{12} \geq w_{22}$ (but not both equalities). In this case there are two equilibria, $p = 0$ and $p = 1$. If p is close to 0, p increases, so the equilibrium at $p = 0$ is unstable. On the other hand, if p is near 1, it moves still closer to 1, so the equilibrium at $p = 1$ is locally stable. Indeed, because p eventually goes to 1 whatever its initial value, the equilibrium at $p = 1$ is also globally stable.

The various types of stability are important in discussing two further cases that can occur when selection involves two alleles of a single gene. A situation called **overdominance** or **heterozygote superiority** occurs when the heterozygous genotype has a greater fitness than either homozygous genotype, or $w_{12} > w_{11}$ and $w_{12} > w_{22}$. With overdominance there is a third equilibrium in addition to $p = 0$ and $p = 1$ because $p(w_{11} - w_{12}) + q(w_{12} - w_{22})$ in Equation 5.7 can equal 0. The third equilibrium $\hat{p}$ can be found by solving $\hat{p}(w_{11} - w_{12}) + (1 - \hat{p})(w_{12} - w_{22}) = 0$, from which a little algebra gives:

$$\hat{p} = \frac{w_{12} - w_{22}}{2w_{12} - w_{11} - w_{22}} \tag{5.19}$$

This can be simplified somewhat by assigning relative fitnesses as $w_{12} = 1$, $w_{11} = 1 - s_{11}$, and $w_{22} = 1 - s_{22}$, where s_{11} and s_{22} are the selection coefficients against A_1A_1 and A_2A_2 respectively. Equation 5.19 then becomes:

$$\hat{p} = \frac{s_{22}}{s_{11} + s_{22}} \tag{5.20}$$

Note that the numerator of Equation 5.19 depends on the selection coefficient against A_2A_2 because the more selection there is against A_2A_2, the greater the equilibrium allele frequency of A_1.

An example of overdominance is shown in Figure 5.8a. Regardless of the initial allele frequency, the frequency of the A_1 allele always converges to its equilibrium, in this example $\hat{p} = 2/3$. Figure 5.8b is a plot of Δp against p showing that, when p is smaller than $\hat{p}$, then $\Delta p > 0$ and p returns to $\hat{p}$; and when p is larger than, then $\Delta p < 0$ and p again returns to $\hat{p}$. The graph shown in Figure 5.8c indicates that the value of $\bar{w}$ is a maximum at the interior equilibrium $\hat{p}$. This is an important point that we will return to in the next section.

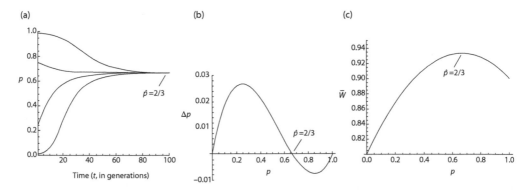

Figure 5.8 Overdominance. (a) The allele frequency converges to the internal equilibrium regardless of its initial value. (b) Graph of Δp showing that the internal equilibrium is stable whereas those at 0 and 1 are unstable. (c) The average fitness $\bar{w}$ is maximal at the internal equilibrium. In this example $w_{11} = 0.9$, $w_{12} = 1$, and $w_{22} = 0.8$.

The local stability of the equilibria with overdominance can be determined analytically by examining the first derivative of Δp with respect to p evaluated at any equilibrium point $\hat{p}$:

$$\Delta p\left(\hat{p}+\varepsilon\right) \approx \left(\frac{d\Delta p}{dp}\bigg|_{\hat{p}}\right)\varepsilon \tag{5.21}$$

Equation 5.21 is a special case of the so-called Taylor series in which ε is some very small deviation from an equilibrium point—so small that terms of order ε^2 and higher power can be ignored. The derivative of Δp from Equation 5.7 is:

$$\frac{d\Delta p}{dp} = \frac{1}{\overline{w}}\left[pq\left(w_{11}-2w_{12}+w_{22}\right)+\left(q-p\right)\left[p\left(w_{11}-w_{12}\right)+q\left(w_{12}-w_{22}\right)\right]\right]$$
$$-\frac{2pq}{\overline{w}^2}\left[p\left(w_{11}-w_{12}\right)+q\left(w_{12}-w_{22}\right)\right]^2 \tag{5.22}$$

When $p=0$, then $d\Delta p/dp=(w_{12}-w_{22})/\overline{w}>0$ and the equilibrium at $p=0$ is unstable; likewise when $p=1$, then $d\Delta p/dt=(w_{12}-w_{11})/\overline{w}>0$ and the equilibrium at $p=1$ is unstable. However, at the interior equilibrium $\hat{p}$ given in Equation 5.19, $d\Delta p/dt=\hat{p}(1-\hat{p})(w_{11}-2w_{12}+w_{22})/\overline{w}<0$, which means that the right-hand side of Equation 5.21 is positive when ε is negative and negative when ε is positive. In other words, the equilibrium at $\hat{p}$ is locally stable. (It is also globally stable, which is clear from Figure 5.8b.)

Although overdominance might seem to be a potent force for maintaining polymorphisms in natural populations, overdominance has been documented in only a few cases. A classic example is the sickle-cell anemia mutation which, when heterozygous, increases resistance to falciparum malaria but which, when homozygous, causes severe anemia. The relative viabilities in high-malaria regions in Africa have been estimated as $w_{11}=0.85$, $w_{12}-1$, and $w_{22}-0$ (Allison 1964), where A_1 represents the nonmutant allele and A_2 the mutant sickle-cell allele. Substitution into Equation 5.19 leads to the predicted equilibrium allele frequencies $\hat{p}=0.87$ and $\hat{q}=0.13$. These are reasonably close to the average allele frequencies observed in West Africa, but there is considerable variation among local populations.

Another classic example of overdominance is an amino acid polymorphism in the alcohol dehydrogenase enzyme of *D. melanogaster* resulting from alleles *Adh-F* and *Adh-S* coding for enzymes that migrate relatively fast or slow under electrophoresis. One of the first amino acid polymorphisms discovered (Ursprung and Leone 1965), the *Adh* polymorphism has been studied in natural populations worldwide. The frequency of the *Adh-F* allele increases as the temperature decreases, and compelling evidence for selection comes from studies of the frequency of *Adh-F* as a function of latitude in Eastern North America (Berry and Kreitman 1993) and Australia (Oakeshott et al. 1982). The data are shown in Figure 5.9, where p on the x-axis represents the allele frequency of *Adh-F* and the inverse sine transformation is used to disentangle the sampling variance of the allele frequency from the mean. The allele frequency of *Adh-F* increases dramatically with distance from the equator, increasing toward the north in the Northern Hemisphere and toward the south in the Southern Hemisphere.

A systematic change in allele frequency across a geographical gradient is known as a **cline**. The major clines shown in Figure 5.9 are matched on a local scale by clines of increasing *Adh-F* with altitude in mountains in both Mexico (Pipkin et al. 1976) and Central Asia (Grossman et al. 1970). The overdominance of the *Adh* alleles may not be due solely to the nonsynonymous mutation in the coding region, as the alleles have single-nucleotide polymorphisms (SNPs) in the intron and synonymous sites that also appear to contribute to enzyme activity (Laurie et al. 1991).

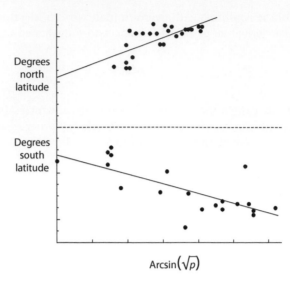

Figure 5.9 Clines in the frequency of *Adh-F* shown as a function of latitude in Eastern North America and Australia. (Data for North America from Berry and Kreitman (1993), those for Australia from Oakeshott et al. (1982).)

Overdominance is not to be confused with heterosis. Overdominance implies that heterozygous genotypes are superior to both homozygous genotypes precisely because they are heterozygous. As we discussed in Chapter 3, heterosis is the hybrid vigor observed in many traits in the progeny of crosses between inbred lines. At one time, overdominance was a viable hypothesis to explain heterosis; however, experimental data overwhelmingly support the alternative hypothesis that heterosis results from favorable dominant alleles inherited from one inbred parent covering harmful recessive alleles inherited from the other inbred parent (Crow 1998). The reason for heterosis is therefore dominance, not overdominance.

The opposite of overdominance is **heterozygote inferiority**, which occurs when the heterozygote is the least fit ($w_{12} < w_{11}$ and $w_{12} < w_{22}$). The mathematics is exactly the same as that for overdominance but some of the signs are changed. In addition to equilibria at 0 and 1, an interior equilibrium also exists and is given by Equation 5.19 but now both the numerator and denominator are negative. An example is depicted in Figure 5.10a. If the initial allele frequency is exactly equal to the equilibrium value (in this example, $\hat{p} = 1/3$), then the allele frequency remains at that value. Otherwise, p goes to 0 or 1 depending on whether the initial allele frequency was less than or greater than the equilibrium value.

In the case of heterozygote inferiority, the interior equilibrium is unstable, whereas the equilibria at $p = 0$ and $p = 1$ are both locally (but not globally) stable. This is evident from Figure 5.10b, which shows that $\Delta p < 0$ when $p < \hat{p}$ and $\Delta p > 0$ when $p > \hat{p}$. Figure 5.10c plots the average fitness in the population, which in this case is a minimum at the interior equilibrium.

Natural examples of heterozygotes inferiority are rare, as one might expect because the polymorphism is unstable. One type of heterozygote inferiority that occurs quite frequently is a reciprocal translocation, in which two nonhomologous chromosomes undergo breakage and reunion. Reciprocal translocations originate as heterozygotes because the interchanged chromosomes are each accompanied by a structurally normal

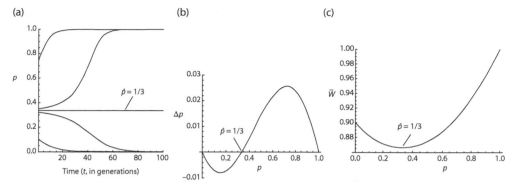

Figure 5.10 Heterozygote inferiority. (a) The allele frequency converges to 0 or 1 depending on its initial value. (b) Graph of Δp showing that the internal equilibrium is unstable whereas those at 0 and 1 are both locally stable. (c) The average fitness $\overline{w}$ is a minimum at the internal equilibrium. In this example $w_{11} = 1$, $w_{12} = 0.8$, and $w_{22} = 0.9$.

homolog. In meiosis, pairing and segregation of the reciprocal translocation and the normal homologous often results in the production of unbalanced gametes carrying large regions that are duplicated or deficient. These aneuploid gametes usually cause death of the gamete in plants or of the zygote in animals, and for this reason translocation heterozygotes are semisterile. The situation is one of heterozygote inferiority in regard to fertility, and because the translocation heterozygote must mate with a normal individual, the translocation-bearing offspring will again be semisterile. Rapid elimination of reciprocal translocations would therefore be expected. Yet the fixation of homozygous reciprocal translocations is commonplace in many lineages. How can this happen? The usual explanation is that in sufficiently small population random genetic drift can increase the frequency of the reciprocal translocation by chance until at some point $p > \hat{p}$, after which p goes to 1 because of selection. One problem with this model is that the chance of fixation is negligible unless the population size is extremely small, of the order of 5–15 individuals (Spirito 1992).

Evolutionary Change in Fitness

Equation 5.7 can be written as:

$$\Delta p = \frac{pq}{2\overline{w}} \frac{d\overline{w}}{dp} \tag{5.23}$$

This is a famous equation that connects the change in allele frequency p with the first derivative of fitness with respect to p (Wright 1937). With directional selection favoring A_1, the maximum average fitness occurs when $p = 1$; and with directional selection favoring A_2, the maximum average fitness occurs when $p = 0$. With overdominance or heterozygote superiority, an interior equilibrium exists and is given by $\hat{p}$ in Equation 5.19, but this equation also implies that $d\overline{w}/dp$ evaluated at $\hat{p}$ equals 0. Hence $\hat{p}$ occurs at either a maximum or a minimum of average fitness depending on whether $d^2\overline{w}/dp^2$ is negative or positive. Since $d^2\overline{w}/dp^2 = 2(w_{11} - 2w_{12} + w_{22})$, the average fitness at $\hat{p}$ is always a maximum with overdominance and a minimum with heterozygote inferiority (Figures 5.8c and 5.10c).

Maximization of average fitness is a frequent outcome of selection in random-mating populations with constant fitnesses, but there are many exceptions to the rule when random mating does not occur, when fitnesses are not constant, or when more than one gene is involved (Ewens 2004). Note particularly that $\overline{w}$ is the average fitness *in* the population, not the average fitness *of* the population. The relative survivorships are relevant only to the differential mortality of the genotypes within a population at any given time. The average of the relative survivorships $\overline{w}$ has no necessary relation to vernacular meanings of "fitness," such as competitive ability, population size, production of biomass, or evolutionary persistence (Haymer and Hartl 1983). The distinction is critical because, if the absolute mortality is high enough, a population can be going extinct even as selection is maximizing the average fitness in the population.

Another famous equation relates the rate of change in fitness to one component of the variance in fitness. The equation is easy to derive: What it means is another matter. To get the equation, let's start by assuming that Δp is small enough that $(\Delta p)^2$ is negligible. Then we can write:

$$\Delta \overline{w} = \overline{w}' - \overline{w} = (p + \Delta p)^2 w_{11} + 2(p + \Delta p)(q - \Delta p) w_{12} + (1 - \Delta p)^2 w_{22}$$

Expanding this expression and ignoring terms in $(\Delta p)^2$ yields:

$$\Delta \overline{w} \approx 2 \Delta p [p(w_{11} - w_{12}) + q(w_{12} - w_{22})]$$

and substituting for Δp from Equation 5.7 leads to:

$$\Delta \overline{w} = 2pq[p(w_{11} - w_{12}) + q(w_{12} - w_{22})]^2 / \overline{w} \tag{5.24}$$

The numerator in Equation 5.24 is known as the **additive genetic variance** in fitness (also called the **genic variance** in fitness). As we shall see in Chapter 8, the additive genetic variance plays a predominant role in predicting the transmission of traits determined by the joint action of multiple genes such as, in this case, fitness. At this point all we need is a symbol to represent the additive genetic variance in fitness, and for convenience let's call it σ_a^2. Equation 5.24 therefore becomes:

$$\Delta \overline{w} = \sigma_a^2 / \overline{w} \tag{5.25}$$

which says that the change in average fitness at any time is equal to the additive genetic variance in fitness at that time. This is the usual statement of what is commonly called the **fundamental theorem of natural selection**, originally due to Fisher (1922). Ewens (1989) provides an illuminating discussion and argues that what Fisher meant by "average fitness" is not the $\overline{w}$ defined in Equation 5.6 but something quite different, nor does the theorem apply to changes across generations but only to changes due to differential survival within a generation. Part of the decades-long misinterpretation of Fisher's principle is due to his own cryptic statement of the principle (Edwards 2002). Ewens (1989) concludes that Fisher's original theorem is mainly a mathematical nicety with little if any biological significance. The interpretation of the theorem in terms of across-generations changes in average fitness (Equation 5.25) is due to Wright (1988). While Equation 5.25 is valid for many simple models of selection, average fitness does not increase in all models, and the equilibrium of average fitness is not always a maximum.

5.3 Mutation–Selection Balance

Outcrossing species typically contain a large amount of hidden genetic variability in the form of recessive or nearly recessive harmful alleles at low frequencies. This situation is to be expected. Selection cannot completely eliminate harmful alleles because of their continual creation through recurrent mutation. In this case, selection and mutation eventually reach a state of balance in which the elimination of harmful alleles by selection is exactly offset by the appearance of new harmful alleles by mutation.

Equilibrium Allele Frequencies for Recessive and Partially Dominant Mutations

Suppose A_2 is a harmful allele and that mutation of A_1 to A_2 occurs at the rate μ per generation. Because q, the allele frequency of A_2, remains small, reverse mutation of A_2 to A_1 can safely be ignored. When the allele frequencies change slowly, the continuous approximation for Δp in Equation 5.8 is still valid, except that a proportion $p\mu$ of A_1 alleles mutate to A_2 in the course of each generation. Therefore:

$$\frac{dp}{dt} = pqs[ph + q(1 - h)] - p\mu \tag{5.26}$$

When selection is balanced by recurrent mutation, there is a globally stable equilibrium at the allele frequencies $\hat{p}$ and $\hat{q}$ that solve Equation 5.26 for $dp/dt = 0$. The two cases of greatest interest are:

1. **When the harmful allele is completely recessive** ($h = 0$). In this case the equilibrium equation becomes $q^2 s = \mu$, and therefore

$$\hat{q} = \sqrt{\frac{\mu}{s}} \tag{5.27}$$

2. **When the harmful allele is partially dominant** ($h > 0$). Here the equilibrium equation is $pqhs + q^2 s(1 - h) = \mu$, and since $\hat{p} \approx 1$ and $\hat{q}^2 \approx 0$,

$$\hat{q} = \frac{\mu}{hs} \tag{5.28}$$

Since Equation 5.27 involves the square root of the mutation rate, whereas Equation 5.28 involves the mutation rate itself, the equilibrium frequency $\hat{q}$ in Equation 5.27 is considerably larger than that in Equation 5.28. In other words, partial dominance even as small as 1 per cent ($h = 0.01$) or 2 per cent ($h = 0.02$) can have a significant effect on decreasing the equilibrium frequency of a harmful recessive allele maintained by recurrent mutation.

To take a specific example, imagine a gene that mutates to a homozygous-lethal allele ($s = 1$) at the rate $\mu = 10^{-6}$ per generation. If the allele were completely recessive ($h = 0$), the equilibrium allele frequency would be $\sqrt{(\mu/s)} = 0.001$. Contrast this case with a situation in which the allele exhibits a small degree of dominance, say $h = 0.01$. Then the equilibrium allele frequency is given by $\mu/(hs) = 0.0001$. The small degree of dominance (the relative fitness of the heterozygous genotype is 0.99) reduces the equilibrium allele frequency by a factor of 10. Why the big difference? Because when an allele is rare, there is a vastly greater number of heterozygous than homozygous genotypes, and even mild selection against so many heterozygotes has a major effect in reducing the equilibrium allele frequency.

Equations 5.27 and 5.28 can be used to determine the long-term effect of mutation–selection balance on the average fitness in the population. This is known as the **mutation load**, and its formal definition is:

$$L = \frac{w_{max} - \overline{w}}{w_{max}} \tag{5.29}$$

where w_{max} is the maximum average fitness that would be achieved if there were no deleterious mutation and $\overline{w}$ is the actual average fitness of a population with deleterious mutation (Crow and Kimura 1970). For a population with no deleterious mutation, $w_{max} = 1$. In the case of a deleterious complete recessive $(h = 0)$ at equilibrium, $\overline{w} = 1 - \hat{q}^2 s = \mu$, and therefore $L = \mu$. On the other hand, in the case of partial dominance, $\overline{w} = 1 - 2\hat{p}\hat{q}hs - \hat{q}^2 s \approx 1 - 2\hat{q}hs = 1 - 2\mu$ because $\hat{p} \approx 1$ and $\hat{q}^2 \approx 0$, and therefore $L = 2\mu$. In other words, as Haldane (1937) first pointed out, the fitness cost of mutation–selection balance depends only on the mutation rate irrespective of the harmfulness of the mutation.

The mutation load is the biological cost of deleterious mutations. In human populations this has been estimated as about 1 per cent (Lynch 2016). Deleterious mutations in humans also have a social cost. Modern medicine's pharmaceuticals, diagnostics, surgical procedures, and other interventions to ameliorate symptoms of genetic disorders will result in relaxed selection against deleterious mutations and a slow increase in the frequency of the underlying genes. The steady increase in these conditions will entail ever-greater amounts of healthcare resources devoted to them (Lynch 2016).

There are many kinds of genetic loads other than the mutation load. These include genetic loads due to migration, heterogeneous environments, the segregation of over-dominant alleles, recombination, meiotic drive, gametic selection, and so forth. They are all defined by analogy with Equation 5.29, where w_{max} is the maximum average fitness in the absence of some process or situation and $\overline{w}$ is the actual average fitness in its presence. Crow and Kimura (1970) give a thorough discussion of the varieties and magnitudes of genetic load.

Degree of Dominance of Severely Versus Mildly Deleterious Mutations

An important principle of population genetics is that mildly deleterious mutations usually exhibit a greater degree of dominance than severely deleterious mutations. In *Drosophila*, for example, the estimated degree of dominance of recessive lethals $(s = 0)$ is in the range $h = 0.01 - 0.03$ (Simmons and Crow 1977), whereas the average degree of dominance of mildly deleterious mutations (average $s = 0.09$) is in the range $h = 0.30 - 0.50$ (Mukai et al. 1972; Ohnishi 1977). The range of h for mildly deleterious mutations means that these mutations are very nearly additive in their effects.

The principle that the degree of dominance increases as the deleteriousness of the mutation decreases at first seems paradoxical, but it is a simple consequence of how genes usually affect phenotypes. In Figures 5.11a and 5.11b, the solid curve traces a pattern of saturation kinetics typical of many biochemical reactions and metabolic pathways in which flux of metabolites through the pathway shows diminishing returns (Hartl et al. 1985). As activity on the x-axis increases, the curve converges asymptotically to a plateau, after which a further increase in enzyme activity results in a negligible increase in flux. Diminishing returns is expected in cases in which flux is limited by substrate availability or in linear metabolic pathways with fixed inputs (Kacser and Burns 1973); however, the model applies not only to enzyme catalysis but also to models of small-molecule adsorption, force generation by motor proteins, gene regulation, transcription, translation, protein modification, chromatin remodeling, and many other biological processes (Wong et al. 2018). Saturation kinetics also provides the framework for the molecular theory of dominance (Wright 1934), and it justifies the assumption that fitness functions are generally concave.

The curves in Figures 5.11a and 5.11b are generic saturation curves, but they serve to illustrate the point we wish to make about degree of dominance. Figure 5.11a shows

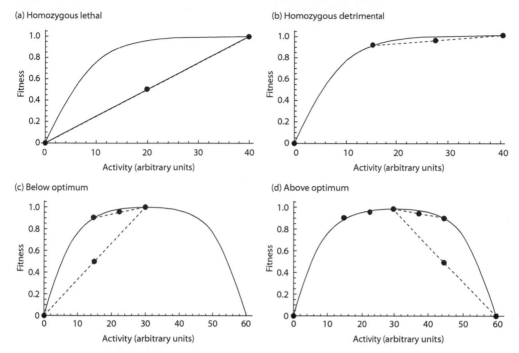

Figure 5.11 Fitness as a concave function of enzyme (or gene) activity. (a) A severely deleterious allele is expected to be largely but not completely recessive. (b) A mildly deleterious allele is expected to be more nearly additive. (c and d) Analogous curves when enzyme or gene activity has an intermediate optimum that maximizes fitness, again showing that mildly deleterious alleles on either side of the optimum are expected to be more nearly additive than severely deleterious alleles.

the expected degree of dominance of a mutation that is severely deleterious when homozygous. The dots show the expected fitnesses of genotypes that are homozygous mutant, heterozygous, and homozygous wildtype, assuming that the mutant allele's effect is additive. The actual fitness is as shown in the saturation curve, and the great distance between the expected and actual fitness is an indication that the severely deleterious allele is largely recessive. Figure 5.11b shows a mildly deleterious allele, where the dots again signify the expected fitnesses with additivity, but in this case the expected fitness of the heterozygous genotype is not so far from that expected with additivity. In other words, given the fitness activity curve in Figures 5.11a and 5.11b, the degree of dominance should be negatively correlated with severity of mutation (Hartl et al. 1985).

Similar reasoning applies when activity has an intermediate optimum, which is usually the case when "activity" refers to level of gene expression (Bedford and Hartl 2009). Figures 5.11c and 5.11d contrast mildly and severely deleterious mutations in this case, and again the degree of dominance is greater for mildly deleterious mutations regardless of whether the mutation decreases gene expression (Figure 5.11c) or increases it (Figure 5.11d).

Background Selection

The level of nucleotide diversity π across the genome of *Drosophila* decreases with the rate of genetic recombination (Begun and Aquadro 1992). This pattern is illustrated in Figure 5.12a for nonoverlapping 500 kb windows across the genome of *D. melanogaster*.

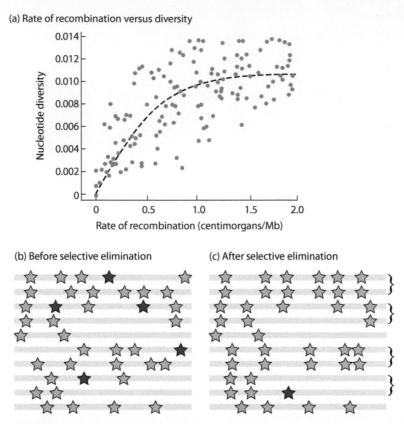

Figure 5.12 Background selection. (a) Genetic diversity is reduced in genomic regions of low recombination, which could result from recurrent selective sweeps or background selection. (b) Haplotypes present in a population with gray stars representing neutral mutations and black stars deleterious mutations. (c) Haplotypes present after those with deleterious mutations have been eliminated, showing increased representation of certain haplotypes and overall reduction in genetic diversity. (Data in (a) from Corbett-Detig et al. (2015).)

Similar patterns are observed across a wide range of taxonomic groups including plants, insects, birds, fish, and mammals (Corbett-Detig et al. 2015). Two diametrically opposed hypotheses have been put forward to explain this finding. One attributes it to successive selective sweeps that reduce genetic variation in a surrounding region of a size inversely proportional to the recombination rate. The other explanation attributes the pattern to deleterious mutations and a process called background selection.

In **background selection**, the low level of polymorphism in regions of tight linkage is attributed to deleterious mutations. Each deleterious mutation that occurs dooms some little region of chromosome in its vicinity to eventual extinction. The lower the rate of recombination, the larger the region of chromosome that is doomed. In effect, each new deleterious mutation reduces by one the number of chromosomes that can contribute to remote future generations.

The effect of background selection is illustrated in Figures 5.12b and 5.12c, where the black and gray stars represent deleterious and neutral mutations, respectively. Figure 5.12b shows a set of haplotypes present in a population prior to the time that the deleterious mutations present are eliminated by selection and Figure 5.12c shows the haplotypes that persist after the deleterious mutations have gone extinct. Note that certain haplotypes (grouped by the braces on the right) become overrepresented merely because they happen to have fewer or less deleterious mutations than other haplotypes. Superficially, the situation resembles what might happen in a soft selective sweep. The black star in Figure 5.12c serves as a reminder that the creation of new deleterious mutations is continuous.

If there is complete linkage (no recombination), as in *Drosophila* chromosome 4, then background selection affects the whole chromosome; but if there is recombination, then only a region around the mutation is affected. In either case, a sufficient density of deleterious mutations will reduce the number of surviving lineages to such an extent that the degree of polymorphism will be smaller than expected, given the actual population size, and the tighter the linkage the greater the disparity (Charlesworth 2009; Charlesworth et al. 1993; Hudson and Kaplan 1995; Stephan 2010). In effect, background selection reduces the effective population number for regions of tight linkage, thereby reducing the nucleotide diversity.

Both selective sweeps and background selection are well documented (Comeron 2017; Stephan 2010); however, disentangling their relative contributions in any particular instance remains difficult. Because outcrossing populations include large numbers of deleterious mutations maintained by mutation–selection balance, as evidenced by inbreeding depression in maize (Figure 3.3) and the effect of whole-chromosome homozygosity in *Drosophila* (Figure 3.4), a case can be made that the background selection is likely pervasive (Comeron 2017) whereas selective sweeps are episodic.

Balance Between Migration and Selection

Just as recurrent mutation can be balanced against selection and result in an equilibrium allele, so can recurrent migration. There are two main differences. First, whereas opposing mutation and selection will almost always lead to a stable equilibrium, opposing migration and selection will often not. Second, whereas with mutation–selection balance the equilibrium frequency of the deleterious allele is usually very small, with migration–selection balance it need not be small.

To illustrate the point we'll consider a simple model of one-way migration from a founder population into a satellite population (Yeaman and Otto 2011). The founder population is often imagined to be on a continental mainland and the satellite population on an island. Suppose that in the satellite population the relative fitnesses of genotypes A_1A_1, A_1A_2, and A_2A_2 are 1, $1-s$, and $(1-s)^2$, respectively, and suppose genotype frequencies in any generation are respectively p^2, $2pq$, and q^2. The allele A_1 is therefore beneficial in the satellite population. Suppose further that in the founder population (and source of migrants) the allele A_2 has a frequency q_f. Then with migration rate m per generation from founder to satellite population, we can follow the logic of Equation 4.38 in Chapter 4 and write:

$$q' = (1-m)\left(\frac{pq(1-s)+q^2(1-s)^2}{p^2+2pq(1-s)+q^2(1-s)^2}\right) + mq_f \qquad (5.30)$$

where the built-up fraction makes use of Equation 5.5 to represent the effects of selection, and the term mq_f represents the effects of recurrent migration.

What motivates this particular choice of fitnesses is that the numerator of the fraction in Equation 5.30 can be written as $q(1-s)[p+q(1-s)]$ and the denominator as $[p+q(1-s)]^2$, which leads to a convenient cancellation. (Note also that when s is small, the fitnesses are approximately $1:1-s:1-2s$, hence the alleles are nearly additive.)

Setting $q' = q$ in Equation 5.30 and solving for q results in quadratic equation with rather messy roots. However, if A_2 is fixed in the founder population, then $q_f = 1$ and the roots are $q = 1$ and $q = m/s$. These tell the story, because if $m >> s$ then A_2 goes to fixation on the island $(q = 1)$ because the selection for A_1 is overwhelmed by migration. At the other extreme, if $m << s$ then $q = m/s$ is a stable equilibrium but q is very small (as in mutation–selection balance). But when m is between these extremes then $q = m/s$ is a stable equilibrium at intermediate frequency maintained by selection balanced against recurrent migration. Moreover, since m can be large relative to the rate of mutation, the equilibrium m/s can in principle be anywhere in the range 0 to 1.

An instructive example of migration–selection balance comes from pigment pattern of water snakes *Natrix sepidon* on islands in Lake Erie off the coast of Ohio (Camin and Ehrlich 1958). The water snakes come in three phenotypes: darkly banded, intermediate, or unbanded. On the Ohio mainland nearest the islands, unbanded phenotypes are virtually absent. On the islands they are more common, most likely because the main predator of water snakes consists of several species of gulls, which can spot banded snakes more readily than unbanded snakes. Random drift seems to play a minor role because the animals are very abundant even on the islands. In the simple model embodied in Equation 5.30, unbanded phenotypes correspond to the genotype A_1A_1. Judging from the difference in abundance of unbanded snakes between newborns and adults, it appears that $s \approx 0.3$ (Camin and Ehrlich 1958). On Kelleys Island, about 5 miles from the mainland, the allele frequency $p = 1 - (m/s) \approx 0.32$, which implies that $m \approx 0.20$, whereas on Middle Island, about 10 miles from the mainland, the allele frequency $p \approx 0.77$, which implies that $m \approx 0.07$. Migration therefore seems to decrease as roughly the square of the distance from the mainland, which might perhaps be expected.

In the absence of any visible phenotype like bandedness in water snakes, how might one detect local adaptation maintained by migration–selection balance? One approach is to exploit the tendency of migration to reduce measures of population differentiation such as F_{ST} as expressed in Equation 4.43 and illustrated in Figure 4.14. In the absence of local adaptation, the F_{ST} values of a large number of SNPs or other genetic markers across the genome will tend to have a homogeneous distribution. F_{ST} values that are outliers in the distribution identify regions of the genome that are candidates for local adaptation with genetic differences among populations maintained by migration–selection balance. This approach of F_{ST} scanning has been used to identify a suite of genes affecting traits associated with adaptation to high altitudes in the speckled teal (*Anas flavirostris*) in the Andes (Graham et al. 2018).

The term **landscape genetics** has come to be used for attempts to explain how geographical and environmental features affect gene flow and population structure (Manel and Holderegger 2013; Richardson et al. 2016). Some authors make a distinction between landscape genetics and **landscape genomics**, using the latter term to mean the use of F_{ST} scanning or other methods to identifying candidate genes for possible local adaptation (Storfer et al. 2018).

5.4 Gametic Selection and Meiotic Drive

To this point we've considered selection in haploid and diploid organisms, but many plants go through a life cycle in which both haploid products of meiosis and the diploid products of fertilization are exposed to selection. In mosses and vascular plants, for example, a diploid organism (the sporophyte) produces spores, each of which germinates to form a haploid organism (the gametophyte) that reproduces asexually by mitosis. The gametophytes give rise to haploid male and female gametes, which undergo fertilization creating a new diploid generation. In mosses, the prominent stage of the life cycle is the gametophyte, whereas in higher plants, the prominent stage is the sporophyte.

Gametic Selection

When the haploid phase of the life cycle is exposed to selection, the selection is called **gametic selection**. As a concrete model, suppose that the relative survivorships of A_1 and A_2 gametophytes (the haploid phase) are given by v_1 and v_2, respectively. In the sporophytes (the diploid phase), the survivorship can be written as before as w_{11}, w_{12}, and w_{22}. If p and q are the allele frequencies of A_1 and A_2 at the beginning of the haploid phase after meiosis, then the allele frequencies after differential haploid mortality has taken place can be written as $p_{ags} = pv_1/\bar{v}$ and $q_{ags} = qv_2/\bar{v}$, where the subscript *ags* means *after gametic selection* and $\bar{v} = pv_1 + qv_2$. These equations are similar to those given in Equation 5.3 for selection in haploids, but in this case we are only part way through the life cycle. With random fertilization among the gametes, the diploid zygotes A_1A_2, A_1A_2, and A_2A_2 are formed in the proportions p_{ags}^2, $2p_{ags}q_{ags}$, and q_{ags}^2, and these survive to sexual maturity in the relative proportions w_{11}, w_{12}, and w_{22}. You may verify for yourself that, at the beginning of the haploid phase of the next generation, the allele frequency p' of A_1 is given by:

$$p' = \frac{p^2 v_1^2 w_{11} + pq v_1 v_2 w_{12}}{p^2 v_1^2 w_{11} + 2pq v_1 v_2 w_{12} + q^2 v_2^2 w_{22}} \tag{5.31}$$

This equation has the same form as the equation for p' in Equation 5.5 except that w_{11} is replaced with $v_1^2 w_{11}$, w_{12} with $v_1 v_2 w_{12}$, and w_{22} with $v_2^2 w_{22}$. The conditions for fixation or for a stable or unstable equilibrium are therefore determined by the relative magnitudes of these composite fitnesses. In particular, an equilibrium analogous to overdominance exists if $v_2 w_{12} > v_1 w_{11}$ and $v_1 w_{12} > v_2 w_{22}$.

Gametic selection is more efficient at changing allele frequencies than is zygotic selection. This principle is shown in Figure 5.13, which compares allele frequency change under gametic viability selection (gray) with that under zygotic viability selection (black)

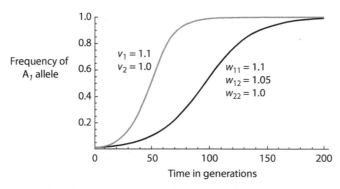

Figure 5.13 Comparison of allele frequency change with gametic selection (gray) or zygotic selection (black). The curve for gametic selection assumes a 10 per cent difference in gametic viability with equal zygotic fitnesses, whereas that for zygotic selection assumes equal gametic fitnesses with a 10 per cent difference in zygotic viability of the homozygous genotypes and additivity.

implied by Equation 5.31. Fixation of the favored allele occurs substantially faster for gametic selection than for zygotic selection.

Meiotic Drive

A situation analogous to, but distinct from, gametic selection takes place when there is non-Mendelian segregation in the heterozygous genotype. In females, unequal recovery of reciprocal products of meiosis can be caused by nonrandom segregation of homologous chromosomes to the functional egg nucleus, which is why non-Mendelian segregation is known generically as **meiotic drive**. In other cases, the cause of the unequal recovery is a gene or genes that act to render gametes carrying the homologous chromosome nonfunctional. Examples include "sperm killers" such as segregation distortion in *D. melanogaster* (Charlesworth and Hartl 1978; Hartl et al. 1967; Larracuente and Presgraves 2012) and the *t* alleles in the house mouse (Lewontin and Dunn 1960; Safronova and Chubykin 2013) as well as "spore killers" described in filamentous fungi (Hammond et al. 2012; Raju 1994).

Because meiotic drive acts only in the heterozygous genotype, its effect is to alter the term pqw_{12} in Equation 5.5 for p'. This term comes from the expression $(1/2) \times 2pqw_{12}$ for the proportion of A_1-bearing gametes from surviving A_1A_2 genotypes, and the $1/2$ is the Mendelian segregation ratio. If the ratio of functional $A_1 : A_2$ gametes from A_1A_2 heterozygotes is $k: 1 - k$ instead of $1/2 : 1/2$, then the expression for p' becomes

$$p' = \frac{p^2 w_{11} + 2kpqw_{12}}{p^2 w_{11} + 2pqw_{12} + q^2 w_{22}} \tag{5.32}$$

which is identical to Equation 5.5 except for the factor $2k$ in the numerator. Because A_1 is the driven allele, $k > 1/2$.

Equation 5.32 is illustrative of meiotic drive even though it requires that the non-Mendelian segregation affect both sexes equally, a case that is not generally found in practice. One implication of the equation is that, unless selection counteracts the meiotic drive, the driven allele goes to fixation. In particular, if the relative viabilities are equal, then $p' = p^2 + 2kpq$ and $\Delta p = pq(2k - 1)$, so that p goes to 1 because $k > 1/2$. As with gametic selection, meiotic drive is very efficient in changing allele frequencies. A comparison of allele frequency change with meiotic drive versus zygotic viability selection is shown in Figure 5.14. The solid curves correspond to meiotic drive and the dashed curves to zygotic viability selection with fitnesses given by $w_{11} = 1 + 2s$, $w_{12} = 1 + s$, and $w_{22} = 1$. With these fitnesses, Equation 5.5 implies that, for $p \approx 0$, $\Delta p \approx pqs$, hence $s \approx 2k - 1$. The comparisons in Figure 5.14 are for $k = 0.60$ and $k = 0.75$.

In most known cases, the driving allele results in some loss in fitness. This may be an artifact of discovery, because in the absence of some counteracting fitness loss, the driving allele would be fixed so fast that it remains undiscovered. In some examples of meiotic drive, including segregation distortion and the *t* alleles, the driven allele is lethal when homozygous (Hartl 1970). Assuming that the lethality is completely recessive, the survivorships are $w_{11} = 0$, $w_{12} = 1$, and $w_{22} = 1$. Equation 5.32 implies that $p' = 2kp/(1 + p)$ and so $\Delta p = p[(2k - 1) - p]/(1 + p)$. There is an interior equilibrium at $\hat{p} = 2k - 1$, which intuition suggests (correctly) is locally stable; it is also globally stable. Note that the equilibrium $\hat{p}$ is between 0 and 1 for any value of k between $1/2$ and 1. More generally, for any set of relative fitnesses, a stable interior equilibrium exists if $1/(2w_{12}) < k < 1 - w_{11}/(2w_{12})$.

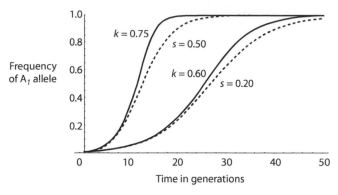

Figure 5.14 Comparison of allele frequency change with meiotic drive (solid curves) or zygotic selection (dashed) for $k = 0.60$ and $k = 0.75$. The value of $s = 2k - 1$.

A gene with alleles affected by meiotic drive can thwart the driver if new alleles can arise by mutations that are insensitive to being driven. An insensitive allele is also called a **suppressor** of the meiotic drive. Suppressor alleles can persist in the population while the driver is displacing the sensitive alleles. If the drive has a fitness cost, suppressor alleles can increase in frequency at the expense of the driver. Along the way, a new driver allele might arise by mutation that can target the suppressor, and this new driver will increase in frequency until still newer suppressors arise, and on and on the system churns in a sort of arms race as successive drivers and suppressors arise and displace their counterparts. Just such an arms race has been reported for the *Segregation Distorter* (*SD*) gene complex in *D. melanogaster* in Africa, where an *SD* driver allele underwent a recent selective sweep that dramatically reduced genetic variation in its vicinity (Presgraves et al. 2009).

Gene Drive

Recent discoveries and technical developments in molecular biology have made it possible to engineer genetic systems that resemble gene conversion and that in populations act like meiotic drive. Chief among these are methods relying on the **CRISPR/Cas9** system for manipulating DNA sequences. CRISPR stands for *clustered regularly interspaced short palindromic repeats*. These repeats were first discovered in bacteria. They form part of the bacterial defense system against viruses and other invading DNA molecules. Bits and pieces of the invading DNA are sequestered in the CRISPR region of the bacterial genome, and these are used as templates for molecules to recognize and destroy the invading DNA. A key discovery was that the Cas9 protein, which cleaves the invading DNA eventually leading to its destruction, is guided to its target by a short sequence of RNA, known as guide RNA, which is complementary in sequence to part of the target (Deltcheva et al. 2011). The CRISPR repeats serve as templates for the guide RNA. When the guide RNA is synthesized and binds with its target sequence, it recruits Cas9 protein that then cleaves the target sequence to initiate its destruction by other nucleases in the cell.

Chemically synthesized guide RNA and Cas9 can be used to engineer precise, predetermined changes in DNA sequence in any organism. This type of genetic engineering, called **DNA editing**, has far-reaching applications in medicine, agriculture, research, and other areas. Applied in such a way as to alter the genotypes of organisms in a natural population, the method is known as **gene drive** (Gantz and Bier 2015; Oberhofer et al. 2019). One version of gene drive using guide RNA and Cas9 protein is outlined in Figure 5.15. This

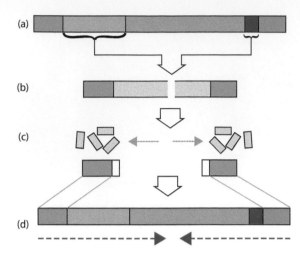

Figure 5.15 Gene drive. (a) Expression of a cassette including sequences for Cas9 (green) and guide RNA (red) catalyze a site-specific cleavage in target DNA (b), whereupon exonucleases widen the gap (c), which is then repaired using the cassette as template (d).

method not only edits genomic DNA, but also it introduces coding sequences for guide RNA and Cas9 at the site of the editing.

The starting point for gene drive is to introduce a DNA cassette like that in Figure 5.15a into an organism. The cassette contains coding sequences for guide RNA (red) and Cas9 (green) along with a template to alter the sequence of some gene in a prescribed way (orange). The cassette also includes genomic sequences flanking these components (blue) that allow the cassette to be used as a template for DNA repair.

When introduced into an organism, the gene-drive cassette produces guide RNA and Cas9, which act together to cleave the targeted gene (gray) at a predetermined site (Figure 5.15b). Note that the target sequence is flanked by genomic regions (blue) that are homologous to those flanking the gene-drive cassette. After cleavage of the target site, nucleases in the cell digest the target DNA working outward from the cleavage site (gray arrows), thereby widening the gap (Figure 5.15c). The gap is repaired by normal repair synthesis, using the gene-drive cassette as a template, first by pairing the homologous sequences at the ends of the target sequence and the cassette and then synthesizing new DNA to close gap (Figure 5.15d, red arrows). The end result is that the engineered DNA in the cassette replaces the target sequence and, in addition, coding sequences for the guide RNA and Cas9 are introduced.

By means of the process in Figure 15.15, a gene-drive cassette can sequentially convert alleles in a natural population into gene-drive cassettes just like itself in a process that has been called the *mutagenic chain reaction* (Gantz and Bier 2015). The system resembles gene conversion in that one allele is used as a template to convert another allele. It also resembles meiotic drive in that the heterozygous genotypes produce an excess of gametes carrying the gene-drive cassette.

To understand the similarity between gene drive and meiotic drive, let A_1 be the gene-drive allele and A_2 be the target with respective frequencies p and q. Suppose the relative viabilities of the genotypes A_1A_1, A_1A_2, and A_2A_2 are given by w_{11}, w_{12}, and w_{22}, respectively, and let c be the probability that, in an A_1A_2 genotype, the A_2 allele is edited and replaced with A_1. For simplicity we assume that the process of gene editing occurs immediately prior to meiosis and gamete formation.

With these stipulations it is straightforward to derive an equation for allele frequency change. For any allele frequency p in gametes, the frequencies of A_1A_1, A_1A_2, and A_2A_2 zygotes are given by p^2, $2pq$, and q^2. These survive in the proportions $w_{11} : w_{12} : w_{22}$, and so, after viability selection, the genotype frequencies are in the ratio $p^2 w_{11} : 2pqw_{12} : q^2 w_{22}$. The A_1A_1 genotypes produce only A_1 gametes and the A_2 genotypes produce only A_2 gametes. In A_1A_2 heterozygotes, a fraction c of the cells are converted to A_1A_1 prior to meiosis, and these produce all A_1 gametes; and a fraction $1 - c$ of the cells remain unconverted, and these produce A_1 and A_2 gametes in the proportions $1/2 : 1/2$. Putting all this together, the expected frequency p' of the gene-drive allele A_1 in gametes forming the next generation is:

$$p' = \frac{p^2 w_{11} + 2pqw_{12}\left[c + (1/2)(1-c)\right]}{p^2 w_{11} + 2pqw_{12} + q^2 w_{22}} = \frac{p^2 w_{11} + (1+c)pqw_{12}}{p^2 w_{11} + 2pqw_{12} + q^2 w_{22}} \qquad (5.33)$$

This is identical to Equation 5.32 except that $2k$ is replaced with $1 + c$, hence gene drive is like meiotic drive with $k = (1 + c)/2$.

As with meiotic drive, fixation of the gene-drive allele can be slowed or stopped if the gene-drive allele has a sufficient fitness cost. A more important impediment is the occurrence of mutant alleles that resist gene editing by the gene-drive cassette. Such mutations can arise in several ways including base substitutions in the target sequence that prevent or inhibit binding of the guide RNA. A more important source of resistant alleles is inherent in the gene-editing machinery itself, because the gap repair process can be short-circuited by joining the digested ends of the target sequence in Figure 5.15. This aborted gap repair process is known as **nonhomologous end joining**, and in deleting the entire sequence to which the guide RNA binds, it creates an allele that is immune to the gene-drive cassette. Such alleles arise at a frequency as high as 1 per cent, and they undermine the potential of gene drive to edit the genomes of natural populations (Unckless et al. 2017). Add to this genetic hurdle the many issues of potential concern in the release of any genetically engineered organism (Esvelt and Gemmell 2017; Gantz and Bier 2015; Gould et al. 2019; Noble et al. 2019), and it becomes unclear how, where, when, or whether gene drive may actually be used to manipulate the genetics of any natural population.

5.5 Other Modes of Selection

So far in this chapter we've considered only the simplest types of selection involving two alleles of a single gene with constant fitnesses determined by differences in zygote viability, gamete viability, or meiotic drive and gene drive. In this section we briefly define some other modes of selection to illustrate the full range and complexity of the issues. Detailed discussion of each of the scenarios is beyond the scope of this book, but in this section we list some cases for the sake of illustration and provide some brief general comments.

Complexities arise when the fitnesses of the genotypes or changes in allele frequency depend on:

- Multiple alleles: despite their analytical complexity, multiple-allele models with constant fitnesses and random mating still maintain the principles that average fitness increases to reach a maximum at equilibrium and that the change in average fitness in any generation is approximately equal to the additive variance in fitness (Ewens 2004). On the other hand, the conditions required for large numbers of alleles to be maintained in a stable equilibrium are quite restrictive (Ewens 2004).

- Multiple loci and gene interaction (epistasis or epistatic selection): among the multiplicity of possible gene interactions and linkage relations, some scenarios contradict findings from one-locus models. Among the surprises are that, unless the loci are in linkage equilibrium, average fitness may decrease for some periods of time, average fitness is not necessarily maximized at equilibrium, and the change in average fitness is not necessarily related to the additive genetic variance in fitness (Ewens 2004).

- Chromosomal location of the gene (X-linked or Y-linked selection): X-linkage has the feature that the allele frequency in male zygotes equals the allele frequency of female gametes in the previous generation, and with random mating the allele frequency in female zygotes equals the average of the allele frequencies in male and female gametes in the previous generation. With selection, the allele frequencies may not be equal in the sexes. For an X-linked recessive lethal, for example, the allele frequency in male gametes equals 0, and all females who carry the allele are heterozygous. The dynamics of Y-linked selection is equivalent to selection in haploids. For X-linked genes the inbreeding effective size equals 3/4 that of autosomal genes, and for Y-linked genes it is 1/4 that of autosomes. In both cases, the smaller effective size implies less genetic variation (Wilson Sayres 2018).

- Cellular location of the gene (mitochondrial or chloroplast selection): selection on these organelles is equivalent to haploid selection, and the inbreeding effective size is 1/4 that of autosomal genes resulting in reduced genetic variation (Wilson Sayres 2018).

- Allele frequencies or genotype frequencies (frequency-dependent selection): it is unclear how important frequency-dependent selection is in nature, however the details of allelic interaction in physiological and neurological systems suggest it could be. One review concludes that "rock-paper-scissors dynamics are common" (Sinervo and Calsbeek 2006). (In a genetic context, rock-paper-scissors dynamics means that allele A_1 is superior to allele A_2, A_2 is superior to A_3, but A_3 is superior to A_1.) There are so many ways that fitnesses can be frequency dependent that one might expect few general principles to emerge; however, with random mating and two alleles of one gene when the genotypic fitnesses are given by polynomials of allele frequency, a simple function of allele frequency can be defined that is nondecreasing and attains a maximum at equilibrium with a rate of change approximately equal to the additive genetic variance in fitness (Curtsinger 1984).

- Rarity of a given allele or genotype (rare-allele advantage, diversifying selection): this is a special but very important class of frequency-dependent selection, and examples are found in genes encoding antigens in parasites and their counterparts in the host immune system (Weedall and Conway 2010), as well as self-incompatibility alleles in plants (Newbigin and Uyenoyama 2005). These are examples of **balancing selection**, a term that encompasses all forms of selection that tend to maintain genetic variation (Fijarczyk and Babik 2015).

- Population size (density-dependent selection): population sizes in nature can differ greatly from one period to the next owing to changes in climate and other factors. Such variation creates situation in which genotypes with superior survival and reproduction at low population density may perform poorly at high population density. This type of selection is often called *r*- **and *K*-selection** after the maximum-growth-rate (*r*) and carrying capacity (*K*) parameters in the standard logistic model of population growth (Equation 1.7). Among many examples is the clutch size of great tits (*Parus major*) in the Netherlands (Saether et al. 2016). In these birds, larger clutch sizes are favored at low population density and smaller clutch sizes at high population density. Furthermore, females with large clutch sizes at low population sizes (conditions favoring *r* selection) show the greatest reduction in fitness when population density increases (conditions favoring K selection) (Saether et al. 2016).

- Age or developmental stage of the organism (overlapping generations): in this case each genotype has its own age-dependent viability and fecundity, which may also depend on sex (Charlesworth 1980).

- The particular genotypes present in each mating pair (fecundity selection): fecundity selection results in equations analogous to those for viability selection provided that the relative fecundity of a mating pair depends on the product of the relative fecundities of the mating genotypes. If this is not the case, then each possible mating pair has its own fecundity, which greatly increases the number of fitness parameters.

- Environmental fluctuations (heterogeneous environments): when the environment varies in space, the average fitness of a genotype equals the arithmetic average of the fitnesses in the different environments; but when the environment fluctuates in time, it is the geometric mean of the fitnesses that matters (Saether and Engen 2015).

- Sex of the organism (sex-limited selection): many genes are expressed differently, or expressed to different degrees, in the two sexes, and some traits are selected in only one sex (e.g., egg production). In these situations, selection impacts the sexes unequally, in some cases in opposite directions leading to genomic conflicts (Patten et al. 2010).

- Success in attracting mates (sexual selection): competition for mates seems to be a kind of playground for evolution that can result in flamboyant behavioral displays and unusual or overgrown ornaments, colors, calls, or odors that attract attention and are present in only one sex. Darwin (1871) first called attention to sexual selection, writing that it depends "on the ardour of love, the courage, and the rivalry of the males, as well as on the powers of perception, the taste, and will of the female." Sexual selection entails the coevolution of traits in males and females, and models of sexual selection include an interesting mix of genetics, behavior, demography, and population dynamics (Kuijper et al. 2012).

- Behaviors enhancing the fitness of genetically related individuals (kin selection): famed geneticist J. B. S. Haldane (1932) averred that he would lay down his life for two siblings, four nephews, or eight cousins, which makes genetic sense because an individual shares 1/2 of his or her genome with a sibling, 1/4 with a niece or nephew, and 1/8 with a cousin. The underlying point was that, in terms of fitness—contribution of genes to future generations—each set of relatives is an equal tradeoff. Taking this argument a step further, one's own genetic contribution to future generations would be enhanced by saving more than two siblings, four nephews, or eight cousins. Just such an argument has been made for the evolution of altruism, behaviors that result in individuals sacrificing themselves to aid in the reproduction their relatives, as observed especially in social insects (Hamilton 1964). How widespread and important kin selection may be in natural populations remains contentious (Liao et al. 2015; Nowak et al. 2010).

- Mechanisms that suppress selfish behavior and enforce cooperation: these mechanisms include rewards and punishment in humans and other vertebrates, policing of behavior in social insects, repression of transposable elements, and uniparental inheritance of mitochondria and plastids (Ågren et al. 2019).

- Differential extinction and recolonization of subpopulations (interdeme selection): interdeme selection is one of the key elements of Wright's shifting balance theory of evolution (Wright 1931, 1965, 1969), in which relatively isolated demes of small effective size undergo random genetic drift. Some demes hit upon a particularly favorable combination of alleles, whereupon the deme expands in size and displaces those nearby, and these in turn do likewise, until the favorable allele combinations have spread through the entire population. The theory has a certain intuitive appeal; however, critics argue that conditions for it to work in the manner Wright envisioned are unlikely to be realized (Coyne et al. 2000).

Problems

5.1 The *gnd* gene in *Escherichia coli* encodes the enzyme 6-phosphogluconate dehydrogenase (6PGD), which is used in the metabolism of gluconate but not ribose. When otherwise genetically identical strains containing the naturally occurring alleles *gnd(RM77C)* or *gnd(RM43A)* were placed in competition on gluconate or ribose, the following data were obtained, in which p_t denotes the frequency of the strain containing *gnd(RM43A)* in generations. Estimate the relative fitnesses of the strains under the two growth conditions.

Genotype of competing strains	Growth medium	p_0	p_{35}
gnd(RM43A) vs *gnd(RM77C)*	Gluconate	0.455	0.898
gnd(RM43A) vs *gnd(RM77C)*	Ribose	0.594	0.587

5.1 ANSWER Equation 5.2 says that $s = m_1 - m_2 = \left[\ln\left(p_{35}/q_{35}\right) - \ln\left(p_0/q_0\right)\right]/35$. For the gluconate data, $s = 0.0673$, hence the relative fitnesses of *gnd(RM43A)* and *gnd(RM77C)* can be written equivalently as $1.0673 : 1$ or $1 : 1 - 0.0673 = 1 : 0.9327$. This indicates very strong selection favoring the strain with the allele *gnd(RM43A)*. For the ribose data, $s = -0.00083$, which is not significantly different from 0, hence the relative fitnesses are $1 : 1$.

5.2 *Industrial melanism* refers to the dark pigmentation due to a dominant mutation that arose in some insects, giving them protective coloration on vegetation darkened by soot in heavily industrialized areas prior to the requirement for smokestack filtration. In this species, the frequency of melanic moths increased from a value of 1 per cent in 1848 to a value of 95 per cent in 1898. The species has one generation per year.

a. Estimate the approximate value of the selection coefficient s against nonmelanics that would be necessary to account for the change in frequency of the melanic forms.

b. How many generations would be required for the same change in frequency of melanic forms in a hypothetical case in which the allele for melanism is recessive, assuming the same value of s against nonmelanics?

5.2 ANSWER Melanic moths result from a dominant allele. Therefore, with random mating, the frequency of the recessive allele can be estimated as the square root of the proportion of nonmelanics. The data given are that $t = 50$ generations (1898–1848), $q_0 = \sqrt{(1 - 0.01)} = 0.9950$, and $q_{50} = \sqrt{(1 - 0.95)} = 0.2236$. **a.** The change in frequency for a favored dominant allele is given by Equation 5.9, hence $st = \ln\left(p_t/q_t\right) + \left(1/q_t\right) - \ln\left(p_0/q_0\right) - \left(1/q_0\right) = \ln\left(0.7764/0.2236\right) + \left(1/0.2236\right) - \ln\left(0.9950/0.0050\right) - \left(1/0.0050\right) = 10.0054$, yielding $s = 0.200$. **b.** If the melanic allele were recessive, we would have $p_0 = \sqrt{0.01} = 0.1000$ and $p_t = \sqrt{0.95} = 0.9747$, with $s = 0.200$ given as the same selection coefficient as above. The change in frequency of a favored recessive allele is given by Equation 5.11, hence we can write $st = \ln\left(p_t/q_t\right) - \left(1/p_t\right) - \ln\left(p_0/q_0\right) + \left(1/p_0\right) = \ln\left(0.9747/0.0253\right) - \left(1/0.9747\right) - \ln\left(0.1000/0.9000\right) + \left(1/0.1000\right) = 14.8226$. Therefore, $t = 74.113$ generations.

5.3 An extensively studied isolated colony of the moth *Panaxia dominula* contained a mutant allele affecting color pattern. The frequency of the mutant allele declined steadily over the period 1939–1968 in agreement with the theoretical prediction for a deleterious additive allele with $s = 0.20$. The species has one generation per year. In 1965, the estimated frequency of the mutant allele was 0.008. Estimate the frequency of the mutant allele in 1950 and in 1940.

5.3 ANSWER A favored additive allele changes according to Equation 5.10, which with slight rearrangement reads $\ln(p_0/q_0) = \ln(p_t/q_t) - st/2$. In this application we set the mutant allele frequency as q, and we are given that $q_t = 0.008$ in 1965. To solve the problem we must infer the initial frequency q_0 from the frequency in 1965. If an initial year of 1950 corresponds to $t = 0$, then 1965 corresponds to $t = 15$ because we are told that that there is one generation per year and that $s = 0.20$. Consequently, $\ln(p_0/q_0) = \ln(0.9920/0.0080) - (0.20)(15)/2 = 3.3203$, from which it follows that $q_0 = 0.0349$. (The observed value in 1950 was 0.037.) An initial year of 1940 makes $t = 25$, hence $\ln(p_0/q_0) = 2.3203$, from which $q_0 = 0.0895$. (The observed value in 1940 was 0.111).

5.4 Show that a random-mating diploid population with relative fitnesses $1 : 1 - s : (1 - s)^2$ for *AA*, *Aa*, and *aa* has the same change in allele frequency as a haploid population with fitnesses $1 : 1 - s$ of *A* and *a*.

5.4 ANSWER From Equation 5.5, $p' = \left[p^2 + pq(1 - s)\right]/\left[p^2 + 2pq(1 - s) + q^2(1 - s)^2\right] = p\left[p + q(1 - s)\right]/\left[p + q(1 - s)\right]^2 = p/\left[p + q(1 - s)\right]$, which is identical to Equation 5.3 for a haploid with $w_1 = 1$ and $w_2 = 1 - s$.

5.5 Warfarin is a blood anticoagulant used for rat control during and after World War II. Initially highly successful, the effectiveness of the rodenticide gradually diminished due to the evolution of resistance among some target populations. Among Norway rats in Great Britain, resistance results from an otherwise harmful mutation, *WARF-R*, in a gene in which the normal, sensitive allele may be denoted *WARF-S*. In the absence of warfarin, the relative fitnesses of *SS*, *SR*, and *RR* genotypes have been estimated as 1.00, 0.77, and 0.46 respectively. In the presence of warfarin, the relative fitnesses have been estimated as 0.68, 1.00, and 0.37, respectively.

a. Calculate the equilibrium frequency $\hat{q}$ of the resistance allele in the presence of warfarin.

b. Noting that, in the absence of warfarin, *WARF-R* and *WARF-S* are very nearly additive in their effects on fitness, estimate the approximate number of generations required for the allele frequency of *WARF-R* to decrease from $\hat{q}$ to 0.01 in the absence of the anticoagulant.

5.5 ANSWER a. In the presence of warfarin, the relative fitnesses of *SS*, *SR*, and *RR* are 0.68, 1.00, and 0.37. When written as $1 - w_{SS}$, 1.00, $1 - w_{RR}$, then $w_{SS} = 0.32$ and $w_{RR} = 0.63$. The stable equilibrium with overdominance is given by Equation 5.20 as $\hat{p} = w_{RR}/(w_{SS} + w_{RR}) = 0.6632$, which is the frequency of the *S* allele. The frequency of the *R* allele is therefore $\hat{q} = 0.3368$. **b.** In the absence of warfarin the relative fitnesses of *SS*, *SR*, and *RR* are given by 1.00, 0.77, and 0.46, which with additive selection are written as 1.00, $1 - s/2$, and $1 - s$. The fitness of the heterozygous genotype yields $s/2 = 0.23$ or $s = 0.46$, whereas the fitness of the homozygous *R* genotype yields $s = 0.54$. This is a reasonable approximation of additive selection when s is taken as the mean, or $s = (0.46 + 0.54)/2 = 0.50$, which yields relative fitnesses of 1.00, 0.75, and 0.50. The change in frequency of an additive allele is given by Equation 5.10; hence $t = \left[\ln(p_t/q_t) - \ln(p_0/q_0)\right]/(s/2) = \left[\ln(0.99/0.01) - \ln(0.6632/0.3368)\right]/(0.25) = 15.67$ generations.

5.6 What is the additive genetic variance due to a gene at equilibrium due to overdominance?

5.6 ANSWER The additive genetic variance is given by Equation 5.24 as $2pq[p(w_{11} - w_{12}) + q(w_{12} - w_{22})]^2$. With overdominance we can write $w_{11} = 1 - s_{11}$, $w_{12} = 1$, and $w_{22} = 1 - s_{22}$, and at the interior equilibrium $p = s_{22}/(s_{11} + s_{22})$ as in Equation 5.19. With these values $p(w_{11} - w_{12}) + q(w_{12} - w_{22}) = [s_{22}/(s_{11} + s_{22})](-s_{11}) + [s_{11}/(s_{11} + s_{22})](s_{22}) = 0$, and hence the additive genetic variance equals 0. In the light of Equation 5.25, this result says that the additive variance in fitness being 0 is the reason for the equilibrium being at the value it is.

5.7 In an experimental population of *Drosophila melanogaster* containing a *Segregation Distorter* chromosome known as *SD*, the equilibrium frequency of the *SD* chromosome was approximately 0.125, and the segregation ratio in *SD* heterozygotes was about $k = 0.75$. The *SD* chromosome is homozygous lethal in both sexes. The equilibrium between viability selection and segregation distortion in this case can be shown to be a wildtype allele frequency of $\hat{p} = 2(k-1)w_{12}/(1 - 2w_{12})$, where w_{12} is the fitness of the heterozygous *SD* genotype relative to homozygous wildtype. Use the equilibrium equation to estimate the approximate value of w_{12} consistent with the value of $\hat{p}$ in the experimental population.

5.7 ANSWER Set $1 - 0.125 = 2(0.75 - 1)w_{12}/(1 - 2w_{12})$ and solve for w_{12}, which evaluates to 0.700.

References

Ågren, J. A., Davies, N. G., and Foster, K. R. (2019), 'Enforcement is central to the evolution of cooperation', *Nat Ecol Evol*, 3 (7), 1018–29.

Allison, A. C. (1964), 'Polymorphism and natural selection in human populations', *Cold Spring Harbor Symp Quant Biol*, 29, 139–49.

Bedford, T. and Hartl, D. L. (2009), 'Optimization of gene expression by natural selection', *Proc Natl Acad Sci U S A*, 106 (4), 1133–8.

Begun, D. and Aquadro, C. F. (1992), 'Levels of naturally occurring DNA polymorphism correlate with recombination rates in D. melanogaster', *Nature*, 356 (6369), 519–20.

Berry, A. and Kreitman, M. (1993), 'Molecular analysis of an allozyme cline: alcohol dehydrogenase in Drosophila melanogaster on the East coast of North America', *Genetics*, 134 (3), 869–93.

Camin, J. H. and Ehrlich, P. R. (1958), 'Natural selection in water snakes (Natrix sipedon L.) on islands in Lake Erie', *Evolution*, 12 (4), 504–11.

Charlesworth, B. (1980), *Evolution in Age-Structured Populations* (Cambridge: Cambridge University Press).

Charlesworth, B. (2009), 'Effective population size and patterns of molecular evolution and variation', *Nat Rev Genet*, 10 (3), 195–205.

Charlesworth, B. and Hartl, D. L. (1978), 'Population dynamics of the segregation distorter polymorphism in Drosophila melanogaster', *Genetics*, 89 (1), 171–92.

Charlesworth, B., Morgan, M. T., and Charlesworth, D. (1993), 'The effect of deleterious mutations on neutral molecular variation', *Genetics*, 134 (4), 1289–303.

Comeron, J. M. (2017), 'Background selection as null hypothesis in population genomics: insights and challenges from Drosophila studies', *Philos Trans R Soc Lond B Biol Sci*, 372 (1736), 20160471.

Corbett-Detig, R. B., Hartl, D. L., and Sackton, T. B. (2015), 'Natural selection constrains neutral diversity across a wide range of species', *PLoS Biol*, 13 (4), e1002112.

Coyne, J. A., Barton, N., and Turelli, M. (2000), 'Is Wright's shifting balance process important in evolution?', *Evolution*, 54 (1), 306–17.

Crow, J. F. (1998), '90 years ago: the beginning of hybrid maize', *Genetics*, 148 (3), 923–8.

Crow, J. F. and Kimura, M. (1970), *Introduction to Population Genetics* (New York: Harper & Row).

Curtsinger, J. W. (1984), 'Evolutionary principles for polynomial models of frequency-dependent selection', *Proc Natl Acad Sci U S A*, 81 (9), 2840–2.

Darwin, C. (1859), *On the Origin of Species* (London: John Murray).

Darwin, C. (1871), *The Descent of Man, and Selection in Relation to Sex* (London: John Murray).

Deltcheva, E., et al. (2011), 'CRISPR RNA maturation by trans-encoded small RNA and host factor RNase III', *Nature*, 471 (7340), 602–7.

Edwards, A. W. F. (2002), 'The fundamental theorem of natural selection', *Theor Popul Biol*, 61 (3), 335–7.

Esvelt, K. M. and Gemmell, N. J. (2017), 'Conservation demands safe gene drive', *PLoS Biol*, 15 (11), e2003850.

Ewens, W. J. (1989), 'An interpretation and proof of the fundamental theorem of natural selection', *Theor Popul Biol*, 36 (2), 167–80.

Ewens, W. J. (2004), *Mathematical Population Genetics* (2 edn.; New York: Springer-Verlag).

Fijarczyk, A. and Babik, W. (2015), 'Detecting balancing selection in genomes: limits and prospects', *Mol Ecol*, 24 (14), 3529–45.

Fisher, R. A. (1922), 'On the dominance ratio', *Proc R Soc Edinburgh*, 42, 321–41.

Gantz, V. M. and Bier, E. (2015), 'Genome editing. The mutagenic chain reaction: a method for converting heterozygous to homozygous mutations', *Science*, 348 (6233), 442–4.

Gould, F., Dhole, S., and Lloyd, A. L. (2019), 'Pest management by genetic addiction', *Proc Natl Acad Sci U S A*, 116 (13), 5849–51.

Graham, A. M., et al. (2018), 'Migration-selection balance drives genetic differentiation in genes associated with high-altitude function in the speckled Teal (Anas flavirostris) in the Andes', *Genome Biol Evol*, 10 (1), 14–32.

Grossman, A. I., Koreneva, L. G., and Ulitscaya, L. E. (1970), 'The variability of the Adh locus in a natural population of Drosophila melanogaster (in Russian)', *Genetika (Moscow)*, 6 (8), 91–5.

Haldane, J. B. S. (1932), *The Causes of Evolution* (London: Longmans, Green).

Haldane, J. B. S. (1937), 'The effect of variation on fitness', *Am Natur*, 71, 337–49.

Hamilton, W. D. (1964), 'The genetical evolution of social behavior. I', *J Theoret Biol*, 7, 1–16.

Hammond, T. M., et al. (2012), 'Molecular dissection of Neurospora Spore killer meiotic drive elements', *Proc Natl Acad Sci U S A*, 109 (30), 12093–8.

Hartl, D. L. (1970), 'A mathematical model for recessive lethal segregation distorters with differential viabilities in the sexes', *Genetics*, 66 (1), 147–63.

Hartl, D. L. and Dykhuizen, D. E. (1981), 'Potential for selection among nearly neutral allozymes of 6-phosphogluconate dehydrogenase in Escherichia coli', *Proc Natl Acad Sci USA*, 78 (10), 6344–8.

Hartl, D. L., Dykhuizen, D. E., and Dean, A. M. (1985), 'Limits of adaptation: the evolution of selective neutrality', *Genetics*, 111 (3), 655–74.

Hartl, D. L., Hiraizumi Y, Crow JF. (1967), 'Evidence for sperm dysfunction as the mechanism of segregation distortion in Drosophila melanogaster', *Proc Natl Acad Sci U S A*, 58 (6), 2240–5.

Haymer, D. S. and Hartl, D. L. (1983), 'The experimental assessment of fitness in Drosophila. II. A comparison of competitive and noncompetitive measures', *Genetics*, 104 (2), 343–52.

Hudson, R. R. and Kaplan, N. L. (1995), 'Deleterious background selection with recombination', *Genetics*, 141 (4), 1605–17.

Jensen, J. D. (2014), 'On the unfounded enthusiasm for soft selective sweeps', *Nat Commun*, 5, 5281.

Kacser, H. and Burns, J. A. (1973), 'The control of flux', *Symp Soc Exp Biol*, 32, 65–104.

Kim, Y. and Stephan, W. (2002), 'Detecting a local signature of genetic hitchhiking along a recombining chromosome', *Genetics*, 160 (2), 765–77.

Kimura, M. (1962), 'On the probability of fixation of mutant genes in a population', *Genetics*, 47, 713–19.

Kuijper, B., Pen, I., and Weissing, F. J. (2012), 'A guide to sexual selection theory', *Annu Rev Ecol Evol Syst*, 43, 287–311.

Larracuente, A. M. and Presgraves, D. C. (2012), 'The selfish segregation distorter gene complex of Drosophila melanogaster', *Genetics*, 192 (1), 33–53.

Laurie, C. C., Bridgham, J. T., and Choudhary, M. (1991), 'Associations between DNA sequence variation and variation in expression of the Adh gene in natural populations of Drosophila melanogaster', *Genetics*, 129 (2), 489–99.

Lewontin, R. C. and Dunn, L. C. (1960), 'The evolutionary dynamics of a polymorphism in the house mouse', *Genetics*, 45, 705–22.

Liao, X., Rong, S., and Queller, D. C. (2015), 'Relatedness, conflict, and the evolution of eusociality', *PLoS Biol*, 13 (3), e1002098.

Lynch, M. (2016), 'Mutation and human exceptionalism: our future genetic load', *Genetics*, 202 (3), 869–75.

Manel, S. and Holderegger, R. (2013), 'Ten years of landscape genetics', *Trends Ecol Evol*, 28 (10), 614–21.

May, R. M. (1985), 'Evolution of pesticide resistance', *Nature*, 315, 12–13.

Maynard Smith, J. and Haigh, J. (1974), 'The hitch-hiking effect of a favourable gene', *Genet Res*, 23 (1), 23–35.

Mukai, T., et al. (1972), 'Mutation rate and dominance of genes affecting viability in Drosophila melanogaster', *Genetics*, 72 (2), 335–55.

Newbigin, E. and Uyenoyama, M. K. (2005), 'The evolutionary dynamics of self-incompatibility systems', *Trends Genet*, 21 (9), 500–5.

Noble, C., et al. (2019), 'Daisy-chain gene drives for the alteration of local populations', *Proc Natl Acad Sci U S A*, 116 (17), 8275–82.

Nowak, M. A., Tarnita, C. E., and Wilson, E. O. (2010), 'The evolution of eusociality', *Nature*, 466 (7310), 1057–62.

Oakeshott, J. G., et al. (1982), 'Alcohol dehydrogenase and glycerol-3-phosphate dehydrogenase clines in Drosophila melanogaster on different continents', *Evolution*, 36 (1), 86–96.

Oberhofer, G., Tobin, I., and Hay, B. A. (2019), 'Cleave and Rescue, a novel selfish genetic element and general strategy for gene drive', *Proc Natl Acad Sci U S A*, 116 (13), 6250–9.

Ohnishi, O. (1977), 'Spontaneous and ethyl methane-sulfonate induced mutations controlling viability in Drosophila melanogaster. III. Heterozygous effect of polygenic mutations', *Genetics*, 87 (3), 547–56.

Orr, H. A. and Betancourt, A. J. (2001), 'Haldane's sieve and adaptation from standing genetic variation', *Genetics*, 157 (2), 875–84.

Patten, M. M., Haig, D., and Úbeda, F. (2010), 'Fitness variation due to sexual antagonism and linkage disequilibrium', *Evolution*, 64 (12), 3638–42.

Pavlidis, P. and Alachiotis, N. (2017), 'A survey of methods and tools to detect recent and strong positive selection', *J Biol Res (Thessalon)*, 24, 7.

Pennings, P. S. and Hermisson, J. (2006), 'Soft sweeps III: the signature of positive selection from recurrent mutation', *PLoS Genet*, 2 (12), e186.

Perlitz, M. and Stephan, W. (1997), 'The mean and variance of the number of segregating sites since the last hitchhiking event', *J Math Biol*, 36 (1), 1–23.

Pipkin, S. B., et al. (1976), 'New studies of the alcohol dehydrogenase cline in D. melanogaster from Mexico', *J Hered*, 67 (5), 258–66.

Presgraves, D. C., et al. (2009), 'Large-scale selective sweep among Segregation Distorter chromosomes in African populations of Drosophila melanogaster', *PLoS Genet*, 5 (5), e1000463.

Raju, N. B. (1994), 'Ascomycete spore killers: chromosomal elements that distort genetic ratios among products of meiosis', *Mycologia*, 86 (4), 461–73.

Richardson, J. L., et al. (2016), 'Navigating the pitfalls and promise of landscape genetics', *Mol Ecol*, 25 (4), 849–63.

Rogers, R. L., et al. (2010), 'Adaptive impact of the chimeric gene Quetzalcoatl in Drosophila melanogaster', *Proc Natl Acad Sci USA*, 107 (24), 10943–8.

Saether, B. E. and Engen, S. (2015), 'The concept of fitness in fluctuating environments', *Trends Ecol Evol*, 30 (5), 273–81.

Saether, B. E., et al. (2016), 'Evidence for r- and K-selection in a wild bird population: a reciprocal link between ecology and evolution', *Proc Biol Sci*, 283 (1829), 20152411.

Safronova, L. D. and Chubykin, V. L. (2013), 'Meiotic drive in mice carrying the *t*-complex in their genome', *Russian J Genet*, 49, 885–97.

Simmons, M. F. and Crow, J. F. (1977), 'Mutations affecting fitness in Drosophila populations', *Ann Rev Genet*, 11, 49–78.

Sinervo, B. and Calsbeek, R. (2006), 'The developmental, physiological, neural, and genetical causes and consequences of frequency-dependent selection in the wild', *Annu Rev Ecol Evol Syst*, 37, 581–610.

Spirito, F. (1992), 'The exact values of the probability of fixation of underdominant chromosomal rearrangements', *Theoret Pop Biol*, 41 (2), 111–20.

Stephan, W. (2010), 'Genetic hitchhiking versus background selection: the controversy and its implications', *Philos Trans R Soc Lond B Biol Sci*, 365 (1544), 1245–53.

Storfer, A., Patton, A., and Fraik, A. K. (2018), 'Navigating the interface between landscape genetics and landscape genomics', *Front Genet*, 9, 68.

Teissier, G. (1942), 'Persistence d'un gène léthal dans une population de Drosophiles', *Compt Rend Acad Sci*, 214, 327–30.

Unckless, R. L., Clark, A. G., and Messer, P. W. (2017), 'Evolution of resistance against CRISPR/Cas9 gene drive', *Genetics*, 205 (2), 827–41.

Ursprung, H. and Leone, J. (1965), 'Alcohol dehydrogenases: a polymorphism in *Drosophila melanogaster*', *J Exp Zool*, 160, 147–54.

Weedall, G. D. and Conway, D. J. (2010), 'Detecting signatures of balancing selection to identify targets of anti-parasite immunity', *Trends Parasitol*, 26 (7), 363–9.

Wilson, B. A., Pennings, P. S., and Petrov, D. A. (2017), 'Soft selective sweeps in evolutionary rescue', *Genetics*, 205 (4), 1573–86.

Wilson Sayres, M. A. (2018), 'Genetic diversity on the sex chromosomes', *Genome Biol Evol*, 10 (4), 1064–78.

Wong, F., et al. (2018), 'Structural conditions on complex networks for the Michaelis-Menten input-output response', *Proc Natl Acad Sci U S A*, 115 (39), 9738–43.

Wright, S. (1931), 'Evolution in Mendelian populations', *Genetics*, 16 (2), 97–159.

Wright, S. (1934), 'Physiological and evolutionary theories of dominance', *Am Nat*, 68 (714), 25–53.

Wright, S. (1937), 'The distribution of gene frequencies in populations', *Proc Natl Acad Sci U S A*, 23 (6), 307–20.

Wright, S. (1965), 'Factor interaction and linkage in evolution', *Proc Roy Soc B*, 162 (986), 80–104.

Wright, S. (1969), *Evolution and the Genetics of Populations, Vol. 2, The Theory of Gene Frequencies* (Chicago, IL: University of Chicago Press).

Wright, S. (1988), 'Surfaces of selective value revisited', *Am Nat*, 131 (1), 115–23.

Yeaman, S. and Otto, S. P. (2011), 'Establishment and maintenance of adaptive genetic divergence under migration, selection, and drift', *Evolution*, 65 (7), 2123–9.

CHAPTER 6

Random Genetic Drift in Small Populations

The term **random genetic drift** refers to fluctuations in allele frequencies that occur by chance as a result of random sampling among gametes, which is particularly important in small populations. In previous chapters we had occasion to examine some of the consequences of random drift, including its effects on linkage disequilibrium, probability of identity by descent, and equilibrium heterozygosity with mutation. We also examined the coalescent genealogy of a set of DNA or amino acid sequences. A **coalescent genealogy** is a principal result of random genetic drift, depicted graphically as tree diagram of the ancestral history of a sample of alleles taken from a population at any given time. Each allele's ancestral history is traced backward in time, and successive internal nodes in the tree are defined by random events in which pairs of alleles coalesce into a single common ancestor. Random genetic drift determines which pairs of alleles coalesce and the timing of their coalescence.

In this chapter we adopt a more formal approach and consider the effects of random genetic drift looking forward in time rather than backward. We focus mainly on the Wright–Fisher model of random drift (Fisher 1930; Wright 1931) owing to its historical importance as well as its dominant role in theoretical population genetics, but we also give attention to alternative models due to Moran (1962) and Cannings (1974).

6.1 Differentiation of Subpopulations Under Random Drift

In the Wright–Fisher model, the adults in an ideal population consisting of N diploid individuals in a species with nonoverlapping generations produces an infinite number of gametes among which exactly $2N$ are chosen at random to form the zygotes of the next generation. An **ideal population** maintains a constant population size, and each individual has the same distribution of offspring number.

In a finite population on a timescale in which mutation can be ignored, two alleles present in any generation may be exact DNA replicas of a single allele present in breeding adults in the previous generation. These exact-replica alleles are **identical by descent**. As time goes forward, the probability of identity by descent increases in a finite population until, eventually, all of the alleles in the population are identical by descent and the population consists exclusively of individuals homozygous for one of the original alleles.

The problem is that what happens in any one population is quite unpredictable. A solution to this problem is to imagine an infinitely large group or ensemble of subpopulations, each identical to one another at the beginning of the random-drift process, and then use a

A Primer of Population Genetics and Genomics. Fourth Edition. Daniel L. Hartl, Oxford University Press (2020). © Daniel L. Hartl.
DOI: 10.1093/oso/9780198862291.003.0006

mathematical model to deduce various statistical attributes of the ensemble as time moves on. For some purposes it is sufficient to know the mean and variance of allele frequencies among the subpopulations, which change as a function of time.

Random Drift in Small Experimental Populations

To examine how the allele frequencies among subpopulations change through time, it is helpful to consider a concrete example from experiments for a PhD thesis carried out many years ago by Buri (1956). It is one of the few experiments of its kind, and we use the data as a showcase to illustrate how various models of random genetic drift can help interpret what features of biological reproduction underlie the observed results.

The goal of Buri's experiment was to track changes in allele frequencies among 107 subpopulations of *Drosophila melanogaster*. Each population was initiated with 8 females and 8 males that were heterozygous for the alleles bw^{75} and bw (bw = brown eyes). These genetic markers are convenient because the genotypes bw^{75}/bw^{75}, bw^{75}/bw, and bw/bw are all distinguishable by eye color. The populations were maintained in separate culture bottles at a constant size of $N = 16$ by randomly choosing 8 males and 8 females as the breeding population in each generation. These lines were maintained and the allele frequencies monitored for 19 consecutive generations.

In an experiment like Buri's, one expects the allele frequencies among the subpopulations to change randomly from one generation to the next, with the differences in allele frequencies among the subpopulations increasing through time. By chance, some subpopulations may become **fixed** (attain an allele frequency of 1) for the bw^{75} allele and others fixed for the bw allele, and once an allele is fixed in a subpopulation it remains fixed.

These expectations are borne out in practice. Buri's paper does not include the generation-by-generation allele frequencies for all subpopulations; however, Figure 6.1 shows the result of a simulation chosen to closely match the overall patterns observed. Each line traces the number of bw^{75} alleles in one subpopulation across all 19 generations of the experiment. There is considerable differentiation among the subpopulations even after one generation, and the dispersion increases with time. As time goes on, an increasing number of subpopulations become fixed for either the bw^{75} allele or the bw allele, and by the end of the experiment (generation 19), 30 subpopulations have lost

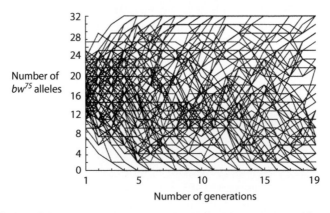

Figure 6.1 Simulation of the Buri experiment resulting in the same number of fixations and losses of the bw^{75} allele as observed in the actual data.

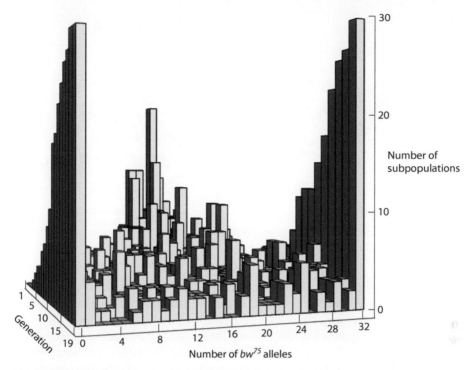

Figure 6.2 Results of random genetic drift in 107 experimental subpopulations of *D. melanogaster*. (Data from Buri (1956).)

bw^{75} (0 bw^{75} alleles) and 28 have become fixed for bw^{75} (32 bw^{75} alleles). Among the remaining 49 **unfixed** subpopulations (those still segregating for bw^{75} and bw), the number of bw^{75} alleles that are present is about equally likely to be any number between 1 and 31 (that is, the distribution of allele frequencies is uniform).

The actual data from Buri's experiment are depicted in the three-dimensional bar graph in Figure 6.2. The height of each bar is proportional to the number of bw^{75} alleles present in each subpopulation. Time proceeds from left to right along the x axis. The initially humped distribution of allele frequency observed in generation 1 gradually becomes broader as subpopulations fixed for bw^{75} or bw begin to pile up. By 19 generations, about half of the subpopulations are fixed for one allele or the other, and among the unfixed populations, the distribution of allele frequencies is essentially flat (uniform).

The Probability Process Underlying the Wright–Fisher Model

The allele frequency trajectories in Figure 6.1 are random trajectories in a Wright–Fisher model in which the probability that a subpopulation with exactly i copies of the bw^{75} allele in generation t has exactly j copies of the allele in generation $t+1$ is given by:

$$p_{ij} = \Pr\left\{p_{t+1} = \frac{j}{2N} \,\middle|\, p_t = \frac{i}{2N}\right\} = \frac{(2N)!}{j!\,(2N-j)!}\left(\frac{i}{2N}\right)^j\left(1-\frac{i}{2N}\right)^{2N-j} \tag{6.1}$$

The probabilities associated with $j = 2N$ or $j = 0$ are the respective probabilities that an allele is fixed or lost in one generation of random drift, starting with an initial allele frequency of $i/(2N)$.

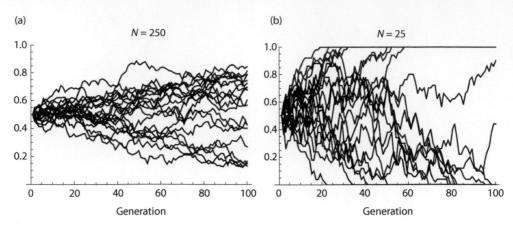

Figure 6.3 Subpopulation divergence under the Wright–Fisher model. Random trajectories of allele frequency for (a) $N = 250$ and (b) $N = 25$.

The extent to which subpopulations accumulate differences in allele frequency under random genetic drift depends on the value of N: the larger N, the less divergence. This principle is illustrated by the trajectories for $N = 250$ and $N = 25$ in Figure 6.3.

The trajectories in Figure 6.1 were obtained by examining independent simulations of 107 subpopulations with $N = 16$ using Equation 6.1 until, by chance, one such simulation yielded a result closely approximating that observed in the Buri (1956) data. As it turns out, the simulation that yielded the result in Figure 6.1 is not typical of random Wright–Fisher trajectories. Instead, these trajectories mimic the behavior of subpopulations of only about half the size in the actual experiment.

Nevertheless, Equation 6.1 is the defining equation of the Wright–Fisher model. The state of a subpopulation at time t corresponds to its allele frequency at that time. Note that the state of a subpopulation at time $t + 1$ depends only on its state at time t and is independent of subpopulation history prior to time t. A random, memory-less process whose future state depends only on its present state is a **Markov process,** named after the Russian mathematician Andrey Markov. The Wright–Fisher process is a special type of Markov process with two **absorbing states** corresponding to the allele frequencies $p_i = 0$ or $p_i = 1$. Each fixation state is called absorbing because, once a subpopulation is fixed, it remains fixed (it is "absorbed"). Except for the absorbing states, the allele frequency can in principle change from any state to any other state in a single generation; however, for large N the changes in allele frequency per generation tend to be small. Eventually, each subpopulation is absorbed at either $p_i = 0$ or $p_i = 1$.

For i and j equal to 1, 2, 3, ... , $2N$, the values of p_{ij} in Equation 6.1 define a square **transition matrix**, **P**, which gives the probability of state i changing to state j in a single generation. The probability that state i changes to state j in exactly t generations is given by the entry p_{ij} in the matrix **P** raised to the power t, $\mathbf{P}^t$. When $2N$ is small or moderate, $\mathbf{P}^t$ can be calculated using computer algorithms. More generally $\mathbf{P}^t$ may be calculated using methods from linear algebra (Ewens 2004) or various shortcuts to deduce quantities of interest (Krukov et al. 2017). In the special case $2N = 32$, corresponding to Buri's experiment, the expected number of bw^{75} alleles (i) among each of 107 subpopulations for values of $t = 1, 2, 3, \ldots, 19$ are shown in Figure 6.4.

At first sight, the match of Figure 6.4 with the actual data in Figure 6.2 looks pretty good, but a close look reveals some important discrepancies. The most obvious is in the number

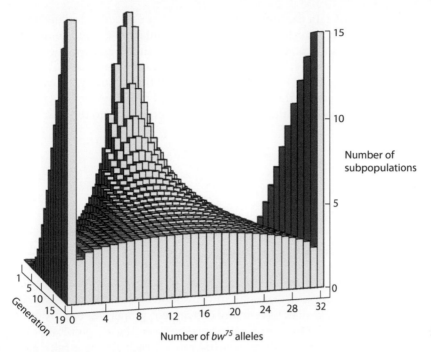

15

10

Number of
subpopulations

5

0

Figure 6.4 Theoretical expectations from the Wright–Fisher model corresponding to Buri's experiment with $2N = 32$.

of fixed populations at the end of 19 generations—30 in Figure 6.4 but 58 in the actual data. The main reason for the discrepancy is that the variance in offspring number in the actual data is substantially larger than that assumed in the model. The Wright–Fisher model assumes that the distribution of offspring is a Poisson distribution with a mean of 1, with the constraint that the total number of offspring in any generation equals N. Because the variance of a Poisson distribution equals the mean, the variance in offspring number is also expected to be equal 1. But the data from Buri's experiment imply that the variance in offspring number is about twice that expected.

A generalization of the Wright–Fisher model that allows for any value for the variance in offspring number was proposed by Cannings (1974) and is discussed in some detail in Ewens (2004). In the Cannings model, the distribution of number of copies that each gene contributes to the next generation is identical for all genes and has mean 1 and variance σ^2. The constraint that the total number of alleles remains constant at $2N$ implies that the covariance in number of copies of any pair of genes equals $-\sigma^2/(2N-1)$ (Ewens 2004). In Buri's experiment, the rate at which the fixed classes increase through time implies that the approximate variance in offspring number is $\sigma^2 \approx 1.8$.

Transition Matrix for the Moran Model

The variance in offspring number of nearly 2 suggests that the Moran model might yield a good fit to the Buri data. This model of random drift, developed by (Moran 1958, 1962), was introduced in Chapter 4 in the context of the coalescent, where we pointed out that many theoreticians like it because it is mathematically somewhat more tractable than the Wright–Fisher model. The underlying distribution of offspring number in the Moran model has a variance of 2. In the Moran model, alleles are chosen at random and replaced one by one. At each step, two alleles are chosen with replacement, and so the same allele may be chosen twice. The first allele chosen reproduces and the second dies. If the alleles are the same, nothing changes. If the first allele is A_1 and the second A_2, the number of A_1 alleles increases by 1; and if the first allele is A_2 and the second A_1, the number of A_1 alleles decreases by 1. Because the number of A_1 alleles can only decrease by 1, increase by 1, or stay the same, the transition probabilities corresponding to those in Equation 6.1 are therefore:

$$p_{i,i-1} = \frac{i(2N-i)}{(2N)^2}$$
$$p_{i,i} = \frac{i^2 + (2N-i)^2}{(2N)^2} \tag{6.2}$$
$$p_{i,i+1} = \frac{i(2N-i)}{(2N)^2}$$

These follow from Equations 4.24 in Chapter 4 by replacing p with $i/(2N)$ and $1-p$ with $(2N-i)/(2N)$.

The transition matrix corresponding to Equations 6.2 consists mostly of zeros, with nonzero entries only along the diagonal and the subdiagonals immediately above and below. Such tridiagonal matrices have many convenient properties, but for our purposes we need not go into detail. One major difference between the Wright–Fisher model and the Moran model is that, in the Wright–Fisher model, each step is a full generation of reproduction, whereas in the Moran model each step is only a fraction—approximately $1/(2N)$—of a full generation because only one allele is replaced. Therefore, $2N$ steps in the Moran model correspond roughly to one generation in the Wright–Fisher model. (We say "roughly" because, in the Moran model, some alleles may not be chosen in the course of $2N$ steps, hence these alleles neither reproduce nor die but carry over.)

Figure 6.5 shows the distribution of allele frequencies in a Moran model for 107 subpopulations each with $2N = 32$. Along the y-axis the word "Generation" is in quotes because each "generation" is actually $2N$ time steps in the model. The fit is considerably improved, particularly in regard to the number of subpopulations that are fixed (60 in the model, 58 in the data). Although the fit is reasonably good, we'll soon see that it can be improved.

Change in Average Allele Frequency Among Subpopulations

One of the implications of random genetic drift for selectively neutral alleles is that the allele frequency, averaged over all subpopulations, remains constant. Although the allele frequency can change willy-nilly in any one subpopulation, a change in one subpopulation that increases the allele frequency is balanced by an equal change in the opposite direction in another subpopulation. With an infinite number of subpopulations, the total number of individuals is also infinite, and therefore the overall average allele frequency must remain the same just as in the Hardy–Weinberg principle.

The Buri data fit the theoretical expectation of a constant average allele frequency very well, even though there are only 107 subpopulations with $N = 16$ in each. The results are illustrated in Figure 6.6. The dashed line is the regression line: its slope is 0.0003 and the P-value testing its difference from 0 is $P = 0.56$. Hence, based on these data, there is no evidence for any significant differences in fitness among genotypes for the bw^{75} and bw alleles. (There could, of course, be very small differences that go undetected in an experiment of this size.)

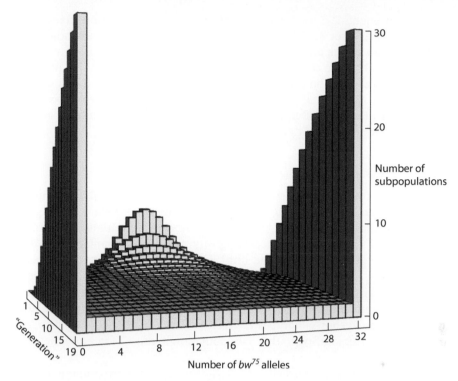

Figure 6.5 Theoretical expectations from the Moran model corresponding to Buri's experiment with $2N = 32$.

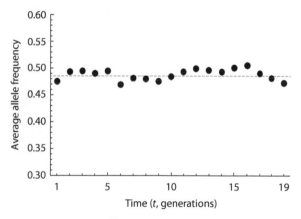

Figure 6.6 Average frequency of the bw^{75} allele across all subpopulations in the Buri experiment. (Data from Buri (1956).)

An important implication emerges from the seemingly simple principle that the allele frequency of neutral alleles, averaged over a large number of subpopulations, remains constant even in the face of random drift. The implication is that, if p_0 and $1 - p_0$ are the average allele frequencies of neutral alleles A_1 and A_2 at time 0, then the probability of ultimate fixation of the A_1 allele must equal p_0. The reason is that, in the absence of new mutation, each of the subpopulations must ultimately become fixed for either A_1 or A_2. In the first set of subpopulations the allele frequency of A_1 equals 1 and in the second set it equals 0. Suppose x is the fraction of subpopulations fixed for A_1 and $1 - x$ the fraction fixed for A_2. Then at a time so advanced that all subpopulations have become fixed, the average allele frequency of A_1 across subpopulations equals $x \times 1 + (1 - x) \times 0 = x$. But because the average allele frequency across subpopulations remains constant at p_0, it must also be that $x = p_0$. Therefore, the probability that a neutral allele at initial frequency p_0 eventually becomes fixed equals p_0. In the special case of a single new mutant allele, the initial frequency $p_0 = 1/(2N)$ and therefore the probability that a new neural mutation eventually becomes fixed equals $1/(2N)$.

Decrease in Average Heterozygosity Among Subpopulations

Although the average allele frequency remains constant for neutral alleles undergoing random drift, the same is not true for the average frequency of heterozygous genotypes. This is clear from Buri's experiment. All subpopulations were initiated with 16 bw^{75}/bw heterozygotes, and so the average heterozygosity among subpopulations at time 0 is $H_0 = 1$. Had the experiment continued until all subpopulations had become fixed for either of the alleles, the average heterozygosity at that time would be $H = 0$ because the fixed populations contain no heterozygotes. So the heterozygosity, averaged across all subpopulations, gradually decreases even though, within any one subpopulation, the proportion of heterozygous genotypes is given by the Hardy–Weinberg principle with allele frequencies corresponding to that subpopulation because of random mating within subpopulations.

The observed decrease in average heterozygosity in Buri's experiment is shown in Figure 6.7. The figure is a bit busy, so let's take it apart piece by piece. First, focus on the blue points. These are the observed values of heterozygosity averaged across all of Buri's subpopulations, including the subpopulations that are fixed. The smooth curves correspond to various models of random drift. In the Wright–Fisher model, we already know from Chapter 3 how average heterozygosity would decrease with random drift. It may not be obvious at first, but if one combines Equation 3.19 with the expression for heterozygosity in Equation 3.5 and sets $F_0 = 0$, one gets:

$$H_t = 2\bar{p}\,\bar{q}\left(1 - \frac{1}{2N}\right)^t \tag{6.3}$$

where H_t is the average heterozygosity in generation t and $\bar{p}$ and $\bar{q}$ are the frequencies of the alleles averaged across all subpopulations. (With an infinite number of subpopulations these allele frequencies are expected to remain constant, but with a finite number of subpopulations, as shown in Figure 6.6, they may fluctuate by chance alone.)

With $N = 16$ in the experimental subpopulations, the expected decrease in H_t in the Wright–Fisher model is indicated by the blue dashed curve. The fit is pretty lousy. Why? As we pointed out earlier, the main reason is that the variance in offspring number is larger than 1 (the Poisson variance), which is assumed in the model. The solid blue curve corresponds to the Wright–Fisher model with $N = 9$. This is a much better fit, and it suggests that the actual variance in offspring number is approximately 1.8. We call $N = 9$ the **effective population number** in the experimental subpopulations, which is usually symbolized as N_e. More precisely, $N_e = 9$ is the **inbreeding effective number** because $N_e = 9$ is the size of theoretically ideal subpopulations that undergo the same rate of increase in the probability of identity by descent F_t (decrease in H_t) as the actual subpopulations. The effective number in the Wright–Fisher model is almost always smaller

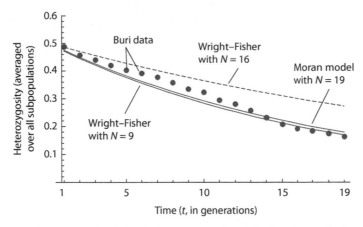

Figure 6.7 Decrease in average heterozygosity among subpopulations in the Buri experiment compared with those expected from various models. (Experimental data from Buri (1956).)

than the actual number. (The inbreeding effective number is only one type of effective number. Other important types of effective number are discussed in Chapter 4.)

Fitting the Buri data to a Moran model, the best fit corresponds to the red curve in Figure 6.7. In this case $N = 19$, which is actually larger than the actual N! What this seeming paradox says is that the variance in offspring number in the actual experiment is a little less than that assumed in the Moran model, and the variance is in fact a bit less—1.8 versus 2.0.

At this point we've exploited the Buri experiment to illustrate some underlying principles of random genetic drift and to show how different models exhibit somewhat different behavior. Which model is correct? As we emphasized in Chapter 1, all models are equally imprecise. The purpose of models is not to mirror the real world, which is bewilderingly complex in its intricacies. In population genetics, the purpose of models is to serve as a guide to interpreting what goes on in real populations and to understand how evolutionary forces play off one against another. In the next section we examine approximations to the Wright–Fisher model that have yielded fundamentally important insights into population genetics.

6.2 Diffusion Approximations

Although Equation 6.1 is exact for the Wright–Fisher model, there is a continuous approximation that is somewhat more tractable (Kimura 1955b, 1957, 1964; Wright 1945). This approach is known as the **diffusion approximation**, and it assumes that random drift disperses allele frequencies among subpopulations in a manner analogous to heat diffusing through a metal rod or tiny particles diffusing under Brownian motion. The main idea is to assume that the subpopulations are large enough that the allele frequencies change smoothly through time, not in large jumps. With this assumption the statistical distribution of allele frequencies at any time is a continuous function that we may denote $\varphi(p, x; t)$, where p is the initial allele frequency among all the subpopulations, x is the current allele frequency ($0 < x < 1$), and t is the generation number. For any time t, the function $\varphi(x; t)$ is a smooth, continuous function approximating the histogram of allele frequencies in Figure 6.4, except that $\varphi(x; t)$ pertains only to the distribution of allele

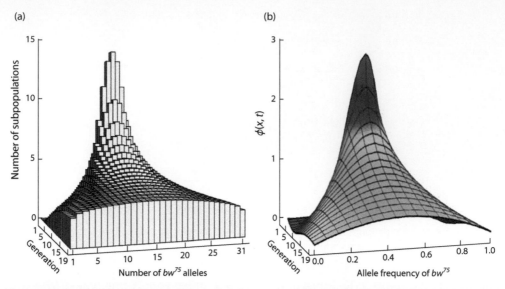

Figure 6.8 Continuous approximation to allele frequency change in the Wright–Fisher model. (a) Discrete model of allele frequency change in the unfixed classes in the Buri experiment. (b) Continuous approximation to the 3D bar graph.

frequencies among unfixed subpopulations (that is, those subpopulations still segregating for both alleles). Figure 6.8 shows the unfixed classes from Figure 6.4 compared with the continuous approximation $\varphi(0.5, x; t)$.

The Forward Equation: An Approach Looking Forward in Time

To explain the meaning of the diffusion equations, let us for the moment interpret p as the number of A_1 alleles present in a subpopulation initially and x as the number present at time t, with $1 < p, x < 2N$. We will also interpret $\varphi(p, x; t)$, as the proportion of subpopulations that have changed from having p copies at time 0 to having x copies at time t. These interpretations are not rigorous from a mathematical standpoint, but we use them merely to explain the diffusion equations rather than to derive them.

To find an equation for $\varphi(p, x; t)$ we may first ask how the distribution of allele frequency changes as we go forward in time for one generation. The bookkeeping is shown in Table 6.1. We assume that the state of each subpopulation changes slowly enough that in any one generation the number of copies of an allele (the "state" of the subpopulation) either stays the same, increases by 1, or decreases by 1. There are two reasons why the state could change. One is random genetic drift (denoted RGD in the table) and the other is a systematic force (denoted SF), which might include mutation, migration, or any type of selection. We will assume that A_1 is the favored allele and define $M(x)$ as the probability that x increases by 1 because of the systematic force. The effect of random drift is measured by $V(x)$, which is the probability that x increases by 1 or decreases by 1 purely because of chance.

Because changes in state are limited to $+1$ or -1, a subpopulation can be in state x at time $t+1$ only if it was in state $x+1$, x, or $x-1$ at time t, and these have probabilities $\varphi(p, x+1; t)$, $\varphi(p, x; t)$, and $\varphi(p, x-1; t)$, respectively. A subpopulation in state $x-1$ can change to state x with probability $M(x-1) + V(x-1)/2$ according to whether it was

Table 6.1 Routes from p to x in $t+1$ generations, looking forward from time t

Possibilities for frequency after t generations	Probability	Possibilities to go to x in next generation	Probability
$x-1$	$\varphi(p,x-1;t)$	$x-1 \to x$ because of SF	$M(x-1)$
	$\varphi(p,x-1;t)$	$x-1 \to x$ because of RGD	$V(x-1)/2$
$x+1$	$\varphi(p,x+1;t)$	$x+1 \to x$ because of RCD	$V(x+1)/2$
x	$\varphi(p,x;t)$	x remains at x	$1-M(x)-V(x)$

$$\varphi(p,\ x;\ t+1) - \varphi(p,\ x;\ t) = -[M(x)\varphi(p,\ x;\ t) - M(x-1)\varphi(p,\ x-1;\ t)]$$
$$+\tfrac{1}{2}\{[V(x+1)\varphi(p,\ x+1;\ t) - V(x)\varphi(p,\ x;\ t)] - [V(x)\varphi(p,\ x;\ t) - V(x-1)\varphi(p,\ x-1;\ t)]\}$$

pushed by the systematic force or changed randomly by random drift. A subpopulation in state $x+1$ can change to state x with probability $V(x+1)/2$ due to random drift. And a subpopulation in state x can remain in state x with probability $1-M(x)-V(x)$. The required function $\varphi(p, x; t)$ is obtained by summing the products of rows 2 and 4 in Table 6.1, yielding the difference equation at the bottom. In this case, the expression relates changes in φ with respect to time to changes in φ with respect to its current state. When p and x are reconsidered as allele frequencies and φ as a statistical distribution, then the terms in the difference equation at the bottom of Table 6.1 converge in the limit to the terms of the Kolmogorov **forward equation**, named after the Russian mathematician Andrey Kolmogorov:

$$\frac{\partial \varphi(p,x;t)}{\partial t} = -\frac{\partial\left[M(x)\varphi(p,x;t)\right]}{\partial x} + \frac{1}{2}\frac{\partial^2\left[V(x)\varphi(p,x;t)\right]}{\partial x^2} \tag{6.4}$$

This equation was first solved by Motoo Kimura for neutral alleles $[M(x)=0]$ (Kimura 1955b). His solution can be written as:

$$\varphi(p,x;t) = \sum_{i=1}^{\infty} p(1-p)\,i(i+1)(2i+1)F(1-i,i+2;2;p)F(1-i,i+2;2;x)\,e^{-\frac{i(i+1)t}{4N}} \tag{6.5}$$

where $F(a, b, c; z)$ is the Gegenbauer polynomial (itself an infinite sum):

$$F(a,b;c;z) = \sum_{k=0}^{\infty} \frac{(a)_k(b)_k}{(c)_k}\frac{z^k}{k!} \tag{6.6}$$

in which $(a)_k = a(a+1)\cdots(a+k-1)$ if $k>0$ and $(a)_k=1$ if $k=0$. The symbol $(a)_k$ has a name; it is called the "rising factorial." And similarly for $(b)_k$ and $(c)_k$.

The outer sum in Equation 6.5 converges very slowly, but with today's computers it is straightforward to calculate $\varphi(p, x; t)$ to any desired degree of accuracy. Figure 6.9 illustrates some examples. Figure 6.9a shows $\varphi(p, x; t)$ for $p=0.50$ at various times t expressed as a multiple of the population size N, and Figure 6.9b shows $\varphi(p, x; t)$ for $p=0.15$ and the same lengths of time. In both cases the allele frequency distributions ultimately become flat (uniform); however, unless the initial allele frequency is near 0.5, a flat distribution is attained so slowly that most subpopulations have already been fixed by the time it happens.

Kimura's publication of the solution in Equation 6.5 (Kimura 1955b) created quite a stir at the time and is credited with revitalizing theoretical population genetics (Ewens 2004). Other approaches and refinements have continued through the years (McKane and Waxman 2007; Zhao et al. 2013).

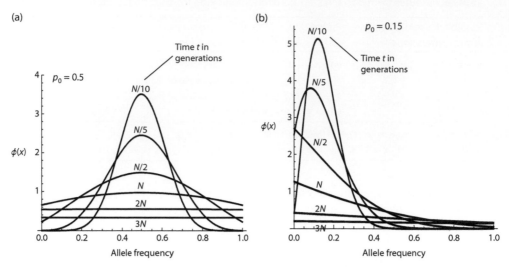

Figure 6.9 Continuous approximations to the distribution of allele frequency among unfixed subpopulations (a) when initially $p = 0.5$, and (b) when initially $p = 0.15$.

Kimura first presented his results in 1955 in a talk entitled "Stochastic processes and distribution of gene frequencies under natural selection" at a symposium at Cold Spring Harbor, Long Island, New York (Kimura 1955a). Hardly anyone understood a word, in part because of his soft voice and heavy Japanese accent, but even more because the math was such tough going. Sewall Wright, who commanded tremendous respect in the field, saved the day by rising to tell the audience that the only people who could truly appreciate Kimura's achievement were those who had tried to solve the problem themselves and failed (Hartl 2004).

As you might expect, there is also a forward diffusion equation for the Moran model. It has the same form as Equation 6.4 except that the factor of 1/2 is not present on the right-hand side (Aalto 1989). The absence of this factor relates to the differing distributions of offspring number in the two models and the more rapid dispersion of allele frequencies in the Moran model.

Although we have explained the underlying logic of the forward diffusion equation, we have not yet specified $M(x)$ or $V(x)$ in terms that have any relation to population genetics. $M(x)$ is another symbol for the change in allele frequency that occurs in one generation due to any systematic force such as mutation, migration, or selection. In previous chapters we've symbolized a one-generation change in allele frequency as $\Delta p = p_{t+1} - p_t$. $M(x)$ can be obtained from any of these equations by replacing p with x. The term $V(x)$ for random drift also has a straightforward biological interpretation. $V(x)$ is the variance in allele frequency after one generation of binomial sampling of $2N$ alleles according to the expression in Equation in 6.1, hence $V(x) = x(1-x)/(2N)$. This if sometimes written as $V(x) = x(1-x)/(2N_e)$, but where N_e is the inbreeding effective population size. This formulation arose at a time when the effective population size was assumed to be singular and unambiguous. We now know this is not the case (Chapter 4), and to avoid unnecessary complexity we will assume an ideal population of size N in which each of the effective population sizes (in reality, possibly different) are all identical.

Table 6.2 Routes from p to x in t generations, looking backward to the first generation

Possibilities for change in first generation	Probability of first-generation change	Probability of changing to x in remaining $t-1$ generations
$p \rightarrow p+1$ because of SF	$M(p)$	$\varphi(p+1,x;t-1)$
$p \rightarrow p+1$ because of RCD	$V(p)/2$	$\varphi(p+1,x;t-1)$
$p \rightarrow p-1$ because of RGD	$V(p)/2$	$\varphi(p-1,x;t-1)$
p remains at p	$1-M(p)-V(p)$	$\varphi(p,x;t-1)$

$$\varphi(p,\ x;\ t) - \varphi(p,\ x;\ t-1) = M(p)\left[\varphi(p+1,\ x;\ t-1) - \varphi(p,\ x;\ t-1)\right]$$
$$+\frac{V(p)}{2}\{[\varphi(p+1,\ x;\ t-1) - \varphi(p,\ x;\ t-1)] - [\varphi(p,\ x;\ t-1) - \varphi(p-1,\ x;\ t-1)]\}$$

The Backward Equation: Musing on the First Step

To find another equation for $\varphi(p, x; t)$ we may also look backward in time to the beginning of the process and consider what may have happened in the very first generation. The bookkeeping in this case is shown in Table 6.2. Again we assume that the state of each subpopulation changes slowly enough that in any one generation the number of copies of an allele (the "state" of the subpopulation) either stays the same, increases by 1, or decreases by 1. With A_1 the favored allele, $M(p)$ is the probability that the number of A_1 alleles increases from p to $p+1$ in the first generation because of the systematic force. Taking into account random genetic drift, the probability that p changes to $p-1$ in the first generation is $V(p)/2$, and the probability that p changes to $p+1$ in the first generation is also $V(p)/2$. Therefore, in the first generation, p can change to $p+1$ with probability $M(p)+V(p)/2$, to $p-1$ with probability $V(p)/2$, or it can remain at p with probability $1-M(p)-V(p)$. For a subpopulation to be in state x in generation t, the state must change from whatever it was after the first generation to the state x in the last $t-1$ generations. The probabilities of changing from $p+1$ to x, or from p to x, or from $p-1$ to x in $t-1$ generations are, respectively, $\varphi(p+1,x;t-1)$, $\varphi(p,x;t-1)$, and $\varphi(p-1,x;t-1)$. This reasoning leads to the difference equation at the bottom of Table 6.2, which suggests the continuous differential equation known as the Kolmogorov **backward equation**,

$$\frac{\partial \varphi(p,x;t)}{\partial t} = M(p)\frac{\partial[\varphi(p,x;t)]}{\partial p} + \frac{V(p)}{2}\frac{\partial^2[\varphi(p,x;t)]}{\partial p^2} \tag{6.7}$$

We'll see how this equation can be used in the following section, but before proceeding we need to say that, in this formulation, the partial derivatives with respect to the initial allele frequency p do not imply that the random variable is p, as is sometimes asserted; the random variable is still the current frequency x, and the backward equation is just an alternative way to get at $\varphi(p,x;t)$ by looking backward to the beginning of the stochastic process. To forestall another potential source of confusion, note also that the backward equation has almost nothing to do with the probabilities of coalescing ancestries in gene genealogies that arise in coalescent theory (Chapter 4).

6.3 Fixation Probabilities and Times to Fixation

Although the forward and backward diffusion equations are different ways of expressing a partial differential equation for $\varphi(p, x; t)$ in random genetic drift, in practice they reveal different facets of this complex process. The forward equation (Equation 6.5) is especially useful for finding the limiting distribution of $\varphi(p, x; t)$ as t goes to infinity in situations with opposing evolutionary forces (e.g., forward mutation versus reverse mutation, or forward mutation versus selection) when polymorphism rather than fixation is the end state of the subpopulations. We will deal with these scenarios later in this chapter. In this section we focus on the utility of the backward equation (Equation 6.7). This way of approaching $\varphi(p, x; t)$ is especially useful in models in which fixation is the ultimate fate of the subpopulations. In such models the backward equation yields expressions for the probability of ultimate fixation of neutral or selected alleles and the average time to fixation or loss.

Probability of Fixation

To obtain the probability of fixation from the backward equation, first set $x = 1$ in Equation 6.7, which reduces $\varphi(p, x; t)$ to a function of only two variables, p and t, yielding a new equation that satisfies:

$$\frac{\partial u(p,t)}{\partial t} = M(p)\frac{\partial u(p,t)}{\partial p} + \frac{V(p)}{2}\frac{\partial^2 u(p,t)}{\partial p^2} \tag{6.8}$$

Because $x = 1$, $u(p, t)$ can be interpreted as the probability that an allele initially at frequency p is fixed at or prior to time t. Now consider a time so advanced that all subpopulations are fixed. Then $u(p, t)$ no longer changes with t and $\partial u(p,t)/\partial t = 0$. This assumption reduces Equation 6.8 to an ordinary differential equation in one variable:

$$0 = M(p)\frac{du(p)}{dp} + \frac{V(p)}{2}\frac{d^2u(p)}{dp^2} \tag{6.9}$$

where $u(p)$ is the probability of ultimate fixation of an allele that has initial frequency p. The desired solution must also satisfy the boundary conditions that $u(0) = 0$ and $u(1) = 1$, which simply say that an allele that doesn't exist in a population cannot be fixed and that an allele already fixed is fixed. The solution to Equation 6.9 that satisfies the boundary conditions is (Crow and Kimura 1970; Ewens 2004):

$$u(p) = \frac{\int_0^p Exp[-2M(x)/V(x)]\,dx}{\int_0^1 Exp[-2M(x)/V(x)]\,dx} \tag{6.10}$$

where $Exp[\bullet]$ is shorthand for $e^{[\bullet]}$.

NEUTRAL ALLELES

For a neutral allele in the Wright–Fisher model, $M(x) = 0$ and $V(x) = x(1-x)/(2N)$, in which case $-2M(x)/V(x) = 0$ and $Exp[-2M(x)/V(x)] = 1$. The probability of ultimate fixation of a neutral allele is therefore $u(p) = p$. For a newly arising neutral mutation $p = 1/(2N)$ and hence the probability of ultimate fixation equals $1/(2N)$. We've already encountered this principle in connection with Figure 6.6 when explaining why the expected average allele frequency of neutral alleles across subpopulations remains constant.

POSITIVELY SELECTED ALLELES

Equation 6.10 also yields the probability of ultimate fixation of alleles subject to selection. The simplest case is that of additive effects using Equation 5.7 and setting $w_{11} = 1 + 2s$, $w_{12} = 1 + s$, and $w_{22} = 1$ with the additional stipulation that s is small enough that $\overline{w} \approx 1$. Then from Equation 5.7, using the symbol x instead of p, $M(x) = \Delta x = x(1-x)s$ and with $V(x) = x(1-x)/(2N)$ then $-2M(x)/V(x) = -4Ns$. Substituting this into Equation 6.10 yields:

$$u(p) = \frac{1 - e^{-4Nsp}}{1 - e^{-4Ns}} \tag{6.11}$$

For a newly arising mutation $p = 1/(2N)$ when s is small and $Ns > 1$, $u(p) \approx 2s$, which says that the probability of ultimate fixation of a beneficial allele is about two times its selective advantage. To take a concrete example, a mutant allele that confers a 1 per cent selective advantage would have to occur independently an average of 50 times for one of the mutants to be fixed. That $u(p) \approx 2s$ for a newly arising mutation agrees with what we deduced earlier in Equation 5.7 under the Wright–Fisher assumption of a Poisson distribution of offspring number.

DELETERIOUS ALLELES

Equation 6.11 makes no stipulation that s must be positive, hence it also applies to deleterious alleles. Note that newly arising deleterious alleles are very unlikely to be fixed. The probability of fixation of a newly arising deleterious allele with $s = -0.01$ in a population of size $N = 50$ is only 0.3 per cent, and in a population of size $N = 100$ it is only 0.03 per cent.

The probability of fixation $u(p)$ given by Equation 6.11 is somewhat easier to interpret in terms of $u(p)/p$, which is the probability of fixation of an allele relative to that of a neutral allele with the same initial frequency. Figure 6.10 shows $u(p)/p$ as a function of the product Ns for various values of p. Look first at the case $p = 1/(2N)$ for a single copy of a newly arising allele. The relative fixation probability for $Ns = -2$ is about 0.003, which serves to emphasize the point made by Alfred Russel Wallace about "the overwhelming odds against the less fit" (Wallace 1892). For new mutants with $Ns = +3$, $u(p)/p$ is about 12. This intensity of selection is really quite small. In *Drosophila*, for example, $Ns = +3$ implies that $s \approx 3 \times 10^{-6}$ (Andolfatto et al. 2011).

The curves for $p = 0.10$ and $p = 0.50$ in Figure 6.10 illustrate another principle about fixation probabilities. At such initially high frequencies, the extent to which the probability of fixation of a weakly selected favorable mutation exceeds that of a neutral mutation is not as great as that for new mutants. The reason is that a new neutral mutation runs its greatest risk of extinction when it is rare, in the first several generations of its existence. If the

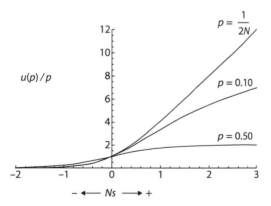

Figure 6.10 Probability of ultimate fixation of a positively or negatively selected allele at initial frequency p relative to that of a neutral allele with the same initial frequency.

mutation can survive long enough to reach an appreciable frequency, then its probability of ultimate fixation is much improved, more so than that of a weakly selected mutation.

Times to Fixation or Loss

Once you know the probability that an allele becomes fixed or lost, it is natural to inquire how long it takes for the allele to be fixed or lost. Here one needs to be careful. If an allele is sometimes fixed and sometimes lost, then the average time to fixation is infinity because, in those instances in which the allele is lost, fixation never occurs. So we will need to think about a *conditional* drift process in which we pay attention only to those trajectories in which an allele is ultimately fixed. To state the question with precision, what we wish to know is this: given that an allele is ultimately fixed, what is the time to fixation? Similar issues arise in considering the time to loss of an allele.

Conditional on ultimate fixation or loss, diffusion approximations to random drift result in equations analogous to the Kolmogoroff forward and backward equations, but the multiplier terms need to be altered (Ewens 2004). These equations lead to the following expressions for the average time to fixation $\bar{t}_{fix}$ of alleles that are ultimately fixed and the average time to loss $\bar{t}_{loss}$ of alleles that are ultimately lost (Crow and Kimura 1970):

$$\bar{t}_{fix} = \int_p^1 \psi(x)u(x)\left[1 - u(x)\right]dx + \frac{1 - u(p)}{u(p)} \int_0^p \psi(x)[u(x)]^2 dx \tag{6.12}$$

$$\bar{t}_{loss} = \frac{u(p)}{1 - u(p)} \int_p^1 \psi(x)[1 - u(x)]^2 dx + \int_0^p \psi(x)u(x)\left[1 - u(x)\right]dx \tag{6.13}$$

where p is the initial allele frequency, $u(x)$ is the probability of ultimate fixation, and:

$$\psi(x) = \frac{2\int_0^1 Exp\left[-\int \frac{2M(x)}{V(x)} dx\right]dx}{V(x)Exp\left[-\int \frac{2M(x)}{V(x)} dx\right]} \tag{6.14}$$

in which $\int \frac{2M(x)}{V(x)} dx$ is the indefinite integral, a function of x. These equations look quite complex, however they reduce nicely in the case of neutrality.

NEUTRAL ALLELES

In this case $M(x) = 0$ and $V(x) = x(1-x)/(2N)$, which means that:

$$Exp\left[-\int_0^x \frac{2M(x)}{V(x)} dx\right] = 1$$

and therefore:

$$\psi(x) = \frac{4N}{x(1-x)}$$

Also $u(x) = x$, and so Equation 6.12 says that:

$$\bar{t}_{fix} = -\frac{4N(1-p)\ln(1-p)}{p} \tag{6.15}$$

Likewise, Equation 6.13 implies that:

$$\bar{t}_{loss} = -\frac{4Np\ln(p)}{1-p} \tag{6.16}$$

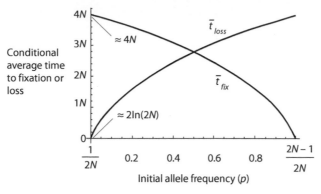

Figure 6.11 The average times to fixation or loss of a neutral allele plotted as a function of initial allele frequency. The average times are for the conditional distributions: $\bar{t}_{fix}$ applies only to those alleles that are ultimately fixed, and $\bar{t}_{loss}$ only to those ultimately lost.

Figure 6.11 illustrates the conditional average times to fixation or loss of neutral mutations as functions of the initial allele frequency running from $p = 1/(2N)$ to $p = (2N-1)/(2N)$. The curves are symmetric and intersect at $p = 0.5$, as a moment's thought will convince you that they must. For newly arising mutations $[p = 1/(2N)]$, the conditional average time to fixation has a mean of $\bar{t}_{fix} \approx 4N$ generations and a median of about $3.5N$ generations. Note that an average time to fixation of $4N$ generations agrees with the average time to the most recent common ancestor looking backward in the coalescent (Equation 4.17 when the sample size n is reasonably large). The conditional average time to loss, $\bar{t}_{loss} \approx 2\ln(2N)$, is very much smaller than the average fixation time. For $N = 1000$, for example, $\bar{t}_{fix} \approx 4000$ generations (in humans about 100 thousand years), whereas $\bar{t}_{loss} \approx 15.2$ generations (about 380 years). The long average fixation time makes one wonder how a neutral allele could ever become fixed in a human population, but the average based on the Wright–Fisher model ignores founder effects (Chapter 4) as well as hitchhiking with beneficial alleles (Chapter 5).

POSITIVELY SELECTED ALLELES

Equation 6.12 also applies to beneficial alleles. In an additive model in which the genotypes A_1A_1, A_1A_2, and A_2A_2 have fitnesses given by $w_{11} = 1 + 2s$, $w_{12} = 1 + s$, and $w_{22} = 1$, the change in allele frequency when s is small is given by Equation 5.7 as $\Delta p \approx sp(1-p)$ because $\bar{w} \approx 1$. In the diffusion approximation we replace p with x to obtain $M(x) = \Delta x = sx(1-x)$.

Unfortunately, there seems to be no closed-form solution to Equation 6.12 even in this simple additive case (Ewens 2004), and one must turn to numerical solutions. Figure 6.12 shows how the conditional time to fixation decreases as a function of Ns. The dashed line shows an approximate solution over the range $Ns = 1.5$ to 10. The approximation is $t = 2\ln(6Ns)/s$, which happens to be the time required in a deterministic model for an additive allele with selective advantage s to increase in frequency from $p_0 = 1/(6Ns)$ to $p_t = 1 - 1/(6Ns)$.

The relatively long fixation times in Figure 6.12 might at first seem in contradiction to the assertion in Chapter 5 that significant pesticide resistance in insect populations typically evolves in 5–50 generations. The main reasons for the apparent discrepancy are that pesticide resistance due to a single allele is often partially or completely dominant, the selection coefficients are orders of magnitude greater than those in Figure 6.12, and

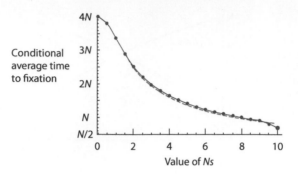

Figure 6.12 The conditional average time to fixation of an additive beneficial allele that confers a selective advantage *s* in heterozygous genotypes. The dashed red line is an approximation based on a deterministic model.

the pesticide becomes ineffectual in practice by the time the resistance frequency reaches a few per cent.

DELETERIOUS ALLELES

The solid curve in Figure 6.12 also applies to an additive model with fitnesses $w_{11} = 1 + 2s$, $w_{12} = 1 + s$, and $w_{22} = 1$ when *s* is negative! This seeming equivalence is counterintuitive, maybe even a little spooky. It has to do with the conditional drift processes being time reversible, which means that the steady-state distribution has the property that the long-run probability of being in state *i* and then transitioning to state *j* is the same as the long-run probability of being in state *j* and then transitioning to state *i* (Ewens 2004). Note, however, that the equality of the conditional time to fixation of a deleterious allele with that of its beneficial counterpart does not imply equality of probabilities of fixation, as Equation 6.11 implies that these are very much different. The equality of conditional average times means only that, in rare cases when a deleterious allele is lucky enough to be fixed, it must be fixed just as fast as its beneficial counterpart.

6.4 Equilibrium Distributions of Allele Frequency

So far in this chapter we've considered only those situations in which random genetic drift results in eventual fixation or loss of one allele or the other, and we considered such issues as the probability of fixation of an allele and the conditional time to fixation. In this section we consider cases in which mutation is an ongoing process, with the result that no allele can be permanently fixed (although temporary fixation can happen). In such cases the allele frequency across a large ensemble of subpopulations attains a **stationary distribution** (also called a **steady-state distribution**), which remains the same through time. In the steady state, allele frequency changes in one subpopulation are balanced by opposite changes in another subpopulation, and hence the overall distribution of allele frequencies remains constant.

An Equation for the Stationary Distribution

In this section we develop an expression for the stationary distribution using the diffusion approximation in Equation 6.4. In later sections we will consider several specific examples in more detail. At steady state, the function $\varphi(p,x;t)$ no longer changes with time, hence the left-hand side of Equation 6.4 equals 0. Furthermore, because permanent fixation cannot occur, each state can be reached from any other state, which implies that the stationary distribution is independent of the starting state p. At steady state, therefore, $\varphi(p,x;t)$ is a function of the single variable x, and Equation 6.4 implies that:

$$0 = -\frac{\partial[M(x)\varphi(x)]}{\partial x} + \frac{1}{2}\frac{\partial^2[V(x)\varphi(x)]}{\partial x^2}$$

Integrating with respect to x yields the first-order differential equation:

$$0 = -M(x)\varphi(x) + \frac{1}{2}\frac{d[V(x)\varphi(x)]}{dx} \tag{6.17}$$

The expression on the right in Equation 6.17 is the net flux in probability across any point x. Table 6.1 gives a rationale for this interpretation based on a discrete-time analogy. The equality in Equation 6.17 states simply that, at steady state, the net probability flux across any given point must equal 0. (If it were anything other than 0, the system would not be in steady state.)

The solution to Equation 6.17 is given by:

$$\varphi(x) = \frac{C}{V(x)}Exp\left[\int\frac{2M(x)}{V(x)}dx\right] \tag{6.18}$$

where C is a constant chosen so that $\int_0^1\varphi(x)dx = 1$.

To verify that Equation 6.18 solves Equation 6.17, note that the derivative

$$\frac{dV(x)\varphi(x)}{dx} = 2C\frac{M(x)}{V(x)}Exp\left[-2\int\frac{M(x)}{V(x)}dx\right] = 2M(x)\varphi(x)$$

can be rearranged in a form to exactly match Equation 6.17.

Reversible Mutation

Consider a scenario of reversible mutation like that in Chapter 4 in which allele A_1 mutates to allele A_2 at a rate μ per generation and A_2 reverse-mutates to A_1 at a rate v per generation. Letting x represent the allele frequency A_1, it follows from Equation 4.3 that, for small changes in allele frequency:

$$M(x) = \Delta x = v(1-x) - \mu x \tag{6.19}$$

where, as usual, $V(x) = x(1-x)/(2N)$.

Substituting $M(x)$ and $V(x)$ into Equation 6.18 yields the steady-state distribution of allele frequencies:

$$\varphi(x) = \frac{\Gamma(4N\mu + 4Nv)}{\Gamma(4N\mu)\Gamma(4Nv)}x^{4Nv-1}(1-x)^{4N\mu-1} \tag{6.20}$$

(Wright 1931), where Γ represents the gamma function, a continuous extension of the factorial function.

Equation 6.20 is that of a beta distribution whose properties are well known. For example, the mean of the distribution is $\hat{x} = 4Nv/(4Nv + 4N\mu) = v/(v+\mu)$, which is perhaps not surprising because it coincides with the equilibrium allele frequency with reversible mutation in an infinite population (Equation 4.6).

The beta distribution can assume a variety of shapes depending on the values of the parameters. Figure 6.13a shows some examples for the case $v = \mu$. The curves are labeled

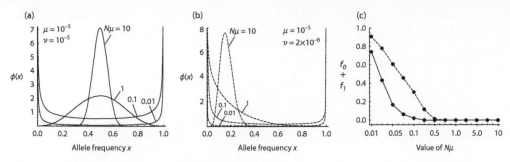

Figure 6.13 Mutation-drift equilibrium with (a) equal forward and reverse mutation rates or (b) unequal forward and reverse mutation rates. (c) Steady-state proportion of populations temporarily fixed for one allele or the other, color-coded according to the parameters in (a) or (b).

according to the value of $N\mu$. When $N\mu$ is large (of the order of 10), the distribution is highly peaked and concentrated around the mean; however, as $N\mu$ decreases, the distribution becomes more flat, and for $N\mu$ less than 1 it becomes U-shaped with little probability mass near the mean. As $N\mu$ approaches 0, the distribution becomes ever more highly concentrated near the values $x = 0$ and $x = 1$. Figure 6.13b shows examples when $v = 0.2\mu$, for which the mean of Equation 6.20 equals 0.17. The same pattern emerges as in Figure 6.13a, except that the mode of the distribution of allele frequencies does not coincide with the mean even for $N\mu = 1$. Once again, as $N\mu$ approaches 0, the distribution becomes more concentrated near the extremes.

Values of x arbitrarily close to 0 or 1 seem to make little sense in a finite population, however they arise naturally in the diffusion limit, which assumes that N goes to infinity and μ goes to 0 in such a way that $N\mu$ remains constant. Moreover, the values of x near 0 and 1 have an interpretation that makes sense even in a finite population. In particular, in a population of size N, values of $x < 1/(2N)$ correspond to subpopulations that are fixed for A_2 and values of $x > 1 - 1/(2N)$ correspond to subpopulations that are fixed for A_1. These values are given explicitly by the equations:

$$f_0 = \int_1^{\frac{1}{2N}} \varphi(x)dx = \frac{\Gamma\left(4N\mu + 4Nv\right)}{\Gamma\left(4Nv + 1\right)\Gamma\left(4N\mu\right)}\left(\frac{1}{2N}\right)^{4Nv} \tag{6.21}$$

$$f_1 = \int_{1-\frac{1}{2N}}^1 \varphi(x)dx = \frac{\Gamma\left(4N\mu + 4Nv\right)}{\Gamma\left(4Nv\right)\Gamma\left(4N\mu + 1\right)}\left(\frac{1}{2N}\right)^{4N\mu} \tag{6.22}$$

(Crow and Kimura 1970), where $\varphi(x)$ is as given in Equation 6.20. The quantities f_0 and f_1 are the fraction of subpopulations that are fixed for allele A_2 or allele A_1 at any time once the ensemble of populations has reached steady state. These are *temporary* fixations, because a population fixed for either allele remains fixed only until a mutation to the other allele takes place within it. It is worth noting that $f_1/\left(f_0 + f_1\right) \approx v/\left(\mu + v\right)$, which means that the average allele frequency in the temporarily fixed classes matches that of the entire distribution.

Figure 6.13c shows the fraction of temporarily fixed subpopulations $\left(f_0 + f_1\right)$ for the examples in Figures 6.13a and 6.13b. When $N\mu$ is small (of the order of 0.01), a substantial fraction of the subpopulations are fixed for one allele or the other even though deterministic theory implies that there is an interior equilibrium. This is true even for moderate values of $N\mu$ (of the order of 0.1) unless the interior equilibrium is near 0.5.

Multiple Alleles and the Ewens Sampling Formula

Multidimensional generalizations of the diffusion approximation are quite complex as you might well imagine (Ewens 2004), but in certain cases the steady-state distribution takes on a relatively simple form. In particular, imagine a single gene with k alleles in which each allele is equally likely to mutate to any other. In this case the steady-state distribution is the **Dirichelet distribution** discussed in Chapter 2 in regard to DNA typing, in which the joint distribution of the allele frequencies $x_1, x_2, \dots, x_{k-1}$ is given by:

$$f(x_1, x_2, \cdots, x_{k-1}) = \frac{\Gamma(k\alpha)}{[\Gamma(\alpha)]^k}(x_1 x_2 \cdots x_k)^{\alpha-1} \tag{6.23}$$

where $x_k = 1 - x_1 - x_2 - \cdots - x_{k-1}$ and $\alpha = 4N\mu/(k-1)$, with μ equal to the total mutation rate for any allele (Ewens 2004).

Suppose now the infinite-alleles model, in which μ is the mutation rate per allele per generation, but each new mutation creates an allele that is unique to the population. A diffusion model for this case is arrived at by letting N go to infinity and μ go to 0 in such a way that $4N\mu = \theta$ remains constant (Ewens 1972, 2004; Karlin and McGregor 1972). An alternative approach is to let k go to infinity in Equation 6.23 (Kingman 1977). The stationary distribution is in any case quite complex; however, the distribution of allele counts in a sample from the stationary distribution can be written in fairly simple form.

Suppose a gene with infinitely many alleles is at steady state with $\theta = 4N\mu$, and a sample of size n is taken from the population with the stipulation that n is much smaller than N. We'll characterize the sample according to the n-tuple $a_1, a_2, \cdots, a_n$, where a_1 is the number of alleles appearing exactly once in the sample, a_2 the number appearing exactly twice, a_3 the number appearing exactly three times, and so forth. Then $\sum_{j=1}^{n} j a_j = n$, where n is the sample size. To help make this symbolism clear, consider the following examples:

- $a_1 = n$ and $a_j = 0$ for $j > 1$. In this extreme case the sample contains n distinct alleles, each represented exactly once.
- $a_n = 1$ and $a_j = 0$ for $j < n$. In this other extreme case the sample contains only one allele, which is represented n times.
- Suppose n is an even number and that $a_{n/2} = 2$ with $a_j = 0$ for $j \neq n/2$. This sample contains two distinct alleles each represented exactly $n/2$ times.

With $a_1, a_2, \cdots, a_n$ defined as above and $\theta = 4N\mu > 0$, the probability of the sample configuration $a_1, a_2, \cdots, a_n$ is given by:

$$\Pr(a_1, a_2, \cdots, a_n; \theta) = \frac{n!\theta^k}{1^{a_1} 2^{a_2} \cdots n^{a_n} a_1! a_2! \cdots a_n! \theta(\theta+1) \cdots (\theta+n-1)} \tag{6.24}$$

where $k = \sum_{j=1}^{n} a_j$ is the number of distinct alleles in the sample of size n (Ewens 1972; Karlin and McGregor 1972).

Equation 6.24 is known as the **Ewens sampling formula**, and it is hard to exaggerate its importance in the history of population genetics (Crane 2016). Until this breakthrough, population genetics theory was largely composed of mathematical models that explored what might happen under certain specified conditions. The Ewens sampling formula turned population genetics on its head. It provided a framework for asking what *had* happened in evolution rather than what *might* happen. It opened the door for estimating population parameters such as $\theta = 4N\mu$ as well as for testing hypotheses such as whether observed data conform to the expectations for neutral alleles. The Ewens sampling formula also came just at the right time, as protein polymorphisms were being studied in many organisms by means of electrophoresis (Lewontin and Hubby 1966; Lewontin 1991; Nevo 1978). As it turned out, protein polymorphisms that differ in electrophoretic mobility are not well suited for analysis by Equation 6.24, owing in part to difficulties in distinguishing among alleles and uncertain patterns of mutation, but by the time DNA sequencing came along about a decade later, the Ewens sampling formula was already in place to analyze the data. The Ewens sampling formula has wide applicability in probability theory (Crane 2016), and it also stimulated additional important research in population genetics that led ultimately to the development of the coalescent (Kingman 2000).

To apply Equation 6.24, let's calculate the steady-state probability $\hat{F}$ that two randomly chosen alleles are identical by descent. In this case $n = 2$, $a_1 = 0$, $a_2 = 1$, and $k = 1$. From Equation 6.24:

$$\hat{F} = \frac{2!\theta^1}{1^0 2^1 0! 1! \theta (\theta + 1)} = \frac{1}{1 + \theta} = \frac{1}{1 + 4N\mu} \tag{6.25}$$

This expression reproduces Equation 4.11 arrived at by different methods. Because of the infinite-alleles assumption, $\hat{F}$ is also the probability that a randomly chosen diploid individual is homozygous. The quantity $1 - \hat{F}$ is therefore the probability of heterozygosity $\hat{H}$ at steady state, which corresponds to two randomly chosen alleles being distinct. In this case $n = 2$, $a_1 = 2$, $a_2 = 0$, and $k = 2$, for which:

$$\hat{H} = 1 - \hat{F} = \frac{2!\theta^2}{1^2 2^0 2! 0! \theta (\theta + 1)} = \frac{\theta}{1 + \theta} = \frac{4N\mu}{1 + 4N\mu} \tag{6.26}$$

This expression reproduces Equation 4.12.

The distribution of allele counts conditional on a given value of k is also of great utility. Suppose that, in sample of size n, exactly k distinct alleles are observed. If the alleles are arranged in any arbitrary order and the allele counts are given as n_1, n_2, ... , n_k, where $\Sigma n_i = n$, then:

$$\Pr\{n_1, n_2, \cdots, n_k | k\} = \frac{n!}{k! n_1 n_2 \cdots n_k \, | S_n^k |} \tag{6.27}$$

where $| S_n^k |$ is the absolute value of a Stirling number of the first kind, which also happens to be the coefficient of θ^k in $\theta (\theta + 1)(\theta + 2) \cdots (\theta + n - 1)$ (Ewens 1972, 2004; Karlin and McGregor 1972).

One of the important features of Equation 6.27 is that the right-hand side is completely independent of θ. This means that, once you know the value of k in a sample, you have no further information about the value of θ irrespective of the allele counts. The allele counts can therefore be tested for their conformity to steady state for neutral alleles, independent of θ. Equation 6.27 also implies that the allele counts in samples are expected to be quite uneven. An example for $k = 2$ in a sample of size $n = 50$ is shown in Figure 6.14a. In this case the allele configurations in the sample range from (1, 49)—that is, 1 copy of the first allele and 49 of the second—to (49, 1). The U-shaped distribution implies that the least frequent allele configuration is (25, 25).

Although the distribution of allele counts in a sample, given k, is independent of θ, the value of k is does depend on θ. In particular, the expected number of alleles in a sample is given by:

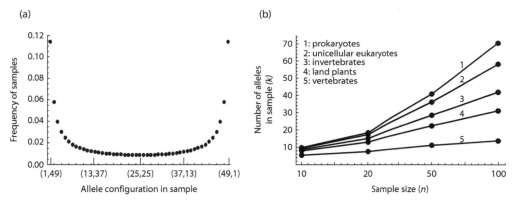

Figure 6.14 Steady-state allele configurations and expected number of alleles in samples under neutrality. (a) Distribution of allele configurations with two alleles in samples of size 50. (b) Expected number of alleles in sample of a 1 kb region of neutrally evolving DNA in various types of organisms.

$$E(k) = 1 + \frac{\theta}{\theta+1} + \frac{\theta}{\theta+2} + \cdots + \frac{\theta}{\theta+n-1} \tag{6.28}$$

The value of θ differs among organisms, decreasing from a mean of 0.1044 per synonymous nucleotide site in prokaryotes to 0.0041 per synonymous nucleotide site in vertebrates, a 25-fold difference with about an order of magnitude variation within each group (Hahn 2019; Lynch 2006). Figure 6.14b shows the expected number of alleles in a 1 kb segment of neutrally evolving DNA for various sample sizes among organisms, taking into account their differing mean θ values. The differences in amount of genetic variation are pronounced, particularly in large samples.

Migration

Like mutation, the effect of migration on allele frequency is linear and therefore the steady-state equations for migration might be expected to resemble those for mutation. For simplicity, we assume two alleles in an island model of migration, in which migrants from all subpopulations are pooled and are equally likely to enter into any subpopulation. The allele frequency change in the deterministic version of this model is given in Equation 4.40, and replacing p in that equation with x yields:

$$M(x) = \Delta x = m(\bar{x} - x) \tag{6.29}$$

where m is the rate of migration per generation and $\bar{x}$ is the average allele frequency among subpopulations, assumed to be constant. As before, $V(x) = x(1-x)/(2N)$. In this case Equation 6.18 yields the stationary distribution:

$$\varphi(x) = \frac{\Gamma(4Nm)}{\Gamma(4Nm\bar{x})\,\Gamma[4Nm(1-\bar{x})]} x^{4Nm\bar{x}-1}(1-x)^{4Nm(1-\bar{x})-1} \tag{6.30}$$

which, like Equation 6.20 for mutation, is a beta distribution with mean $\bar{x}$.

Figure 6.15 shows some examples of the stationary distribution for $\bar{x} = 0.50$ (Figure 6.15a) and $\bar{x} = 0.15$ (Figure 6.15b). These look similar to those for reversible mutation, as might be expected, but a major difference between the situations is that, unless Nm is very small, migration results in a much faster approach to the stationary state than mutation. The smaller values of Nm in Figure 6.15 also imply that a significant proportion of

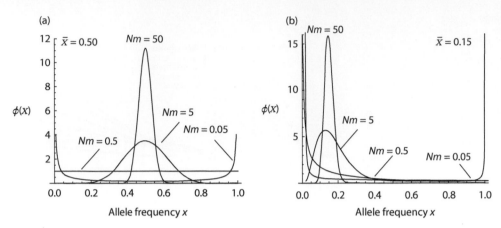

Figure 6.15 Stationary distributions of allele frequency in the island model of migration. (a) $\bar{x} = 0.50$. (b) $\bar{x} = 0.15$.

subpopulations will be temporarily fixed for one allele or the other. The proportions can be estimated by integrations analogous to those in Equations 6.21 and 6.22, but the smaller Nm the greater the proportions of temporarily fixed subpopulations. Don't be misled by the curve for $Nm = 0.05$. At first glance it seems symmetrical; however, it rises much more abruptly near $x = 1$ than near $x = 0$. Because the mean of the entire distribution remains constant at $\bar{x}$, this means that there are many more temporary fixations at $x < 1/(2N)$ then at $x > 1 - 1/(2N)$.

You may recall from Chapter 4 that, in the island model, the steady-state fixation index $\hat{F}_{ST}$ decreases with Nm according to $\hat{F}_{ST} = 1/(1 + 4Nm)$ (Equation 4.43). Applying this equation to the values of Nm in Figure 6.14 results in $\hat{F}_{ST} = 0.005$ for $Nm = 50$ (little subpopulation divergence, according to Figure 4.14), $\hat{F}_{ST} = 0.05$ for $Nm = 5$ (moderate subpopulation divergence), $\hat{F}_{ST} = 0.33$ for $Nm = 0.5$ (very great subpopulation divergence), and $\hat{F}_{ST} = 0.83$ for $Nm = 0.05$ (exceptional subpopulation divergence).

Because mutation and migration are linear forces on allele frequency, it is easy to incorporate both. In the island model of migration, with μ the forward mutation rate and v the reverse mutation rate, Equation 6.30 yields the steady-state distribution if you replace m with $m + \mu + v$ and $m\bar{x}$ with $m\bar{x} + v$.

Mutation–Selection Balance

As we've seen in Chapter 5 there is almost no end to the variety of selection regimes that one can imagine, and if you add reversible mutation into the mix many of these regimes result in a well-defined stationary distribution. In this section we focus on a simple but important model in which selection against an allele is balanced by reverse mutation. To be specific, suppose the relative fitnesses of genotypes A_1A_1, A_1A_2, and A_2A_2 are given by $w_{11} = 1 - 2s$, $w_{12} = 1 - s$, and $w_{22} = 1$, respectively, where s is positive. A_1 is therefore a deleterious allele, and its effects are additive. In this case, $\Delta x = -sx(1 - x)/\bar{w}$, and if s is small enough that $\bar{w} \approx 1$, then $\Delta x = -sx(1 - x)$. The supposition that s is small also makes the assumption about additivity more realistic as discussed earlier in connection with Figure 5.11. If the mutation rate from A_1 to A_2 is μ and that of A_2 to A_1 is v, then:

$$M(x) = \Delta x = -sx(1 - x) - x\mu + (1 - x)v \tag{6.31}$$

As before, $V(x) = x(1-x)/(2N)$. In this case:

$$\int \frac{2M(x)}{V(x)}dx = -4Nsx + 4N\mu\ln(1-x) + 4Nv\ln(x)$$

and the stationary distribution in Equation 6.18 becomes:

$$\varphi(x) = Ce^{-4Ns}(1-x)^{4N\mu-1}x^{4Nv-1} \tag{6.32}$$

where C is again a constant chosen so that $\int_0^1 \varphi(x)dx = 1$.

Figure 6.16 shows some examples of the stationary distribution of allele frequencies for various values of Ns. In an infinite population, the equilibrium value of x is given by the positive root of the quadratic in Equation 6.31, and if mutation is weak with respect to selection, the equilibrium is given approximately by $\hat{x} = 2v/s$, which is in accord with Equation 5.27 for an additive model $(h = 1/2)$. A value of $Ns = -5$ may be considered rather strong selection, and in this case the negative selection dominates the process. As s becomes smaller relative to N, the average allele frequency becomes drawn toward the deterministic mutational equilibrium of $v/(v+\mu)$ in Equation 4.6, which is why the mode of the distribution in Figure 6.16 shifts to the right as Ns approaches 0.

Protein Polymorphisms

The use of protein electrophoresis to identify amino acid polymorphisms was a crucial first step in the development of molecular population genetics, as it opened large parts of the protein-coding genome to investigation (Harris 1966; Lewontin 1991; Lewontin and Hubby 1966; Smithies 1955, 2012). Protein electrophoresis identifies a subset of all nonsynonymous nucleotide substitutions, because protein molecules of the same size that differ in charge due to an amino acid replacement can be separated. Polymorphisms of this type are called **allozymes**. Because the detection of allozyme polymorphisms requires a difference in amino acid sequence, there are fewer protein polymorphisms than DNA polymorphisms.

Protein polymorphisms are usually more difficult to interpret than DNA polymorphisms, because a change in amino acid sequence of a protein might very well be expected to have some effect on the survival or reproduction of the organism. Some allozyme polymorphisms are maintained because the heterozygous genotype has the

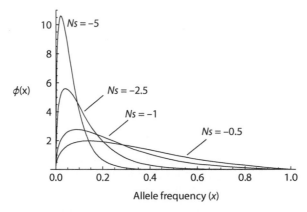

Figure 6.16 Stationary distributions of a deleterious allele with additive fitness effects for various values of Ns. In these examples the mutation rates are such that $v = 0.75\mu$.

highest average reproductive fitness. One well-established case is that of sickle-cell anemia in tropical Africa, where the malaria resistance of the heterozygotes offsets the near-lethality of the sickle-cell homozygotes (Allison 1954). Another example is an allozyme polymorphism for alcohol dehydrogenase in *D. melanogaster* (Kreitman and Hudson 1991), but in this case the physiological basis of the heterozygote superiority is unclear although correlations with latitude are clear (Figure 5.9). On the other hand, heterozygote superiority cannot explain the high incidence of allozyme polymorphisms in certain bacterial populations, such as the intestinal bacterium *Escherichia coli* (Whittam et al. 1983), since these organisms are haploid and therefore heterozygous genotypes do not exist. In this organism many of the amino acid polymorphisms appear to be slightly detrimental (Sawyer et al. 1987).

Polymorphism for allozymes is demonstrated in Figure 6.17, which summarizes the results of electrophoretic surveys from 14 to 71 enzymes (mostly around 20) in populations of 242 species. The numbers in parentheses are the number of species examined in each type of organism. The vertical axis refers to the observed proportion of genes that are polymorphic based on the frequency of the major (most frequent) allele being less than 0.95, and the horizontal axis refers to the proportion of protein-coding genes that are heterozygous in an average individual. Each point is the average of many different species in each category, and the errors bars (not shown) are rather large.

Because of the large variability in polymorphism and heterozygosity found within each group of organisms, Figure 6.17 admits of no simple summary. On the whole, there is a positive relationship between amount of polymorphism and degree of heterozygosity, which is as expected because the greater the fraction of polymorphic genes in a population, the more genes that are expected to be heterozygous in an average individual. Organisms also differ in their average values of $\theta = 4N\mu$ (Figure 6.14b).

The overall mean polymorphism in Figure 6.17 is 0.26 ± 0.15, and the mean heterozygosity is 0.07 ± 0.05. Vertebrates have the lowest average amount of genetic vari-

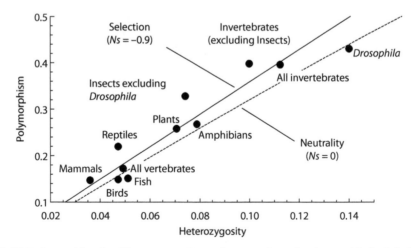

Figure 6.17 Estimated levels of heterozygosity and proportion of polymorphic loci derived from allozyme studies of various groups of plants and animals. The lines are based on a model of neutral alleles (dashed) or one of mutation–selection balance (solid). (Data from Nevo (1978).)

ation among the groups, land plants come next, and invertebrates have the highest. *Drosophila* is the most genetically variable group of higher organisms so far studied, and mammals the least variable. Humans are fairly typical of large mammals. Extensive electrophoretic surveys among human populations yield an average allozyme polymorphism of about 0.30 and an average allozyme heterozygosity of about 0.05 (Harris and Hopkinson 1972).

The lines in Figure 6.17 illustrate theoretical results calculated using two of the models discussed earlier in this chapter. The dashed line shows the expected relation between polymorphism and heterozygosity based on the Dirichlet distribution for neutral alleles (Equation 6.23), where for simplicity the minor (least frequent) alleles are grouped into a single category. The fit is satisfactory but not impressive. For any observed level of polymorphism, the predicted level of heterozygosity is about 10 per cent greater than that actually observed. The solid line is based on the same mutation rates but assuming mutation–selection balance (Equation 6.32). The fit is much better, in fact very close to the regression line of polymorphism against heterozygosity. On the other hand, the fit ought to be better because it includes Ns as an additional parameter. The intensity of selection in this model is $Ns = -0.9$. This represents rather mild negative selection and is in the range that Ohta has called **nearly neutral** (Ohta 1992; Akashi et al. 2012). In the nearly neutral model, an allele is neither neutral ($Ns = 0$) nor strongly selected $|Ns| >> 1$. Nearly neutral alleles are those subject to such weak selective forces that their fate is determined by the joint effects of random genetic drift and weak selection.

Although Figure 6.17 suggests that most amino acid polymorphisms are slightly deleterious, the analysis needs to be hedged with a number of caveats. First, in many eukaryotes, only a minority of the genome codes for protein. This is particularly true of eukaryotes with large genomes including mammals and grassy plants. In humans, only about 1.5 per cent of the nucleotides actually codes for the amino acids in proteins, and perhaps another 8 per cent accounts for the regulatory sequences that determine when and in which cells the protein-coding genes are expressed. Any conclusion about evolutionary forces operating on proteins ignores the large amount of genomic DNA that has other functions (or perhaps, as some population geneticists might argue, no function).

Second, any analysis of average polymorphism and average heterozygosity across dozens of genes in hundreds of species necessarily conceals what may be critical biological variation. The organisms in Figure 6.17 are highly diverse with multifarious habitats, lifestyles, and demographic histories. The averages smooth over all of these important differences. Likewise the genes that are included have different physiological and metabolic functions. Each gene may have its own configuration of alleles that are neutral, nearly neutral, or selected in any of a myriad of ways.

Third, the theoretical curves in Figure 6.17 assume demographic equilibrium and a stationary state of allele frequency distribution. This assumption is problematic in light of how long it takes populations to reach a steady-state distribution of allele frequency and the sometimes rapid changes in population size that actually occur in populations due to local and varying climate conditions.

On the other hand, the inference that most amino acid polymorphisms within populations are slightly deleterious is supported by the much more sophisticated analyses made possible by DNA sequencing and the methods of molecular population genetics (Hahn 2019). These are the subjects of Chapter 7.

Problems

6.1 Suppose random genetic drift occurs according to the Wright–Fisher model in a large ensemble of genetically isolated diploid subpopulations each of size N. Each subpopulation initially includes two neutral alleles A_1 and A_2 at frequencies p_1 and p_2 ($p_1 + p_2 = 1$). After t generations have passed and some subpopulation are fixed for A_1 and others for A_2:

a. Among the subpopulations that are fixed, what fraction are fixed for A_1 and what fraction are fixed for A_2?

b. What is the average allele frequency across all subpopulations?

c. What is the average heterozygosity across all subpopulations in generation t?

d. If t is very large, what can you say about the allele frequencies among the unfixed subpopulations (i.e., those still segregating for A_1 and A_2)?

6.1 ANSWER a. Because the alleles are neutral, among the fixed subpopulations a fraction p_1 is expected to be fixed for A_1 and p_2 to be fixed for A_2. **b.** The average allele frequencies across all subpopulations are expected to remain constant at p_1 and p_2. **c.** The average heterozygosity among the subpopulations decreases according to Equation 6.3, and hence in generation t is expected to be $H_t = 2p_1p_2[1 - 1/(2N)]^t$. **d.** As t increases the distribution of allele frequencies among unfixed populations becomes progressively more flat, hence the distribution of allele frequencies across unfixed subpopulations is eventually a uniform distribution.

6.2 Equation 6.1 defines a matrix of elements $\mathbf{P} = [p_{ij}]$ in which p_{ij} is the probability that an allele goes from i copies to j copies after one generation of random genetic drift in the Wright–Fisher model, where $i,j = 0,1,2,\dots,2N$. The $\mathbf{P}$ matrix for $N = 2$ is shown below. (Note: the Wright–Fisher model in a diploid population of size $N = 2$ would seem to be equivalent to sib mating, but it is not quite equivalent because the model allows for self-fertilization.)

$$\begin{pmatrix} 1 & 0 & 0 & 0 & 0 \\ 81/256 & 27/64 & 27/128 & 3/64 & 1/256 \\ 1/16 & 1/4 & 3/8 & 1/4 & 1/16 \\ 1/256 & 3/64 & 27/128 & 27/64 & 81/256 \\ 0 & 0 & 0 & 0 & 1 \end{pmatrix}$$

a. What does each entry represent?

b. What does each row represent?

c. What does each column represent?

d. Why do the entries of each row sum to 1?

e. Why are the entries at the upper left and lower right equal to 1?

6.2 ANSWER a. Each entry in $\mathbf{P}$ represents p_{ij}, the probability that an alleles goes from i copies to j copies in one generation of random drift. **b.** Each row represents the outcome of one generation of drift given that the starting number of alleles is $i = 0,1,2,\dots$, where i is the row number. **c.** Each column represents the proportion of populations with j copies of the allele where $j = 0,1,2,\dots$ is the column number and the initial condition is the row number $i = 0,1,2,\dots$. **d.** The entries of each row sum to 1 because the probabilities of all possible j values for any given i must sum to 1. **e.** The entry at the upper left equals 1 because an allele that is lost is lost for good, and that at the lower right equals 1 because an allele that is fixed is fixed for good (i.e., no mutation is allowed in this version of the model).

6.3 Suppose a neutral allele A_1 is present at i copies in a Moran model of a population of size $2N$. Then its allele frequency is $p = i/(2N)$. Let a and b be integers with values 0, 1, 2, ..., $2N$ such that $a < i < b$. Then it can be shown that the probability that i visits the state a before it visits the state b equals:

$$\Pr\{a \text{ before } b|i\} = \frac{i-a}{b-a}$$

a. What is the probability $\Pr\{b \text{ before } a\}$?
b. What is the probability that A_1 is lost? What is the probability that A_1 is fixed?
c. If $p = 0.4\,(i = 0.8N)$, what is the probability that $p = 0.6$ before $p = 0$?
d. If $i = 1$ and $2N = 50$, what is the largest value of $b < 100$ that is attained with probability ≥ 0.05 before A_1 is lost from the population?

6.3 ANSWER a. $\Pr\{b \text{ before } a\} = 1 - \Pr\{a \text{ before } b\} = (b-i)/(b-a)$. **b.** Here $a = 0$, $b = 2N$, and i is given. Hence $\Pr\{0 \text{ before } 1\} = i/(2N)$, $\Pr\{1 \text{ before } 0\} = 1 - i/(2N)$. **c.** In this case $a = 0$, $b = 0.6 \times (2N) = 1.2N$ and $i = 0.8N$ is given, and therefore $\Pr\{0 \text{ before } 1.6N\} = 0.8N/1.2N = 2/3$. **d.** Solve $(1-0)/(b-0) \geq 0.05$ for b, which yields $b \leq 20$.

6.4 Taking into account the probabilities of fixation or loss of a neutral allele with initial frequency p, what is the average time to fixation or loss implied by Equations 6.15 and 6.16? In a diploid population of size $N = 100$, what is the average time to fixation or loss for an allele at frequency $p = 0.005$? At frequency $p = 0.5$?

6.4 ANSWER Since the probability of fixation equals p and the probability of loss equals $1-p$ the average time to fixation or loss equals $p\bar{t}_{fix} + (1-p)\bar{t}_{loss}$, and this in turn equals $-4N[p\ln(p) + (1-p)\ln(1-p)]$. For $N = 100$ and $p = 0.005$ this equals 12.59, and for $N = 100$ and $p = 0.5$, it equals 277.26.

6.5 In a Moran model with $2N$ haploid individuals the conditional average times to fixation are $\bar{t}_{fix} = -(2N)^2(1-p)\ln(1-p)/p$ and $\bar{t}_{loss} = -(2N)^2p\ln(p)/(1-p)$, where time is measured in Moran-model units and $2N$ is large. Convert these expressions so that time is measured in units comparable to generations in the Wright–Fisher model and compare with Equations 6.15 and 6.16.

6.5 ANSWER $2N$ time units in the Moran model correspond approximately to one generation in the Wright–Fisher model; hence, the expressions given need to be divided by $2N$. The result is that $\bar{t}_{fix} = -2N(1-p)\ln(1-p)/p$ and $\bar{t}_{loss} = -2Np\ln(p)/(1-p)$. These are half as large as those in Equations 6.15 and 6.16, which means that drift is speeded up in the Moran model. The factor of 2 reflects the fact that for large $2N$ in the Moran model the variance in offspring number equals 2 (versus 1 in the Wright–Fisher model).

6.6 Use the Ewens sampling formula in Equation 6.24 to deduce the probability that 3, 4, or 5 randomly chosen alleles are identical by descent (IBD), assuming an equilibrium population in which all alleles are selectively neutral. From these results conjecture a general expression for the probability that n randomly chosen alleles are IBD.

6.6 ANSWER For $n = 3$, the vector $\alpha = (0, 0, 1)$ and $\Pr\{IBD\} = 6/[(\theta+1)(\theta+2)]$, for $n = 4$ the vector $\alpha = (0, 0, 0, 1)$ and $\Pr\{IBD\} = 24/[(\theta+1)(\theta+2)(\theta+3)]$, and for $n = 5$ the vector $\alpha = (0, 0, 0, 0, 1)$ and $\Pr\{IBD\} = 120/[(\theta+1)(\theta+2)(\theta+3))(\theta+4)]$ These results support the conjecture that, for n alleles, $\Pr\{IBD\} = n!/\prod_{i=1}^{n-1}(\theta+i)$, which is in fact correct.

6.7 Equation 6.18 for the steady state with reversible mutation is a beta distribution, which has a variance of $\alpha\beta/[(\alpha+\beta)^2(\alpha+\beta+1)]$ where $\alpha = 4N\mu$ and $\beta = 4N\nu$. How much does the variance decrease if the population size is reduced by half? What is the decrease in the specific case of $N = 10^6$ and $\mu = \nu = 10^{-6}$?

6.7   **ANSWER** $Var(N) = \mu v/\{(\mu+v)^2[4N(\mu+v)-1]\}$ and $Var(N/2) = \mu v/\{(\mu+v)^2$ $[2N(\mu+v)-1]\}$.   The   ratio   is $[2N(\mu+v)-1]/[4N(\mu+v)-1] = \{2+1/[N(\mu+v)]\}/$ $\{4+1/[N(\mu+v)]\}$, which is a little greater than one-half. For the specific values given, the ratio of the variances equals $5/9 = 0.555$. The conclusion is that the variance scales almost with the population size, but not quite.

6.8 In a model of mutation–selection balance in which the allele A_1 is a recessive lethal, we may assume that x is restricted to very small positive values. This restriction means that we can assume $1-x \approx 1$ and that forward mutation of A_1 to A_2 can be ignored. With these stipulations we can write $M(x) = v - x^2$ and $V(x) = x/(2N)$. Find the steady-state distribution $\phi(x)$ in the Wright–Fisher model. (Hint: this has to be solved from first principles because $V(x) \neq x(1-x)/(2N)$. Note that the indefinite integral of $1/x$ is $\ln(x)$ and that of x is $x^2/2$.)

6.8 ANSWER $2M(x)/V(x) = 4Nv/x - 4Nx$. The integral of the first term is $4Nv\ln(x)$ and that of the second is $-2Nx^2$. Therefore $\mathrm{Exp}(2[M(x)/V(x)]dx) = x^{4Nv}e^{-2Nx^2}$. Setting $V(x) = x/(2N)$, Equation 6.18 implies that $\phi(x) = Cx^{4Nv-1}e^{-2Nx^2}$.

References

Aalto, E. (1989), 'The Moran model and validity of the diffusion approximation in population genetics', *J Theor Biol*, 140 (3), 317–26.

Akashi, H., Osada, N., and Ohta, T. (2012), 'Weak selection and protein evolution', *Genetics*, 192 (1), 15–31.

Allison, A. C. (1954), 'Protection afforded by the sickle-cell trait against subtertian malaria infection', *Br Med J*, 1 (4857), 290–4.

Andolfatto, P., Wong, K. M., and Bachtrog, D. (2011), 'Effective population size and the efficacy of selection on the X chromosomes of two closely related Drosophila species', *Genome Biol Evol*, 3, 114–28.

Buri, P. (1956), 'Gene frequency in small populations of mutant Drosophila', *Evolution*, 10 (4), 367–402.

Cannings, C. (1974), 'The latent roots of certain Markov chains arising in genetics: a new approach. I. Haploid models', *Adv Appl Prob*, 6 (2), 260–90.

Crane, H. (2016), 'The ubiquitous Ewens sampling formula', *Statist Sci*, 31 (1), 1–19.

Crow, J. F. and Kimura, M. (1970), *Introduction to Population Genetics* (New York: Harper & Row).

Ewens, W. J. (1972), 'The sampling theory of selectively neutral alleles', *Theor Popul Biol*, 3, 87–112.

Ewens, W. J. (2004), *Mathematical Population Genetics* (2 edn.; New York: Springer-Verlag).

Fisher, R. A. (1930), *The Genetical Theory of Natural Selection* (2 edn.; Oxford: Clarendon).

Hahn, M. W. (2019), *Molecular Population Genetics* (New York: Oxford University Press).

Harris, H. (1966), 'Enzyme polymorphisms in man', *Proc R Soc Lond B*, 164, 298–310.

Harris, H. and Hopkinson, D. A. (1972), 'Average heterozygosity per locus in man: an estimate based on the incidence of enzyme polymorphisms', *Ann Hum Genet*, 36 (1), 9–20.

Hartl, D. (2004). Talking with Jim Crow (Digital Versatile Disc (DVD) recording, Genetics Society of America).

Karlin, S. and McGregor, J. (1972), 'Addendum to a paper of W. Ewens', *Theor Popul Biol*, 3 (1), 113–16.

Kimura, M. (1955a), 'Stochastic processes and distribution of gene frequencies under natural selection', *Cold Spring Harbor Symp Quant Biol*, 20, 33–53.

Kimura, M. (1955b), 'Solution of a process of random genetic drift with a continuous model', *Proc Natl Acad Sci U S A*, 41 (3), 144–50.

Kimura, M. (1957), 'Some problems of stochastic processes in genetics', *Ann Math Stat*, 28 (4), 882–901.

Kimura, M. (1964), 'Diffusion models in population genetics', *J Appl Prob*, 1 (2), 177–232.

Kingman, J. F. C. (1977), 'The population structure associated with the Ewens sampling formula', *Theor Popul Biol*, 11 (2), 274–83.

Kingman, J. F. C. (2000), 'Origins of the coalescent 1974–1982', *Genetics*, 156 (4), 1461–3.

Kreitman, M. and Hudson, R. R. (1991), 'Inferring the evolutionary histories of the Adh and Adh-dup loci in Drosophila melanogaster from patterns of polymorphism and divergence', *Genetics*, 127 (3), 565–82.

Krukov, I., de Sanctis, B., and de Koning, A. P. J. (2017), 'Wright-Fisher exact solver (WFES): scalable analysis of population genetic models without simulation or diffusion theory', *Bioinformatics*, 33 (9), 1416–17.

Lewontin, R. C. (1991), 'Electrophoresis in the development of evolutionary genetics: milestone or millstone?', *Genetics*, 128, 657–62.

Lewontin, R. C. and Hubby, J. L. (1966), 'A molecular approach to the study of genic heterozygosity in natural populations II. Amount of variation and degree of heterozygosity in natural populations of Drosophila pseudoobscura', *Genetics*, 54, 595–609.

Lynch, M. (2006), 'The origins of eukaryotic gene structure', *Mol Biol Evol*, 23, 450–68.

McKane, A. J. and Waxman, D. (2007), 'Singular solutions of the diffusion equation of population genetics', *J Theor Biol*, 247 (4), 849–58.

Moran, P. A. P. (1958), 'Random processes in genetics', *Proc Cambridge Phil Soc*, 54 (1), 60–71.

Moran, P. A. P. (1962), *Statistical Processes of Evolutionary Theory* (Oxford: Clarendon Press).

Nevo, E. (1978), 'Genetic variation in natural populations: patterns and theory', *Theor Popul Biol*, 13 (1), 121–77.

Ohta, T. (1992), 'The nearly neutral theory of molecular evolution', *Annu Rev Ecol Evol Syst*, 23 (1), 263–86.

Sawyer, S. A., Dykhuizen, D. E., and Hartl, D. L. (1987), 'A confidence interval for the number of selectively neutral amino acid polymorphisms', *Proc Natl Acad Sci U S A*, 84, 6225–8.

Smithies, O. (1955), 'Zone electrophoresis in starch gels: group variations in the serum proteins of normal human adults', *Biochem J*, 61 (4), 629–41.

Smithies, O. (2012), 'How it all began: a personal history of gel electrophoresis', in B. Kurien and R. Scofield (eds.), *Protein Electrophoresis: Methods in Molecular Biology* (Methods and Protocols) (Totowa, NJ: Humana Press), 1–21

Wallace, A. R. (1892), 'Note on sexual selection', *Nat Sci*, 1, 749–50.

Whittam, T. S., Ochman, H., and Selander, R. K. (1983), 'Multilocus structure in natural populations of Escherichia coli', *Proc Natl Acad Sci U S A*, 80 (6), 1751–5.

Wright, S. (1931), 'Evolution in Mendelian populations', *Genetics*, 16 (2), 97–159.

Wright, S. (1945), 'The differential equation for the distribution of allele frequencies', *Proc Natl Acad Sci U S A*, 31 (12), 382–9.

Zhao, L., Yue, X., and Waxman, D. (2013), 'Complete numerical solution of the diffusion equation of random genetic drift', *Genetics*, 194 (4), 973–85.

CHAPTER 7

Molecular Population Genetics

The origins of theoretical population genetics date back to William E. Castle (1903), who was Sewall Wright's PhD mentor, and even earlier to Gregor Mendel who worked out the genetic consequences of repeated self-fertilization in a population of plants (Mendel 1866). (See Abbott and Fairbanks (2016) for a modern translation.) For much of its history, population genetics was mainly theoretical. Its focus was on the relations between mating systems, population structure, mutation, migration, selection, and random genetic drift, insofar as these could be deduced a priori from Mendelian inheritance and Darwinian processes. The essence of this theory is contained in the previous chapters. Its fundamental variables are allele frequencies. The focus on theory reflected the ongoing lack of experimental methods of general utility to detect allelic differences between organisms present in natural populations. Apart from a handful of special cases, such as human blood groups that could be detected by antigen–antibody reactions and chromosomal inversions in populations of *Drosophila* that could be studied cytologically, there were almost no allele frequency data to which the theory could be applied.

This situation was turned upside down by refocusing the theory on the types of information about evolutionary processes that could be inferred from samples taken from natural populations (Ewens 1972). The samples were initially of the sort obtained from protein electrophoresis of allozyme polymorphisms (Harris 1966; Lewontin and Hubby 1966), but a major technical advance came in the 1970s with the development of DNA sequencing and then again in the 1990s with the creation of automated high-throughput DNA sequencing machines that permitted relatively inexpensive sequencing of the complete genomes of multiple individuals from natural populations. The abundance of molecular polymorphisms within species and differences among species has brought about unprecedented opportunities for inferring mechanisms of evolutionary change and for testing evolutionary hypotheses. Just as some surnames imply a lot about their bearers' ethnic heritage, so do nucleotide and amino acid sequences contain information about their evolutionary heritage.

An ongoing problem is to be able to interpret this information. Which brings us back to population genetics theory, both classical theory and modern sampling theory, because the theory provides the framework upon which any inferences must be based. Nevertheless, the focus of population genetics has changed absolutely—from inquiring what deductions can be made about the evolutionary process from the abstract principles of Mendelian inheritance and Darwinian selection, to testing hypotheses and making inferences about the evolutionary process from the analysis of sequences of genes and genomes sampled from actual evolving populations. This is the subject of **molecular population genetics**. This chapter affords a concise overview of this broad and vigorously growing field. A more advanced treatment may be found in Hahn (2019).

A Primer of Population Genetics and Genomics. Fourth Edition. Daniel L. Hartl, Oxford University Press (2020). © Daniel L. Hartl.
DOI: 10.1093/oso/9780198862291.003.0007

7.1 Rates of Nucleotide Substitution

Our starting point is a selectively neutral mutation. By *selectively neutral* we mean a mutation that has a negligible effect on the organism's ability to survive and reproduce. Neutrality is a convenient starting point because, in terms of selection, there is only one way that a mutation can be neutral ($s = 0$) and many ways in which it can be nonneutral ($s \neq 0$). Note that the term *neutral* applies to a mutation or to a mutant allele. A gene can mutate in myriad ways, some of which may be neutral and others nonneutral, or some of which may be neutral but only in certain genetic backgrounds or under particular environmental conditions. Then, too, there are also many ways in which a mutant allele can be neutral without necessarily having its fate determined solely by random genetic drift. Examples include neutral mutations that are genetically linked to deleterious alleles (background selection), those linked to beneficial alleles (selective sweep), or those that occur in populations that are not at demographic equilibrium.

For present purposes we assume the simplest possible case of a neutral mutation that has a constant value of $s = 0$ and remains neutral through time and over all genetic backgrounds and environments. In this case an important principle is that, for neutral mutations, the rate of nucleotide substitution equals the rate at which neutral mutations take place (Kimura 1968). To understand why, suppose that nucleotide sites undergo new mutations at a rate of μ per generation and that the fraction of these that are selectively neutral is f_0. The overall rate at which new neutral mutations occur is therefore $f_0\mu$. Equation 6.10 says that the probability of ultimate fixation of a new neutral mutation is $u(p) = p$, where p is the initial frequency of the new mutation. In a diploid population of size N, $p = 1/(2N)$. Hence the number of new neutral mutations that take place in each generation in a diploid population of size N equals $2N \times f_0\mu$, and since each has a probability of fixation of $1/(2N)$, at steady state the rate of fixation of neutral mutations is given by:

$$2Nf_0\mu \times \frac{1}{2N} = f_0\mu \qquad (7.1)$$

In words, Equation 7.1 says that the steady-state rate of fixation of neutral alleles equals the rate of mutation to neutral alleles. To say the same thing in a slightly different way, the average time between successive fixations of neutral mutations equals $1/(f_0\mu)$ generations. In the next section we shall see that the maximum rate of molecular evolution observed in mammals is approximately 4 nucleotide substitutions per nucleotide site per 10^9 years. Assuming that the maximum rate is observed for nucleotide mutations that are selectively neutral (for these nucleotides $f_0 = 1$), then the rate of nucleotide mutation would be approximately $\mu = 1/(4 \times 10^9) = 2.5 \times 10^{-10}$ per nucleotide site per year. For humans, with a generation time of 20–30 years, this implies a rate of nucleotide mutation of about 5.0×10^{-9} to 7.5×10^{-9} mutations per nucleotide site per generation, an estimate that is in reasonable agreement with whole-genome sequencing studies of mother–father–offspring trios from Iceland (Kong et al. 2012).

Nucleotide Substitutions in Noncoding DNA

The rate of nucleotide substitution can differ across different regions of the genome, from one gene to the next, and even from one part of a gene to another (Figure 7.1). The most rapidly evolving nucleotide sites are found in **pseudogenes**, which are

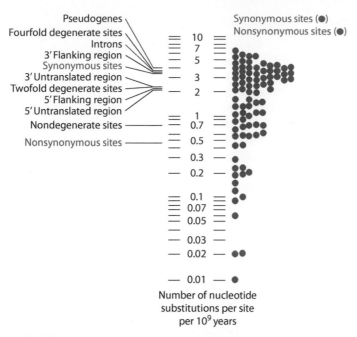

Figure 7.1 Rates of nucleotide substitution in mammalian genes. On the left are the average rates for different types of sequences. On the right are rates for synonymous and nonsynonymous sites in a sample of 43 genes. (Data from Li (1997).)

homologous to known genes but which have undergone one or more frameshift or chain-termination mutations that obliterate their ability to function. Pseudogenes are thought to be completely nonfunctional relics of mutational inactivation. This view is supported by their rapid rate of nucleotide substitution, averaging about 4×10^{-9} per site per year, which is shown in comparison with other types of nucleotide sites in Figure 7.1. The data in Figure 7.1 pertain to substitution rates in mammals. The average is about 3.5×10^{-9} substitutions per site per year, which is much slower than the 15.6×10^{-9} per year observed in *Drosophila* (Li 1997) or the 12.1×10^{-9} per year estimated in *Escherichia coli* (Reid et al. 1999; Whittam et al. 1998).

Note that the scale in Figure 7.1 is logarithmic, which means that there is a 1000-fold difference in substitution rate from the bottom to the top. The most rapidly evolving nucleotide sites are shown on the left, headed by those in pseudogenes, and also including sites found in introns and the 3′ regions that flank protein-coding genes. The 5′ regions that flank protein-coding genes evolve more slowly, as do the 5′ and 3′ regions that are transcribed but not translated, presumably because these regions contain sequences that function in the regulation of transcription, translation, or mRNA stability. What this means is that f_0 in Equation 7.1 is smaller than that observed for nucleotide sites in pseudogenes, which is a more formal way of saying that the need to conserve certain nucleotide identities in order to preserve function acts as a **selective constraint** on the likelihood of a nucleotide undergoing a substitution. Although the per-nucleotide mutation rate varies across the genome, most of the differences in the rate of nucleotide substitution within a species result from selective constraints rather than mutation pressure.

Synonymous and Nonsynonymous Substitutions

Figure 7.1 also includes data on nucleotide sites in coding regions. The classification of sites is based on the standard genetic code shown in Table 1.1. The codons for eight amino acids can have any nucleotide (U, C, A, or G) in their third position, which is therefore known as a **fourfold degenerate site**. The codons for seven amino acids can terminate in either U or C, and those for five amino acids can terminate in either A or G; these third-position sites are called **twofold degenerate sites**. A site is **nondegenerate** if any nucleotide substitution changes the amino acid encoded. Because of the codon degeneracies, nucleotides in a coding sequence can change without affecting the amino acid sequence. These changes are called **synonymous substitutions**. As might be expected, fourfold degenerate sites have the highest rate of nucleotide substitution in mammalian coding regions (Figure 7.1). The average rate of synonymous substitution is 3.4×10^{-9} synonymous substitutions per synonymous site per year. The number of synonymous sites corresponds roughly to the number of fourfold degenerate sites plus 1/3 the number of twofold degenerate sites; the fraction 1/3 come in because, with random mutation at twofold degenerate sites, 1/3 of all nucleotide substitutions are expected to result in a synonymous codon. This definition makes certain simplifying assumptions about patterns of nucleotide mutations that more complex models do not make (Hahn 2019).

Nucleotide substitutions that do change amino acids are **nonsynonymous substitutions** or **amino acid replacements**. The overall average rate of nonsynonymous substitutions in mammals is 0.46×10^{-9} nonsynonymous substitutions per nonsynonymous site per year (Figure 7.1), where the number of nonsynonymous sites corresponds roughly to the number of nondegenerate sites plus 2/3 the number of twofold degenerate sites. (This approximation also makes simplifying assumptions about patterns of nucleotide mutation.)

Especially among nonsynonymous substitutions, there is great variation from one gene to the next. This is shown by the blue dots on the right in Figure 7.1. The large variation in rates is attributed to selective constraints on amino acid replacements that do not operate as strongly on synonymous nucleotide substitutions (red dots). In this case the selective constraint limits the tendency for amino acids at particular sites to change because of natural selection for optimal protein function. Not just any amino acid will serve at a particular position in a protein molecule, because each amino acid must participate in the chemical interactions that fold the molecule into its three-dimensional shape and give the molecule its specificity and ability to function. The need for proper chemical interactions and folding constrains the acceptable amino acids that can occupy each site. Although some amino acid replacements may be functionally equivalent or nearly equivalent, many more are expected to impair protein function to such an extent that they reduce the fitness of the organisms that contain them. We've already seen evidence for weak selection against most amino acid polymorphisms in Figure 6.17, and further evidence is presented later in this chapter.

Nucleotide Divergence Between Species

The rate of nucleotide substitution in properly aligned sequences between two species is estimated as the number of fixed nucleotide differences per nucleotide site between them, divided by the time since the species diverged from a common ancestor. The number of fixed differences per nucleotide site is a measure of genetic **distance (*d*)** or divergence between the species. If the species are too closely related, then d must be corrected by the magnitude of polymorphism within each species (Hahn 2019), and if the species are too distantly related, then d must be corrected for nucleotides that have undergone multiple mutational hits of the same nucleotide. Ideally, the time since species divergence is estimated from the fossil record; however, estimates can also be based on rates of protein evolution, as we'll see later.

In two properly aligned nucleotide sequences, the number of synonymous differences per synonymous site is typically denoted d_S, and the number of nonsynonymous differences per nonsynonymous site is typically denoted d_N. Values of d_N are usually much smaller than those of d_S, as illustrated in Figure 7.2a by comparison of 11,492 orthologous proteins between *Arabidopsis thaliana* and its close relative *A. lyrata* (Yang and Gault 2011). (Homologous genes in different species are said to be **orthologous** if they descend from a common ancestral gene and retain the same function in both species.) As Figure 7.2a indicates, there are much stronger selective constraints on nonsynonymous (amino acid changing) sites than there are on synonymous sites. Some genes have $d_N = 0$ because the selection constrains nearly every amino acid; for example, the 103 amino acids of histone H4 are identical between these two species.

The comparison can be taken a step further, which is to compare the ratio of d_N to d_S (d_N/d_S) for each individual coding sequence. The data for the *A. thaliana–A. lyrata* ortholog are shown in Figure 7.2b. The mean d_N/d_S is estimated as 0.114 ± 0.014, which implies that about 90 per cent of the nonsynonymous mutations that occur are sufficiently detrimental that they have little chance of being fixed. The proportion of severely deleterious mutations could well be greater because any beneficial mutations that become fixed contribute disproportionately to divergence.

At one time d_N/d_S was thought to be an efficient way to detect positive selection, especially looking for genes for which d_N/d_S is greater than 1. As Figure 7.2b clearly shows, such genes would be extreme outliers but they do occur in certain rapidly evolving genes in the immune and reproductive systems of animals. On the other hand, d_N/d_S ratios cannot distinguish between negative selection against synonymous mutations reducing d_S and positive selection for amino acid replacements increasing d_N (Spielman and Wilke 2015). A good discussion of these and other pitfalls of interpreting d_N/d_S ratios may be found in Hahn (2019).

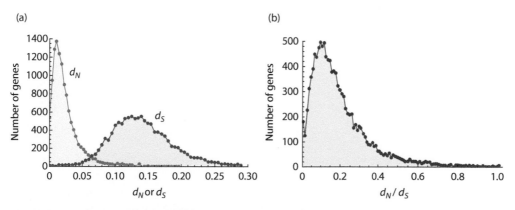

Figure 7.2 Sequence divergence between *A. thaliana* and *A. lyrata*. (a) Distributions of d_N and d_S. (b) Distribution of d_N/d_S. (Data from Yang and Gault (2011).)

Correction for Multiple Mutational Hits

When nucleotide sequences are compared between species that have diverged sufficiently long that individual nucleotides may have experienced two or more substitutions, the observed genetic distance between the sequences will be an underestimate of the true distance and must be corrected. Figure 7.3 shows why. As multiple substitutions occur at an increasing proportion of nucleotide sites, the sequences become nearly randomized with the result that any two aligned nucleotide have a 1/4 chance of being identical and a 3/4 chance of being different. In Figure 7.3, the rate of nucleotide substitution at synonymous sites (solid curve) is five times faster than that of nonsynonymous sites (dashed), which is similar to that for *Arabidopsis* in Figure 7.2a. Saturation therefore occurs much more rapidly at synonymous sites. When the species are recently diverged, the rate of divergence increases linearly and d_N/d_S can be estimated without correction. As time goes on, the synonymous sites begin to saturate at 0.75 divergence whereas the nonsynonymous sites continue to diverge almost linearly. The result is that, without correction, d_S is underestimated and d_N/d_S overestimated.

The simplest model of correction for multiple hits is the **one-parameter model** of Jukes and Cantor (1969), which assumes that any nucleotide is equally likely to be substituted by any other nucleotide. Letting $d_0(t)$ denote the observed proportion of nucleotide differences between two aligned sequences, we can write:

$$1 - d_0(t) = [1 - d_0(t-1)](1 - 2\lambda) + d_0(t-1)(2\lambda/3) \tag{7.2}$$

where λ is the probability of a nucleotide substitution per unit time. Equation 7.2 says that the proportion of nucleotide sites that are identical at time t is equal to whatever it was at time $t-1$ multiplied by the probability that neither sequence underwent a substitution, which is $1 - 2\lambda$ (the 2 is needed because a substitution might have occurred in either of the two lineages); plus the proportion of sites that were different at time $t-1$ multiplied by the probability that either of the lineages underwent a substitution to a nucleotide matching that in the other lineage, which is $2\lambda/3$ (the 1/3 comes from the assumption that, when a nucleotide is substituted, the new nucleotide is equally likely to be any of the other three). The units of time are chosen such that $\lambda^2 \approx 0$, which assigns a negligible probability to a substitution occurring in both lineages in a single time unit.

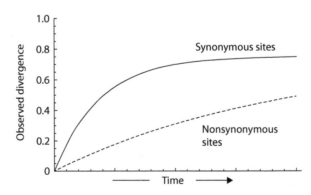

Figure 7.3 Multiple nucleotide substitutions at the same sites ultimately result in sequence divergence plateauing at 75 per cent. This happens much more rapidly at synonymous sites than at nonsynonymous sites.

Equation 7.2 can be written as:

$$d_0(t) - \frac{3}{4} = \left(d_0(t-1) - \frac{3}{4}\right)(1 - 8\lambda/3) \tag{7.3}$$

which has the solution:

$$d_0(t) - \frac{3}{4} = \left(d_0(0) - \frac{3}{4}\right)(1 - 8\lambda/3)^t \approx -\frac{3}{4}e^{-8\lambda t/3} \tag{7.4}$$

because $d_0(0) = 0$.

Although the observed number of substitutions per site at time t equals $d_0(t)$, the true number of substitutions per site d is expected to be $2\lambda t$ (that is, λt in each lineage). Replacing $2\lambda t$ in Equation 7.3 with d and then solving for d yields the Jukes–Cantor estimate of the number of mutations per nucleotide site corrected for multiple hits (Jukes and Cantor 1969):

$$d = -\frac{3}{4}\ln\left[1 - \frac{4}{3}d_0\right] \tag{7.5}$$

where d_0 is the observed proportion of nucleotide differences at time t. As is evident Figure 7.4, the correction in Equation 7.5 is reasonable as long as d_0 is not too close to its maximum of 3/4. As d_0 approaches its maximum, the corrected d blows up, and any tiny difference in d_0 due to sampling error or any other cause results in a huge change in d. Where to draw the line is a matter of judgment, however for the one-parameter model the correction begins to lose accuracy for $d_0 > 0.5$. The variance of d is estimated as $Var(d) = (1/L)\, d_0\, (1 - d_0)\, /[1 - (4/3)\, d_0]^2$, where L is the total number of nucleotides in the sequence.

Because multiple hits are potentially a serious problem for analyzing sequences that are very divergent, considerable effort has gone into corrections based on more realistic models than that in which all nucleotide substitutions are equally likely. There is the **two-parameter model**, which ascribes different mutation rates to transitions and transversions (Kimura 1980; H. Wang et al. 2008). A **transition** is a change from one pyrimidine nucleotide (T or C) into the other, or from one purine nucleotide (A or G) into the other; a **transversion** is a change from a pyrimidine to a purine or the other

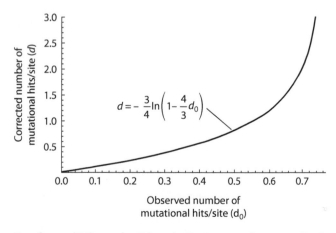

Figure 7.4 Correction for multiple nucleotide substitutions at the same site based on the one-parameter model of Jukes and Cantor (1969).

way around. Models with 4, 6, 9, and 12 parameters have also been investigated. (The 12-parameter model assigns a different substitution rate to each of the four nucleotides changing into any of the other 3.) Mathematical details of the models can be found in Zharkikh (1994) and a good general discussion in Arenas (2015). In each case the substitution parameters are estimated from the data themselves. Each model has its own maximum value of d_0, and regardless of the model, as d_0 approaches its maximum the estimate of the true number of mutational hits per nucleotide site becomes increasingly unreliable.

Amino Acid Divergence Between Species

It is less necessary to correct for multiple amino acid replacements per amino acid site than for nucleotide substitutions per nucleotide site because there are twenty amino acids and only four nucleotides. The amino acid analog of the one-parameter correction in Equation 7.5 is:

$$d = -\frac{19}{20}\ln\left[1 - \frac{20}{19}d_0\right] \tag{7.6}$$

The one-parameter model for amino acids is somewhat at odds with the genetic code in Table 1.1 in assuming that each amino acid can change with equal likelihood to any other, nevertheless the model is useful for many purposes and clearly shows that the maximum d_0 for amino acid replacements is $19/20 = 0.95$. By the time two proteins approach this level of divergence, they are likely to be unalignable. The variance of d in Equation 7.6 is estimated as $Var(d) = (1/L)d_0(1 - d_0)/[1 - (20/19)d_0]^2$, where L is the total number of amino acids in the sequence.

Rates of amino acid replacement in a sample of mammalian proteins are shown in Figure 7.5. Among the most slowly evolving proteins are the histones that form the core of the nucleosome particles into which DNA is packaged; among the most rapidly evolving proteins are the interferons that contribute to host defense against viral infection. A number of other well-known proteins are also identified. The variation in rate of evolution of mammalian proteins is huge—almost a factor of a thousand. Some of the variation may be due to differences in mutation rate, which is correlated with level of transcription among other factors (Thornlow et al. 2018), but most of the variation is undoubtedly due to differing degrees of selective constraint on the molecules. For example, histone

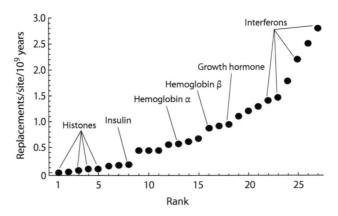

Figure 7.5 Rates of amino acid replacement across 27 proteins in mammals. (Data from Graur (1985).)

proteins are subject to strong selective constraints because they must interact with one another, with negatively charged DNA, and with various proteins that add or remove chemical groups to promote or impede gene expression. Interferons, on the other hand, are small soluble signaling molecules released by infected cells. More than 20 structural or functional correlates of the evolutionary rate of proteins have been identified (Zhang and Yang 2015), but all of them relate one way or another to raising or relaxing selective constraints.

Molecular Clockwork

In spite of variation in rates of evolution among different proteins, for a specified protein the average rate of molecular evolution often manifests an apparent uniformity throughout long periods of evolutionary time. Such uniformity is known as a **molecular clock**. As an example we'll look at hemoglobin α, which has a fairly typical rate of amino acid replacement. The observed percentage of differences, d_0, in hemoglobin α between pairs of vertebrate species is shown above the diagonal in Table 7.1. The observed differences are rather large relative to the protein's length (142 amino acids in the human lineage), and below the diagonal are the estimated number of amino acid replacements per site d corrected for multiple hits according to Equation 7.6. (Correction by more sophisticated methods makes very little difference in this example.)

For example, the α-globins of dog and human differ in 23 of 142 amino acid sites, or $d_0 = 16.2$ per cent, which yields $d = -(19/20)\ln[1-(20/19)d_0] = 0.177$, rounded off in the table to 0.18. The percentages exclude differences that result from the insertion or deletion of amino acids, which are called **gaps** in sequence comparisons. For example, the comparison between human and shark α-globin is based on 139 aligned amino acids and excludes gaps that span 11 amino acid sites. Although gaps are not typically used in phylogenetic analysis, they can contain a great deal of information about the number and size distribution of insertions and deletions and the evolution of genome size (Petrov et al. 1996, 2000).

Table 7.1 also includes the average value of d in all comparisons with humans and the estimated divergence times based on paleontological data. The average number of hits per

Table 7.1 Rate of evolution in the a-globin gene

	Shark	Carp	Newt	Chicken	Echidna	Kangaroo	Dog	Human
Shark		59.4	61.4	59.7	60.4	55.4	56.8	53.2
Carp	0.93		53.2	51.4	53.6	50.7	47.9	48.6
Newt	0.99	0.78		44.7	50.4	47.5	46.1	44.0
Chicken	0.94	0.74	0.60		34.0	29.1	31.2	24.8
Echidna	0.96	0.79	0.72	0.42		34.8	29.8	26.2
Kangaroo	0.83	0.72	0.66	0.35	0.43		23.4	19.1
Dog	0.86	0.67	0.63	0.38	0.36	0.27		16.2
Human	0.78	0.68	0.59	0.29	0.31	0.21	0.18	
Mean d	0.90	0.73	0.64	0.36	0.37	0.24	0.18	
Time (MY)	450	410	360	290	225	135	80	

Note: values above the diagonal are the observed per cent amino acid differences d_0 per pair-wise comparison. (Data from Kimura (1983).) Values below the diagonal are the values d corrected for multiple hits using a one-parameter model. The average values of d and estimated times of divergence (in millions of years) are given at the bottom.

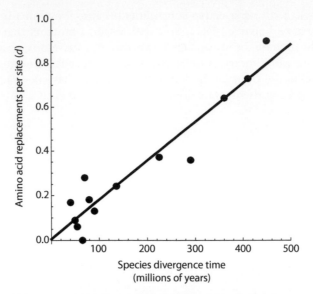

Figure 7.6 Relation between estimated number of amino acid substitutions per amino acid site in the α-globin gene between pairs of the vertebrate species, plotted against estimated time of species divergence. The straight line is expected based on a uniform rate of amino acid substitution during the entire period.

site is plotted against divergence time in Figure 7.6. The reasonable fit to a straight line is evident. Since the divergence time between species is half of the total time available for evolution (because each divergence creates two independently evolving branches), the rate of evolution λ can be estimated as one-half times the slope of the line in Figure 7.6. For these data, the slope is 1.8×10^{-9}, and therefore $\lambda = 0.9 \times 10^{-9}$ amino acid substitutions per amino acid site per year. The reasonable fit of the points to a straight line indicates that the average rate of α-globin evolution has been approximately constant for the past 450 million years, but different genes, and different parts of genes, evolve at different rates (Figure 7.5). The α-globin gene is near the middle of the range.

Strictly speaking, the molecular "clock" is not very clocklike. The rate of gene substitutions differs in different lineages as well as in different proteins and in different regions of a protein molecule (Spielman and Kosakovsky Pond 2018). It is at least twofold faster in rodents than in higher primates, and about 30 per cent faster in the Old World monkey lineage than in the human lineage (Li 1997). These differences are thought to be due in part to differences in mutation rate per year as a function of generation time. Furthermore, a molecular "clock" for the evolution of a molecule can have a variance in substitution rate very much larger than that of the Poisson distribution expected from Equation 7.1, which predicts that the mean and variance should be equal. Gillespie (1989) found that nonsynonymous substitutions in mammals have an average variance that is 7.75 times greater than the mean, whereas for synonymous substitutions the ratio is about 3.3. The high variance-to-mean ratio for nonsynonymous substitutions supports the hypothesis that protein evolution undergoes evolutionary episodes in which the rate of amino acid replacement is accelerated. Some current models try to account for such episodes by explicitly including constraints due to protein folding or epistatic interactions between amino acids; however, application of these models often entails complex computational methods based on Bayesian inference in which the key parameters are assumed a priori to have some specified underlying distribution (Arenas 2015). The much smaller variance-to-mean ratio of 3.3 observed for synonymous substitutions (Gillespie 1989) suggests that these are more nearly neutral than nonsynonymous substitutions.

7.2 Analysis of the Site Frequency Spectrum

The observed distribution of allele frequencies in a sample can in some cases reveal a great deal about the evolutionary history of a population. Such inferences are possible because the hypothesis that a population is at steady state of mutation drift for neutral alleles makes specific, quantitative predictions of the relative frequencies of alleles. Demographic or evolutionary processes that depart from the steady-state neutral model result in predictable deviations from the relative frequencies of alleles expected under neutrality. The manner in which a sample of alleles may deviate from neutral expectation therefore supports an alternative model that invokes a type of demographic or evolutionary process that could result in just such a deviation. In this section we discuss the principles underlying such analyses.

The Unfolded Site Frequency Spectrum

To set the stage, suppose you sample n alleles from a population, sequence L nucleotides from each allele, and identify S nucleotide sites that are polymorphic. Table 7.2 shows a specific example in which $n = 7$, $S = 10$, and $L = 400$. (The 390 nucleotide sites that are monomorphic are not listed.) Each nucleotide at each polymorphic site is classified as to whether it is ancestral (black font in Table 7.2) or derived (red font). Ordinarily, one does not know which nucleotide is ancestral and which derived; however, the ancestral nucleotide can often be deduced from the identity of the corresponding nucleotide observed in a closely related species (an **outgroup**). The **site frequency spectrum (SFS)** in the sample refers to the relative frequencies of alternative nucleotides at each polymorphic site; the SFS is also referred to as the **allele frequency spectrum (ASF)**. When the ancestral and derived nucleotides can be distinguished, the SFS is called the **unfolded SFS**.

Formally, the unfolded SFS is defined as the vector $\xi = (\xi_1, \xi_2, \ldots, \xi_{n-1})$ in which ξ_i equals the number of polymorphic nucleotide sites that have exactly i derived nucleotides in the sample. In Table 7.2, for example, the values of ξ_i can be obtained by summing the entries for $i = 1, 2, \ldots, 6$ at the bottom of the table. In this example $\xi = (4, 2, 1, 1, 1, 1)$. Note that

Table 7.2 Unfolded site-frequency spectrum

Sample number ($n = 7$)	Polymorphic nucleotide site ($S = 10$)									
1	C	G	A	C	A	T	G	T	G	T
2	C	A	T	T	A	T	T	C	G	C
3	C	G	T	T	A	C	T	C	G	T
4	A	A	A	T	G	T	T	C	T	T
5	C	A	A	T	A	T	T	T	G	C
6	C	A	A	T	A	T	T	C	G	T
7	C	A	A	T	G	T	T	T	G	C
Number of derived alleles	1	5	2	1	2	1	6	4	1	3

the sum $\sum_{i=1}^{n-1} \xi_i = S$, which corresponds to the number of polymorphic nucleotide sites in the sample.

The SFS in a sample depends on the demographic history of the population, which means its history of increases, decreases, and bottlenecks of population size. The SFS also depends on mutation, recombination, random drift, natural selection, and other factors including population structure and migration. Except for the simplest models, simulating the coalescent process can be difficult and time consuming, especially if the sample size is large or if the polymorphic nucleotide sites are as numerous as typically found in genome-wide sequences. For this reason, inferences from the SFS are usually based on summary statistics, which are straightforward to calculate and often robust to the retails of the underlying model (Ferretti et al. 2017).

The most widely used summary statistics for analysis of the SFS are various estimators of $\theta = 4N\mu$, where N is the inbreeding effective population size and μ is the mutation rate per nucleotide site per generation. (In a haploid population, $\theta = 2N\mu$.) For a population at steady state for neutral mutations with frequencies governed by random genetic drift, the expected number of polymorphic sites with derived allele frequency $\xi_i = i/(n-1)$ equals:

$$E(\xi_i) = \theta/i \tag{7.7}$$

(Y.-X. Fu 1995; Griffiths and Tavaré 1998; Tajima 1983). Under the steady-state neutral model, any of the ξ_i values in the SFS, or any linear combination of ξ_i values, can be used to formulate an estimator of θ, although different estimators have different variances due to sampling.

Two of the most common estimators of θ have been foreshadowed in Chapter 4. An estimator corresponding to Equation 4.18 is:

$$\theta_W = (1/a_n) \sum_{i=1}^{n-1} \xi_i \quad \text{where } a_n = \sum_{i=1}^{n-1} 1/i \tag{7.8}$$

θ_W has become known as Watterson's estimator after its originator (Watterson 1975). Equation 7.8 estimates the true value of θ for the entire set of L nucleotides analyzed. To obtain the corresponding estimate of θ *per nucleotide site*, divide by L.

An alternative estimator of θ, due to Tajima (1983), corresponds to Equation 4.21, namely:

$$\theta_\pi = \frac{2}{n(n-1)} \sum_{i=1}^{n-1} i(n-i)\xi_i \tag{7.9}$$

The subscript π in θ_π is a reminder that the estimator in Equation 7.9 is based on the average number of nucleotide mismatches between the sequences taken in all pairwise combinations. Again, the estimator in Equation 7.9 pertains to the entire set of nucleotide sites sequenced, and the per-nucleotide-site estimate of θ is θ_π/L.

As Equation 7.7 implies, many other estimators of θ can be devised based on the SFS (Ferretti et al. 2017; Y.-X. Fu and Li 1993; Zeng et al. 2006), but for present purposes we need to consider only one more, namely:

$$\theta_H = \frac{2}{n(n-1)} \sum_{i=1}^{n-1} i^2 \xi_i \tag{7.10}$$

The estimator θ_H gives increasing weight to ξ_i values with a high frequency of derived alleles, and it was devised to detect whether an SFS is consistent with positive directional selection on derived nucleotides (Fay and Wu 2000). As before, θ_H estimates θ for the entire set of nucleotides, and to convert it to an estimate of θ per nucleotide site, divide by L.

The estimators θ_W, θ_π, and θ_H in Equations 7.8–7.10 are all estimators of the true value of θ for a gene at mutation drift equilibrium for neutral mutations, and hence in such a population their values should be within sampling variation of each other. For the simulated neutral data in Table 7.2, the values of these estimators are shown in Example 1 of Table 7.3. They differ slightly, but only because the expected values in the SFS have been rounded to the nearest integer, with the result that ξ_5 and ξ_6 in the SFS are slightly too large. Examples 2–6 show values of θ_W, θ_π, and θ_H for SFS that depart in various ways from neutral expectation, but these are most easily interpreted in terms of the differences D, H, and L in the last three columns.

The difference D, H, and L are statistics often used to detect specific types of departure of the SFS from the steady-state neutral expectation. The differences:

$$D = \theta_\pi - \theta_W$$
$$H = \theta_\pi - \theta_H \qquad (7.11)$$
$$L = \theta_W - \theta_H$$

are the test statistics. D is known as **Tajima's D** (Tajima 1983), H as **Fay and Wu's H** (Fay and Wu 2000), and L is from Ferretti et al. (2017).

In Table 7.3, let's first focus on Tajima's D. Example 1 is the steady-state neutral case, and D is close to 0 as expected. Example 2 has an excess of **singletons** (derived sites present only once in the sample), and D is negative. Such a pattern might be expected of a recent selective sweep in or near the gene itself, with the excess of singletons arising from recent mutations. A negative Tajima's D can also result from recent population growth, as an increasing population size results in a greater number of new mutations. To understand why population growth results in $D < 0$, recall from Equation 4.15 that the average time to coalescence from i to $i-1$ alleles equals $4N/[i(i-1)]$. In a growing population, therefore, the average time to coalescence lengthens as N increases. The longer coalescence times mean that a growing population has longer branches at the tips and shorter branches at the root than expected in a population of constant size. This pattern of coalescence is illustrated in Figure 7.7a.

Example 3 in Table 7.3 has a deficiency of rare alleles. This pattern can result from any kind of balancing selection that maintains alleles at intermediate frequency, or it can

Table 7.3 Unfolded site frequency spectrum (SFS)[a]

Example	SFS	θ_W	θ_π	θ_H	$D = \theta_\pi - \theta_W$	$H = \theta_\pi - \theta_H$	$L = \theta_W - \theta_H$
1	{4, 2, 1, 1, 1, 1}[b]	4.082	4.000	4.667	−0.082	−0.667	−0.585
2	{8, 1, 1, 0, 0, 0}[c]	4.082	3.333	1.000	−0.748	2.333	3.082
3	{1, 2, 4, 3, 0, 0}[d]	4.082	5.238	4.429	1.156	0.809	−0.347
4	{0, 0, 1, 2, 3, 4}[e]	4.082	4.286	12.380	0.204	−8.095	−8.299
5	{1, 2, 2, 2, 2, 1}[f]	4.082	4.762	6.905	0.680	−2.143	−2.823
6	{0, 2, 3, 3, 2, 0}[g]	4.082	5.333	6.333	1.252	−1.000	−2.252

[a] In these examples, $n = 7$ and $S = 10$.
[b] Neutral expectation.
[c] Too many low-frequency alleles.
[d] Too few low-frequency alleles.
[e] Too many high-frequency derived alleles.
[f] Too uniformly distributed.
[g] Too many intermediate-frequency alleles.

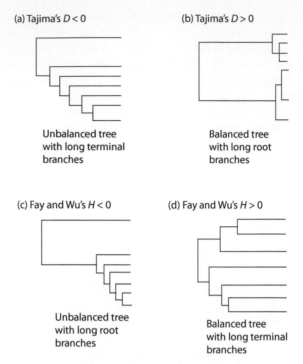

(a) Tajima's $D < 0$

Unbalanced tree
with long terminal
branches

(b) Tajima's $D > 0$

Balanced tree
with long root
branches

(c) Fay and Wu's $H < 0$

Unbalanced tree
with long root
branches

(d) Fay and Wu's $H > 0$

Balanced tree
with long terminal
branches

Figure 7.7 Coalescent trees exemplifying departures from steady-state neutrality detected by means of Tajima's D (a and b) or Fay and Wu's H (c and d).

result from a recent reduction in population size. A key point is that deviations from a steady-state neutral SFS can be caused by either selection or demographic changes, which enjoins caution in drawing conclusions. A coalescent tree exemplifying $D > 0$ is shown in Figure 7.7b.

Fay and Wu's H was designed specifically to detect an excess of high-frequency derived alleles, and Example 4 in Table 7.3 shows that it does quite a good job in yielding a strongly negative value with this type of SFS. This pattern yields a coalescent tree resembling that in Figure 7.7c and would be expected of a recent selective sweep, although the pattern can be mimicked by various demographic scenarios. A strongly positive value of H would indicate a deficiency of derived alleles with moderate or high frequency, yielding a coalescent tree such as that in Figure 7.7d.

The statistic $L = \theta_W - \theta_H$ (Ferretti et al. 2017) is analogous to H in yielding large negative values when there is an excess of high-frequency derived alleles, but it is somewhat better in detecting SFS with alleles that are too uniformly distributed or that have too many alleles at intermediate frequency (Examples 5 and 6 in Table 7.3). These are patterns that would be expected with various types of balancing selection that maintains alleles at intermediate frequencies, but as in earlier examples, demography can also play a role.

Many other test statistics to detect departures from a steady-state neutral SFS have also been proposed (Ferretti et al. 2017; Fu and Li 1993; Li 2011; Yang et al. 2018), but those in Table 7.3 are illustrative. There is also the important issue of how one decides when a test statistic differs enough from 0 to be regarded as statistically significant. There is no perfect answer. Tajima (1983) suggested a test for Tajima's $D = \theta_\pi - \theta_W$ by dividing D by $\sqrt{Var(\theta_\pi - \theta_W)}$, where:

$$Var(\theta_\pi - \theta_W) = Var(\theta_\pi) + Var(\theta_W) - 2Cov(\theta_\pi, \theta_W) \qquad (7.12)$$

Under the null hypothesis of equilibrium and selective neutrality, the mean and variance of the normalized value of D are approximately 0 and 1. In Equation 7.12, the values of $Var(\theta_W)$ and $Var(\theta_\pi)$ are given by Equations 4.20 and 4.23 and $Cov(\theta_\pi, \theta_W)$ is given by Tajima (1983) as:

$$Cov(\theta_\pi, \theta_W) = \frac{S}{a_n^2} + \frac{(n+2)S^2}{2na_n^3}$$

Alternative approaches are to establish critical values of the test statistics using simulated models with known values of θ and n (Fu and Li 1993), or else to carry out a grid search over values of θ assumed to be plausible (Simonsen et al. 1995). For large values of L one can calculate the test statistic for each set of nucleotides in a sliding window across the sequences and to designate extreme outliers as statistically significant, but his approach risks false positives due to multiple testing (Chapter 2).

Apart from issues of statistical significance are those of interpretation. For example, Cvijovíc et al. (2018) have found that, especially in large samples, strong purifying selection can result in counterintuitive distortions of the SFS mimicking the effects of population expansion or allowing some deleterious alleles to reach anomalously high frequencies that resemble selective sweeps even though the alleles are doomed to eventual fixation. These authors also give explicit formulas analogous to Equation 7.7 for various ranges of the selection coefficient and population size, allowing biases in the test statistics in Equation 7.11 to be estimated.

As noted earlier, problems also arise when the population size changes. When the population size changes willy-nilly, the expected SFS may depend on the entire history of changes in population size through time, and although there are mathematical methods for smoothing the effects (Rosen et al. 2018), the number of segregating sites required to estimate ancient population size can be very large.

The Folded Site Frequency Spectrum

Distinguishing derived from ancestral alleles in a SFS is, in many cases, not possible. This is especially true in prokaryotes and eukaryotic microbes for which no suitable outgroup can be identified. In such cases the unfolded SFS must be formulated and analyzed as a **folded** SFS, in which the minor allele frequencies across sites are combined. The **minor allele frequency (MAF)** at a site is the frequency of the least-frequent allele or nucleotide. In a sample of size n, the folded SFS is given by the vector ξ where:

$$\begin{aligned} \xi &= \left(\xi_1 + \xi_{n-1}, \xi_2 + \xi_{n-2}, \ldots, \xi_{(n-1)/2} + \xi_{(n+1)/2}\right) \text{ if } n \text{ is odd, or} \\ \xi &= \left(\xi_1 + \xi_{n-1}, \xi_2 + \xi_{n-2}, \ldots, \xi_{n/2}\right) \text{ if } n \text{ is even} \end{aligned} \qquad (7.13)$$

Defining ξ in this way yields a vector of minor allele frequencies without the need to specifying whether the minor allele is ancestral or derived.

Although the folded SFS contains less information than the unfolded SFS, it nevertheless contains a great deal of information. For example, Tajima's $D = \theta_\pi - \theta_W$ is still a useful statistic because it primarily tests for too many or too few rare alleles, although in this case Equations 7.8 and 7.9 should be interpreted with the summations taken over all elements in the folded SFS. The statistics H and L are not so useful with the folded SFS because they were devised specifically to test for departures from the expected allele frequencies of ancestral versus derived alleles. In the following, we will also show how the folded SFS can be analyzed to make inferences about the intensity of natural selection acting on amino acid polymorphisms as well as on the nonrandom use of synonymous codons. Before speeding on to see the details of the analysis, however, we need to take a short detour to discuss codon usage bias.

Codon Usage Bias

The synonymous substitutions in Figure 7.1 (red) occupy a relatively narrow range of rates, but some synonymous substitutions are also constrained. Some constraints reflect selection to maintain secondary structures that form in certain RNA molecules due to internal base pairing (Parsch et al. 1998; Kudla et al. 2009). Others are due to strategically placed rare codons that slow translation to allow a protein to fold properly as it is being translated (Jacobs and Shakhnovich 2017). Nevertheless, the major source of constraint on codon usage appears to be translational speed and accuracy (Plotkin and Kudla 2011). Such constraints are correlated with the relative abundance of tRNA molecules used to translate different synonymous codons. Nonrandom use of synonymous codons is known as **codon usage bias**. In bacteria and yeast, highly expressed proteins tend to use synonymous codons for tRNA molecules that are abundant in the cell, whereas proteins produced in small amounts do not show such codon usage bias (Ikemura 1985; Eyre-Walker 1996). Natural selection also acts to optimize codon usage in such organisms as *Drosophila* (Akashi 1993, 1995), but in vertebrates codon usage bias is confounded with large-scale differences in base composition (G + C content) and patterns of nucleotide substitution across different regions of the genome (Bernardi 2000; Eyre-Walker and Hurst 2001; Plotkin and Kudla 2011).

Nonrandomness in synonymous codon usage can be measured in a variety of ways. The most commonly encountered measures are the following:

- The **effective number of codons (ENC)** is a measure of departure from equal codon usage that is independent of gene length, amino acid composition, and any reference set of genes (Wright 1990). For each amino acid, the ENC is the number, which if used equally frequently, would yield the same sum of squares of the actual codon frequencies used for that amino acid. The ENC for the entire gene is the sum of the ENCs for each amino acid. The ENC has the convenient interpretation that it measures the inverse of the probability that two randomly chosen synonymous codons are identical. A low ENC corresponds to high codon usage bias, a high ENC to low codon usage bias. The minimum value of 20 occurs when one codon is used exclusively for each amino acid, and the maximum is 61 when synonymous codons are used equally. The typical range of ENC is 25–55.

- The **scaled chi-square** (χ^2/L) is another measure of deviation from equal codon usage (Shields et al. 1988). For each synonymous codon group, a χ^2 value is calculated according to Equation 2.3. Calculate the scaled χ^2 value by summing these values for each group of synonymous codons and dividing by the total number (L) of codons in the gene. The scaled chi-square theoretically ranges from 0–1, and the observed range is almost as wide, but most genes are in the range 0.1–0.6.

- The **codon adaptation index (CAI)** estimates the extent to which codon usage is biased toward codons used in a set of highly expressed reference genes in the same species (Sharp and Li 1987). Each codon is assigned a relative "adaptiveness" value according to its frequency of use in the reference genes. The CAI for any gene is then calculated as the geometric mean of the relative adaptiveness of its codons, divided by the geometric mean of the relative adaptiveness if each codon used were optimal. (The geometric mean is the Lth root of the product of L numbers, where in this case L is the number of codons in the gene.) A CAI near 1 means high codon usage bias; near 0 means that the gene has an extensive concentration of otherwise rarely used codons. In yeast the CAI ranges from about 0.1–0.9 and in *E. co*li from about 0.2–0.8, with highly expressed genes having the greater CAI.

Selection for Optimal Codons and Amino Acids

The pattern of codon usage bias in a gene exemplifies the interplay between mutation, random genetic drift, and weak selection. Each codon in a gene represents a sort of replicate experiment that, at any point in time, may be fixed for an optimal codon, fixed for a nonoptimal codon, or be polymorphic. To analyze this situation quantitatively, consider a coding sequence containing some number of synonymous sites evolving in a haploid organism such as yeast or *E. coli*. In any generation, any of the sites can undergo mutation. Let us assume that each site evolves independently, which means that the rate of recombination between sites is at least on the order of the mutation rate per site. Let us assume furthermore that, once a site mutates, its evolutionary fate of fixation or loss is resolved before the same site mutates again.

The properties of independence with at most two nucleotides at any polymorphic site define the **Poisson random field (PRF)** model of molecular evolution (Sawyer and Hartl 1992). Alternative models that allow multiple nucleotides at a site or linkage between sites are discussed in Ewens (1972) and Watterson (1985). In the PRF model, each new mutation initiates an independent process of random genetic drift with selection that can be modeled as a diffusion process of the type discussed in Chapter 6.

Suppose now that n sequences are sampled from the population. Using the PRF model it can be shown that, at a steady state flux of new mutations whose fate is governed by mutation, selection, and random genetic drift, the expected number of polymorphic sites with exactly r derived and $n-r$ ancestral nucleotides $(r = 1, 2, 3, \ldots, n-1)$ has a Poisson distribution with mean:

$$\lambda_r = 2\mu \int_0^1 \frac{1}{x(1-x)} \times \frac{n!}{r!(n-r)!} x^r (1-x)^{n-r} dx \tag{7.14}$$

for neutral alleles $(s = 0)$ and

$$\lambda_r = 2\mu \int_0^1 \frac{1}{x(1-x)} \frac{1-e^{-2Ns(1-x)}}{1-e^{-2Ns}} \times \frac{n!}{r!(n-r)!} x^r (1-x)^{n-r} dx \tag{7.15}$$

when $s \neq 0$ (Sawyer and Hartl 1992; Hartl et al. 1994). In both equations, μ is the total mutation rate across the set of nucleotides being analyzed, scaled by the effective population size (that is, μ equals the per-site mutation rate times N times L, where L is the number of sites). The factor preceding the times sign in these equations is the limiting probability distribution of derived nucleotides with frequencies in the range $(x, x+dx)$, and the factor following the times sign is the binomial sampling distribution. Under the PRF model, the expected numbers given in Equations 7.14 and 7.15 are independent for $1 \le r \le n-1$. Equations 7.14 and 7.15 therefore yield the expected unfolded SFS. When the ancestral and derived alleles cannot be distinguished, the SFS must be folded according to Equation 7.13. The elements of the folded SFS there refers to sites consisting either of r ancestral nucleotides and $n-r$ derived nucleotides or else of r derived nucleotides and $n-r$ ancestral nucleotides, where $r = 1, 2, \ldots (n-1)/2$ when n is odd and $r = 1, 2, \ldots, n/2$ when n is even.

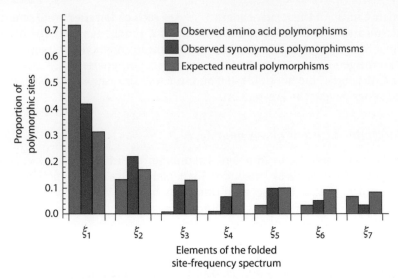

Figure 7.8 Unfolded site frequency spectra expected with neutrality versus those observed for synonymous and nonsynonymous polymorphisms in the *gnd* gene encoding 6-phosphogluconate dehydrogenase in *E. coli*. (Data from Hartl et al. (1994).)

An application of the PRF sampling theory to actual data is shown in Figure 7.8, where the sample consists of $n = 14$ sequences of the *E. coli* gene *gnd* encoding 6-phosphogluconate dehydrogenase (Hartl et al. 1994). The total number of nucleotides sequenced per sample was 1407, which included 367 sites with potential synonymous polymorphisms and 439 codons with potential nonsynonymous polymorphisms. Among these sequences, 60 sites exhibited a synonymous polymorphism and 22 codons exhibited a nonsynonymous polymorphism. Figure 7.8 shows the folded SFS expected for neutral polymorphisms along with those observed for synonymous and nonsynonymous polymorphisms. The bar chart shows the folded SFS where, for each type of polymorphism, the elements of the folded SFS have been normalized to sum to 1. The folded SFS for synonymous and nonsynonymous polymorphisms both differ significantly from that expected with neutrality, and they also differ significantly from each other. The principal difference between the distributions is that both the synonymous polymorphisms and nonsynonymous polymorphisms exhibit an excess of singletons and a deficiency of non-singletons (also called **multitons** (Hahn 2019)).

A deficiency of singleton polymorphisms would be expected of negative selection acting to hinder polymorphisms from increasing in frequency by chance alone. Both the synonymous and nonsynonymous SFS in Figure 7.8 are consistent with this hypothesis. Expected numbers for both classes of sites can be obtained by choosing values of μ and Ns in Equations 7.14 and 7.15 so as to maximize the probability of obtaining the observed data under these values (Hartl et al. 1994). The sample configurations for synonymous polymorphisms fit the expectations of a folded SFS with $Ns = -1.34$, where s is the average selection coefficient against nonoptimal synonymous codons. The nonsynonymous acid polymorphisms fit the expectations of Equation 7.15 with $Ns = -3.66$, which implies that there is about a 2.7-fold greater selection against an average deleterious amino acid replacement than against an average nonoptimal codon in this gene. We can take this analysis a step further by assuming that the per-nucleotide mutation rate is 5×10^{-10} (Drake 1991), which implies an effective population number of 1.8×10^8. If we use this value of N and the previous estimates of Ns, the average selection coefficient against a nonoptimal synonymous codon is 7.4×10^{-9} and the average selection coefficient against a deleterious polymorphic amino acid replacement is 2.0×10^{-8}. These selection coefficients are very small, but in a population so large the selection coefficient against a deleterious polymorphism must be small or else the mutation would not be polymorphic. For the sake of convenience we have assumed that the selection rate is the same for all new mutations in a class (synonymous or nonsynonymous), or at least for all new mutations that have any chance of becoming polymorphic in the population. Our estimates of s are therefore an average taken across different mutable sites. More realistic models could incorporate a probability distribution for the selection coefficient, but this is unnecessary for present purposes. (For the record, we should also point out that an excess of singleton polymorphisms would also be expected in an expanding population; however, if demography alone were responsible for the differences in Figure 7.8, it is unclear why synonymous sites should have a smaller excess of singletons than nonsynonymous sites.)

7.3 Polymorphism and Divergence

In this section, we examine the relation between molecular polymorphism within species and genetic divergence between species. This is an important issue because the essence of Darwinism is sometimes asserted to be that polymorphism within species is transformed into divergence between species. While true as far as it goes, the assertion is all too easily misinterpreted. It does not assert, for example, that the totality of genetic variation within species becomes transformed into genetic differences between species. Nor does it claim that genetic divergence between species must mirror the types and frequencies of genetic polymorphisms within species. For Darwin, speciation was above all a process of adaptation by means of natural selection. But it is only in the absence of selection that there is a smooth, unbiased transformation of polymorphism into divergence.

The McDonald–Kreitman Test

The **McDonald–Kreitman (MK)** test is an approach of commendable simplicity and robustness that assesses whether levels of polymorphism within species and divergence between species are consistent with those expected under mutation and random genetic drift of neutral alleles (McDonald and Kreitman 1991). The layout of the test is shown in Figure 7.9a. The raw data consist of a set of aligned protein-coding sequences from multiple individuals of a species and at least one individual from a closely related species. All variable sites are in the alignment are first classified as either synonymous (if the nucleotide substitution results in a synonymous codon), or else as replacement (if the nucleotide substitution results in an amino acid replacement in the protein). Each variable site is then classified as to whether it is polymorphic in one or both species, and monomorphic sites that differ between the species are classified as divergent. This dual classification allows each variable site to be placed into one of four classes in a 2×2 table, as shown in Figure 7.9a. The value of D_s is the number of synonymous fixed differences between the species and that of P_s is the number of synonymous sites that are polymorphic in one or both species. The value of D_a is the number of amino acid sites that are fixed differences between the species and that of P_a is the number of amino acid sites that are polymorphic in one or both species.

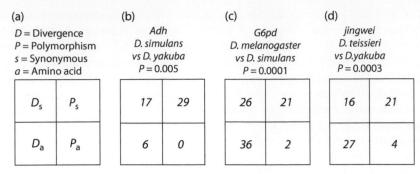

Figure 7.9 Examples of MK tests. (a) The general layout: counts of synonymous nucleotide and amino acid polymorphisms (P_s and P_a) and those of synonymous nucleotide and amino acid divergence (D_s and D_a) are arrayed in the form of a 2×2 table and tested for independence. Example data are (b) the gene encoding alcohol dehydrogenase, (c) the gene encoding glucose-6-phosphate dehydrogenase, and (d) a gene encoding a newly evolved protein of unknown function. (Data for (b) from McDonald and Kreitman (1991), (c) from Eanes et al. (1993), and (d) from Long and Langley (1993).)

McDonald and Kreitman realized that if polymorphisms within species are faithfully transformed into divergence between species, then the cells in the 2×2 table should be independent. This means that the ratio of divergence to polymorphism should be the same for both synonymous and nonsynonymous sites. The best test for independence is Fisher's exact test; however, if the numbers are sufficiently large, a chi-square test for independence is adequate. The point of a significant P-value is that the hypothesis of independence can be rejected, and in this case the usual interpretation is either that there is too much divergence at nonsynonymous sites (suggesting positive selection) or too much polymorphism at nonsynonymous sites (suggesting mildly deleterious mutations or perhaps balancing selection).

Figure 7.9b shows the data comparing the gene *Adh* for alcohol dehydrogenase in *D. simulans* and *D. yakuba* (McDonald and Kreitman 1991). Some of the cells have very low counts but the P-value from Fisher's exact test is highly significant. For this gene there seems to be an excess of divergent over polymorphic replacement sites, relative to what would be expected from the ratio of divergent to polymorphic synonymous sites. If synonymous sites are at most weakly selected, then the excess of divergent replacement sites can most likely be attributed to the result of adaptive evolution incorporating beneficial amino acid replacements into the protein.

The other data in Figure 7.9 are for the genes *G6pd* (Eanes et al. 1993), which encodes the metabolic enzyme glucose-6-phosphate dehydrogenase, and *jingwei* (Long and Langley 1993), which is a gene of unknown function that evolved relatively recently in the *D. teissieri*/*D. yakuba* lineage after coopting some of the exons of *Adh*. These genes also show the signature of adaptive amino acid replacements in having an excess of nonsynonymous divergence over nonsynonymous polymorphisms, relative to that seen for synonymous sites. For both genes the large number of replacement sites implies that selection is pervasive across the molecule and not restricted to one or a few sites. This is a pattern that would be expected if much of the selection were based on protein stability rather than functional capability (DePristo et al. 2005).

Not all genes show the patterns exhibited in Figure 7.9. For example, analogous 2×2 tables for the *Drosophila* genes *period* (Kliman and Hey 1993), *zeste* and *yolk protein 2* (Hey and Kliman 1993), and *bride of sevenless* (Ayala et al. 1993) do not show any significant departure from independence. This suggests that not all genes are undergoing adaptive evolution all of the time. The examples in Figure 7.9 are somewhat special, too, because there were a priori reasons to expect adaptive protein evolution. The *Adh* gene is a well-known target of selection in *D. melanogaster* (David et al. 1986; Kreitman and Hudson 1991; Stam and Laurie 1996; van Delden et al. 1978). In addition, *Adh* shows a geographical gradient, or **cline**, of allozyme frequencies strongly suggestive of selection (see Figure 5.9 in Chapter 5), as does *G6pd* (Cogni et al. 2017), while *jingwei* is a new gene still likely to be acquiring its optimal function (W. Wang et al. 2000). For additional discussion of MK tests see Smith and Eyre-Walker (2002) and Hahn (2019).

Modern sequencing technology has enabled the MK test to be applied on a genomic scale. For example, Bustamante et al. (2005) studied variation in 11,000 protein-coding genes among 39 humans and compared these in MK tests with orthologous sequences from chimpanzees. About 5000 of these genes were informative, and among these about 9 per cent gave evidence of rapid amino acid evolution whereas about 14 per cent showed excess amino acid polymorphisms in humans suggesting mildly deleterious effects or, in some cases, balancing selection. Transcription factors were among the most significant fast-evolving genes, whereas conserved cytoskeletal proteins were among those showing the strongest evidence for deleterious polymorphisms. Balancing selection is the likely explanation for the pattern observed in genes associated with the immune response.

Why does the MK test work? It works because alleles that are fixed by positive selection spend less time as polymorphisms in the population than neutral alleles fixed by chance. For a specified value of $Ns > 0$, the proportion of all fixations due to a fraction f of positively selected alleles is given by:

$$pFix = \frac{4fNs}{1 + 4fNs - e^{-4Ns}} \tag{7.16}$$

whereas the proportional contribution to polymorphisms is given by:

$$pHet = \frac{f\left(4Ns - 1 + e^{-4Ns}\right)}{f\left(4Ns - 1 + e^{-4Ns}\right) + 2Ns\left(1 - e^{-4Ns}\right)} \tag{7.17}$$

These expressions follow from Kimura (1983) (his equations 3.14 and 3.18), and note that the exponential term goes quickly to 0 as Ns increases.

Plots of Equations 7.16 and 7.17 are shown in Figure 7.10 for $f = 0.05$. As Ns increases, new favorable mutations contribute much more to fixations than they do

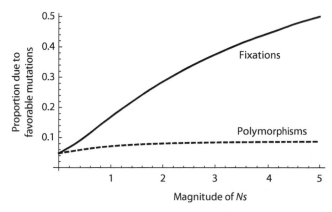

Figure 7.10 Relative contributions of positive selection to fixations versus polymorphisms for various values of *Ns*. Here $f = 0.05$ is the assumed fraction of new mutations that are positively selected.

to polymorphisms. For example, when $Ns = 5$, favorable mutations account for 50 per cent of the fixations but only 9 per cent of the polymorphisms.

In addition to the statistical power implied by Figure 7.10, the MK test is also surprisingly robust (Hahn 2019). It is relatively insensitive to whether sampling is from one subpopulation or many, to the presence of nonneutral amino acid polymorphisms, to weak selection for codon usage bias, to the rate of recombination among sites, and even to recent changes in population size if not too drastic. The most important assumption is that the average genealogical history of synonymous and nonsynonymous sites must be the same. The reason is that, if the most recent common ancestor of the two types of sites differs, then the timescale of their evolution also differs, and the sites in the MK table will not be independent even under neutrality. On the other hand, because the synonymous and nonsynonymous sites across a gene are interspersed, the average genealogies are unlikely to differ by much if at all.

Large changes in population size through time can also affect the outcome of MK tests. The reason is that most fixations have occurred in times past while most polymorphisms are contemporary. Hence, a chronically small population size in the past followed by a recent expansion results in fewer selectively driven fixations and more polymorphisms than expected, whereas a large population size in the past followed by a recent contraction results in more selectively driven fixations and fewer polymorphisms than expected. In these kinds of scenarios MK tests can be misleading.

Refinements of the McDonald–Kreitman Test

The accumulation of slightly deleterious amino acid polymorphisms can obscure positive selection in MK tests. One solution is to make a separate category for singleton polymorphisms on grounds that slightly deleterious mutations are likely to be singletons (Templeton 1996). The differences among the aligned sequences are then classified as singleton polymorphisms, multiton polymorphisms, or fixed differences, and the resulting 2×3 table is then tested for independence. Alternatively, the singleton polymorphisms can just be ignored and multiton polymorphisms versus fixed differences tested in a 2×2 table (Fay et al. 2001). Taking this a step further, Fay et al. (2001) suggest ignoring all polymorphisms with allele frequencies smaller than 0.15; however, Charlesworth and Eyre-Walker (2008) show that while this reduces the bias in estimating the prevalence of adaptive evolution, many instances of positive selection will still remain undetected.

There is also an issue of how to combine MK tables to obtain an overview of how much adaptive evolution occurs among a large set of genes, in principle all of the genes in the genome. For a single gene (say, the ith gene) among a set of genes, Smith and Eyre-Walker (2002) argue that the fraction of fixations due to positive selection should be estimated as $\alpha_i = D_{ai} - (D_{si}/P_{si})P_{ai}$ because D_{si}/P_{si} estimates the proportion of polymorphic genes destined to fix assuming neutrality. Dividing through by D_{ai} yields:

$$\alpha_i = 1 - \frac{D_{si}P_{ai}}{D_{ai}P_{si}} \tag{7.18}$$

Merely averaging across the α_i to obtain the mean of α for all genes introduces a bias; however, an unbiased estimate of the mean is given by:

$$\bar{\alpha} = 1 - \frac{\sum_i D_{si}P_{ai}/(P_{si}+D_{si})}{\sum_i P_{si}D_{ai}/(P_{si}+D_{si})} \tag{7.19}$$

(Stoletzki and Eyre-Walker 2011). These authors also estimate that the mean proportion of fixations driven by positive selection in *Drosophila* is about 40 per cent. Another approach to analyzing MK tables for multiple genes is discussed in the next section.

Polymorphism and Divergence as a Poisson Random Field

The MK test is disarming in its simplicity, yet many parameters affecting the outcome are hidden beneath the surface. These parameters include:

- μ_s, the mutation rate per site at synonymous sites, including monomorphic synonymous sites.
- μ_r, the mutation rate per site at replacement (nonsynonymous) sites, including monomorphic replacement sites.
- N, the effective population size, assumed here to be the same in both species.
- s_s, which is negative, the selection coefficient against nonoptimal synonymous codons.
- s_r, the selection coefficient for (if positive) or against (if negative) amino acid replacements.
- T, which equals $t_{div}/(2N)$, the divergence time between the species, in units of $2N$ generations.

The expected number in each cell in the MK table can be expressed in terms of these parameters by invoking the PRF theory discussed earlier in the context of codon usage bias (Equations 7.14 and 7.15). As noted, the main assumptions of the theory are that each nucleotide site can be segregating for at most two nucleotides at any given time, and that the sites are independent. Under the PRF model, Sawyer and Hartl (1992) showed that the expected numbers in the MK table are given by the formulas in Figure 7.11, where m is the number of sequences from species A and n is the number of sequences from species B. The U's are composite parameters defined as follows:

- $U_s = \mu_s \times L_s \times N$, where L_s is the number of synonymous sites, including monomorphic synonymous sites.
- $U_r = \mu_r \times L_r \times N$, where L_r is the number of replacement (nonsynonymous) sites, including monomorphic replacement sites.

Note that μ_s will in general differ from μ_r even though both are mutation rates per nucleotide site, because the rates include only those mutations that have a chance of becoming polymorphic or divergent in the sample. For example, many amino acid replacements are so deleterious in their effects on fitness that they could not possibly contribute to polymorphism or divergence, and these mutations are not included in μ_r.

	Divergence	Polymorphism
Synonymous	$2U_s \dfrac{2\gamma_s}{1-e^{-2\gamma_s}} [T + G(m) + G(n)]$	$2U_s \dfrac{2\gamma_s}{1-e^{-2\gamma_s}} [F(m) + F(n)]$
Amino acid	$2U_r \dfrac{2\gamma_r}{1-e^{-2\gamma_r}} [T + G(m) + G(n)]$	$2U_r \dfrac{2\gamma_r}{1-e^{-2\gamma_r}} [F(m) + F(n)]$

Figure 7.11 Expected numbers for the entries in an MK table, based on the Poisson random field model. (Formulas from Sawyer and Hartl (1992).)

Based on the very small number of amino acid replacements in the *Adh* data in Figure 7.9b, Sawyer and Hartl (1992) estimated that, at any one time, only approximately 6 of the 256 amino acids in the *Adh* protein are susceptible to a favorable amino acid replacement. The other composite parameters in Figure 7.11 are:

- $\gamma_s = 2Ns_s$, $2N$ times the selection coefficient s_s for synonymous sites
- $\gamma_r = 2Ns_r$, $2N$ times the selection coefficient s_r for replacement sites

The functions G and F in Figure 7.11 are defined as:

$$G(m) = \int_0^1 (1-x)^{m-1} \frac{1-e^{-2\gamma x}}{2\gamma x} dx \tag{7.20}$$

and

$$F(m) = \int_0^1 \frac{1-x^m-(1-x)^m}{1-x} \frac{1-e^{-2\gamma x}}{2\gamma x} dx \tag{7.21}$$

where γ should be replaced with γ_s in the upper two formulas and with γ_r in the lower two.

For any specified values of the parameters, these integrals can be evaluated numerically using any of a number of mathematical analysis programs. If one is willing to assume that $\gamma_s = 0$ (no selection for optimal codon usage), then the formulas in the top row in Figure 7.11 should be replaced with:

$$2U_s \left(T + \frac{1}{m} + \frac{1}{n} \right) \quad \text{and} \quad 2U_s \left[\sum_{i=1}^{m-1} \frac{1}{i} + \sum_{j=1}^{n-1} \frac{1}{j} \right]$$

When the expected numbers are shown in the form of a MK table as in Figure 7.11, the model appears overspecified, as there are five parameters (U_s, U_r, γ_s, γ_r, and T) but only four observations. This is deceptive, however, because the SFS of the synonymous sites allows independent estimation of U_s and γ_s using Equation 7.15. The SFS also allows independent estimation of T (Sawyer and Hartl 1992), but T is also often available from independent datasets. Further detail and extensions of the PRF model can be found in Nielsen et al. (2005), Desai and Plotkin (2008), Sethupathy and Hannenhalli (2008), and Ortega-Del Vecchyo et al. (2016).

What intensities of selection are implied by the results in Figure 7.9? From the PRF model the estimates of Ns for synonymous sites are -0.5, and -0.4 for *Adh* and *jingwei*, indicating weak selection against nonoptimal synonymous codons in these genes. This result is consistent with findings of Akashi (1995) using a different approach. The estimates of Ns for nonsynonymous sites are much larger: $+5$ for *Adh*, $+10$ for *G6pd*, and $+3$ for *jingwei*. These estimates should be interpreted as averages, since the PRF model makes the questionable assumption that all amino acid replacements that are candidates for polymorphism or divergence have the same selective effect. All of these Ns values are much smaller than what could be detected in laboratory experiments. The reason is that the effective population size of these *Drosophila* species is estimated to be on the order of 10^6 (Akashi 1995). If $Ns = +5$, this implies that $s = 0.000005$. Darwin (1859) envisaged adaptive evolution as the gradual accumulation of very slight improvements over eons of geological time, and the selection coefficients so far estimated from molecular population genetics support this point of view.

Underlining the finding of small selection coefficients is an analysis of Sawyer et al. (2007), who used the PRF framework to analyze coding sequences of up to 12 alleles of each of 91 genes in *D. melanogaster* sampled near Lake Kariba, Zimbabwe, compared with divergence from a highly inbred line of *D. simulans* sampled in Chapel Hill, North Carolina. Estimates of U_s, U_r, γ_s, γ_r, and T were obtained using a rather complex algorithm, assuming T (the species divergence time) is the same for all genes in the data.

According to this analysis, about 95 per cent of new nonsynonymous mutations are deleterious, an estimate that excludes those mutations that are so deleterious that they have virtually no chance of ever becoming polymorphic. Among the new mutants that become polymorphic, about 75 per cent are deleterious with an average $Ns \approx -2$, and the 25 per cent that are beneficial have an average $Ns \approx +1$. Among the nonsynonymous mutations that eventually become fixed, the inferred distribution of Ns values is as shown in Figure 7.12.

About half the fixations are either very slightly deleterious or only mildly beneficial, with about 5 per cent having $Ns < 0$, about 20 per cent having $0 < Ns < +1$, and about 25 per cent having $+1 < Ns < +2$. Only about 1 per cent of fixations have $Ns > +7$. A mean value of $\alpha = 0.40$ as in Stoletzki and Eyre-Walker (2011) corresponds to fixations that have $Ns < +1.75$.

Wilson et al. (2011) have analyzed *Drosophila* data with an algorithm that uses allele frequency data but also works its way probabilistically through all possible ancestral coding sequences. The method estimates about the same proportion of mutations fixed by positive selection as does the PRF model; however, it yields more extreme values of Ns both positive and negative. The model imposes no constraints on the distribution of Ns, with the curious result that the distribution of Ns among new mutations is almost uniform except for a large proportion of highly deleterious mutations, and the large Ns values inferred for fixed differences have wide credible intervals. On the other hand, the PRF assumption that Ns is normally distributed may constrain the selection coefficients too narrowly.

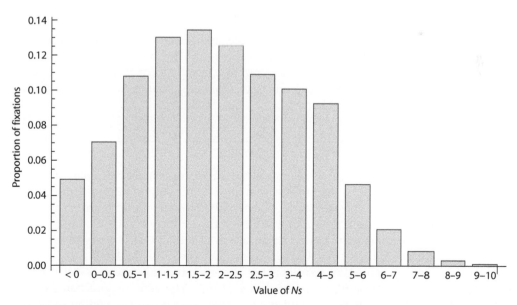

Figure 7.12 Distribution of Ns values among nonsynonymous mutations in *Drosophila* that ultimately become fixed. (Data from Sawyer et al. (2007).)

The Hudson–Kreitman–Aguadé Test

How fast does divergence due to neutral mutations accumulate between reproductively isolated species? We know from Equation 4.21 that, with two sequences chosen at random from within a species, the expected number of pairwise differences equals $\theta = 4N\mu$, where N is the effective population size and μ the per-site mutation rate. Because this equation applies to sequences within a species, it is the minimum divergence we should expect between sequences from different species. We also know from Equation 7.1 that the steady-state rate of fixation of neutral mutations is given by $f_0\mu$, where μ is again the per-site mutation rate and f_0 is the fraction of new mutations that are neutral. In a purely neutral model $f_0 = 1$. In this case the expected proportion of pairwise differences $E(D)$ between two homologous sequences, one from each of two species, is:

$$E(D) = 2\mu t_{div} + \theta = \frac{4N\mu t_{div}}{2N} + \theta = \theta(T+1) \tag{7.22}$$

where t_{div} is the divergence time in generations and N is the effective population number in each species. The factor of 2 after the first equal sign comes from the fact that, for each species, t_{div} is the number of generations back to the common ancestor, and so the total time elapsed in both branches of the phylogenetic tree is $2t_{div}$. The simplification on the right comes from defining the divergence time T in units of $2N$ generations, so that $T = t_{div}/(2N)$. The variance in the proportion of pairwise differences is also known (Li 1977; Gillespie and Langley 1979), namely:

$$Var(D) = \theta(T+1) + \theta^2 \tag{7.23}$$

The second term in this equation measures the contribution to the variance made by polymorphisms present in the ancestral population (Hahn 2019).

Hudson et al. (1987) devised a statistical test for the selective neutrality of within-species polymorphism and between-species divergence based on Equations 7.22 and 7.23. The rationale of the test is that, with no selection, the equilibrium polymorphism depends on θ but the divergence depends on θ and T. Both parameters can therefore be estimated by combining data on polymorphism and divergence, and these estimates can be used in a test for goodness of fit to the hypothesis of no selection. The data for the test consist of a sample of each r unlinked genes (or gene regions) from each of two species A and B, with $r \geq 2$. For illustration we take $r = 2$, and in this case the test statistic for the **Hudson–Kreitman–Aguadé (HKA)** test is:

$$\chi^2 = \sum_{i=1}^{2} \frac{\left[S_i^A - E\left(S_i^A\right)\right]^2}{Var\left(S_i^A\right)} + \frac{\left[S_i^B - E\left(S_i^B\right)\right]^2}{Var\left(S_i^B\right)} + \frac{[D_i - E(D_i)]^2}{Var(D_i)} \tag{7.24}$$

which under the null hypothesis of neutrality has an approximately chi-square distribution (Hudson et al. 1987). For r regions, the summation in Equation 7.24 goes from 1 to r. The test statistic is conservative in rejecting the hypothesis of no selection, since it assumes complete linkage within each gene and free recombination between genes. Estimates of $E\left(S_i^A\right)$ and $Var\left(S_i^A\right)$ can be derived from Equations 4.19 and 4.20, and likewise for $E\left(S_i^B\right)$ and $Var\left(S_i^B\right)$; estimates of $E(D_i)$ and $Var(D_i)$ follow from Equations 7.22 and 7.23.

The first application of the HKA test was to two regions in the *Adh* gene of *D. melanogaster* and its sibling species *D. sechellia* (Hudson et al. 1987). The gene regions consisted of the 5′ flanking sequence (region 1) and the synonymous sites in the coding sequence (region 2). For *D. melanogaster* (species A), 81 sequences were examined and yielded $S_1^A = 9$ polymorphisms among 414 sites in region 1 and $S_2^A = 8$ polymorphisms among

79 synonymous sites in region 2. In *D. sechellia* only one sequence was examined, and so the terms involving S_1^B and S_2^B in Equation 7.24 must be dropped. The average number of pairwise differences was $D_1 = 210$ among 4052 sites in region 1 and $D_2 = 18$ among 324 synonymous sites in region 2. Based on these data the estimates of θ were $\theta_1 = 0.0066$ and $\theta_2 = 0.0090$ per nucleotide site, and the estimated divergence time was $T = 6.73$. The expected values and variances substituted into Equation 7.24 yield:

$$\chi^2 = \frac{[9 - 13.57]^2}{25.76} + \frac{[8 - 3.53]^2}{4.36} + \frac{[210 - 206.72]^2}{921.92} + \frac{[18 - 22.54]^2}{31.04} = 6.07$$

This chi-square has one degree of freedom (since there are four observations and three parameters estimated from the data), and $P = 0.014$. The hypothesis of neutrality can therefore be rejected with some confidence. But what does this mean? Among the possibilities:

- Too little polymorphism in the 5′ flanking region.
- Too much polymorphism at the synonymous sites.
- Too little divergence at the synonymous sites.
- Too much divergence in the 5′ flanking region.

Hudson et al. (1987) give arguments why they prefer the second explanation, but they also emphasize that the HKA test is an omnibus test that cannot, by itself, identify which of the possibilities, or which combination of possibilities, is the true underlying cause. The test has been extended to allow for an explicit test of selection at individual loci (Wright and Charlesworth 2004).

The HKA test can also be applied on a genome-wide scale (Langley et al. 2012) but the results are difficult to interpret. The main problem is the requirement that the regions being compared are unlinked. Even more so than the SFS, the HKA test is sensitive to differing genealogies among the regions being compared. The sites in the SFS are at least tightly linked, which implies that their genealogies are likely to be similar, whereas the unlinked regions in the HKA test will often have differing genealogies. These and other issues with applications of the HKA test are well discussed in Hahn (2019).

Neutrality Versus Selection: An Emerging Consensus

It is understandable if by this point you surmise that recent research in molecular population genetics has been preoccupied with statistical tests that can reject the hypothesis of neutrality. To a large extent this is true. The principal reason is that the neutral theory is a well-defined null hypothesis that leads to precise predictions about steady-state levels of polymorphism and divergence. There is only one way that a mutation can be neutral ($s = 0$) but myriad ways that $s \neq 0$. Hence the steady-state neutral theory is the go-to null hypothesis even in understanding the somatic evolution of cancer (Cannataro and Townsend 2018).

Another reason for the preoccupation with finding evidence for selection relates to the neutralist–selectionist controversy that dominated population genetics for more than a generation starting in about 1970 (Lewontin 1991; Hey 1999). This sometimes acrimonious debate has largely subsided. It is now widely assumed that many nucleotide substitutions have negligible effects on organismal fitness. This is especially true of substitutions in repetitive elements and noncoding DNA, and it is often assumed to be true of synonymous substitutions in the MK test and its generalizations and refinements.

But there's more to the neutral theory than the mere existence of neutral mutations. In its broad sense, the neutral theory asserts not only that neutral mutation exist *but also* that their fate is determined exclusively by random genetic drift. It is the "but also" part of this assertion that admits of reasonable doubt. The ultimate fate of a mutation that is neutral in and of itself may be determined not by random drift alone but also by background selection due to linked deleterious mutations or by hard or soft selective sweeps due to linked favorable mutations or by any of various forms of balancing selection acting on nearby sites. The extent to which these and other factors influence the fate of a neutral allele may differ from one region of the genome to the next owing to variation in local recombination rate, gene density, genealogical history, and other complications.

7.4 Demographic History

Natural populations are subject to many forces that affect their genetic variation and evolution. Mutation is the origin of genetic variation, but its rates and patterns can vary in time and across the genome. Migration influences the distribution of genetic variation among subpopulations; however, it can change rapidly in time and directionality among subpopulations. Recombination affects the extent of linkage disequilibrium, but linkage disequilibrium also depends on population structure, patterns of migration, and demographic history. Selection shapes genetic variation in many ways according to the selective effects of individual mutations, background selection, hard sweeps, soft sweeps, and patterns of epistasis among selected loci. Population structure, population size, and demographic history also have effects on the amount and distribution of genetic variation.

A major challenge in making inferences from gene or genomic sequences sampled from natural populations is that the effects of some of these forces can mimic the effects of others. For example, background selection in regions of low recombination constrains levels of selectively neutral genetic variation maintained in these regions, which mimics the effects of a reduction in population size (Corbett-Detig et al. 2015). As another example, strong purifying selection can in some cases distort the SFS and mimic the pattern expected from a selective sweep (Cvijovíc et al. 2018). Among the major factors that can confound inferences from population data is that of **demographic history**, by which we mean changes in population size through time or the fusion or splitting of subpopulations. Much current research is therefore directed at making inferences about demographic history from the amount and distribution of genetic variation (Barrandeguy et al. 2017; Hahn 2019).

Changes in Population Size Through Time

The simplest scenarios of demographic history are illustrated in Figure 7.13. Part (a) is the typical assumption in population models, in which the population size remains fixed for a long period of time. This scenario has only one parameter to be estimated, namely the effective population size. Parts (b) and (c) add one level of complexity, which is to assume that at some time t the population size either increases as in (b) or decreases as in (c). These models have three parameters to be estimated: the population size before and after the change and the time at which the change took place. Most models assume that the change in size is instantaneous. Models with more gradual chance in size are possible, but require additional parameters.

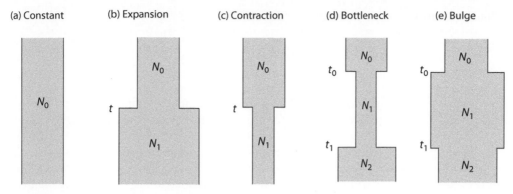

Figure 7.13 Models for the demographic history of a single population. In (d) and (e), N_2 need not be equal to N_0; also growth/decline may be exponential (or some other form) rather than being instantaneous.

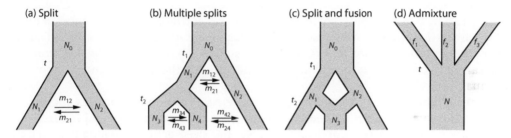

Figure 7.14 Models for the demographic history of populations that split or merge.

Parts (d) and (e) add still more complexity in that the population size is assumed to change abruptly at one time, either shrinking (d) or expanding (e), and then changing in size again at a later time. The reduction in population size in (d) is called a **bottleneck**, however the contraction in population size in (c) is also a type of bottleneck. The models in (d) and (e) each feature five parameters to be estimated: three effective population sizes and two times when changes in size took place.

Population Splits and Fusions

The illustrations in Figure 7.14 show how quickly the number of parameters increase when the model of demographic history includes population splits and fusions. Part (a) is the simplest case in which a population splits into two subpopulations with reciprocal migration between them. In this case there are six parameters to be estimated: three effective population sizes, two migration rates, and the time of the population split.

A three-way split as in part (b) entails estimating four effective population sizes, eight migration rates (migration m_{32} from the N_3 subpopulation into the N_2 subpopulation and m_{23} for the reverse is not shown), and two splitting times.

Simple models of population fusion or admixture are shown in parts (c) and (d). If the ancestral subpopulations are known, as in (c), then the goal is to estimate the effective population sizes and the times of splitting and fusion. If the ancestral subpopulation as not known, as in (d), then the goal is to identify the ancestral subpopulations and to estimate the fractional contribution of each to the admixture (f_1, f_2, and f_3 in this case).

Estimating Parameters in Demographic Models

As input data for estimating the parameters in demographic models, many methods deal with the **multipopulation site- frequency spectrum** or **mpSFS**. For two subpopulations, the mpSFS is a matrix in which the rows record the number of single-nucleotide polymorphisms (SNPs) in which the derived allele was observed. The number of rows equals the largest number of haplotypes sampled from any of the subpopulations. Each column in the mpSFS records the number of instances in which the derived allele was observed in each subpopulation. To use an example from Gutenkunst et al. (2009), if 20 haplotypes were sequenced from subpopulation 1 and 10 from subpopulation 2, then the mpSFS would be a 21-by-11 data matrix $\mathbf{D}$ in which the entry in row i and column j ($d[i, j]$) is the number of cases in which derived allele was observed in the first subpopulation i times and in the second subpopulation j times. (Many of the $d[i, j] = 0$). In this example, the entry $d[2, 0]$ is the number of cases in which the derived allele was present in 2 haplotypes in subpopulation 1 but in 0 haplotypes in subpopulation 2; $d[20, 5]$ is the number of cases in which the derived allele was observed in all 20 haplotypes in subpopulation 1 and 5 haplotypes in subpopulation 2; and $d[0, 10]$ is the number of cases in which the derived allele was observed in no haplotypes in subpopulation 1 and all 10 haplotypes in subpopulation 2. As Gutenkunst et al. (2009) note, when the polymorphic sites can have only two states and are independent, then the mpSFS contains all of the information in the data.

As you might imagine, estimating the parameters of the more complex models in Figures 7.13 and 7.14 from population data is challenging for theory and intensive in computation. Nevertheless, impressive progress has been achieved (Gutenkunst et al. 2009; Jouganous et al. 2017; Nielsen et al. 2009). For complex, multipopulation scenarios, the mpSFS is a multidimensional matrix, and in these cases the following two approaches seem promising.

One powerful approach is approximate Bayesian computation (Beaumont et al. 2002; Navascués et al. 2017; Solar et al. 2019). In typical Bayesian analysis, the goal is to evaluate

$$\Pr\{Parameters|Data\} = \frac{\Pr\{Data|Parameters\}\Pr\{Parameters\}}{\Pr\{Data\}}$$

in which the conditional probability on the left-hand side is the likelihood of a particular set of parameters given the data and the conditional probability on the right-hand is the likelihood of observing the data given the particular set of parameters. One problem is that, in models of sufficient complexity, it is computationally too intensive, and in some cases impossible, to evaluate $\Pr\{Data|Parameters\}$. This is the situation for many complex demographic models.

In **approximate Bayesian computation** (**ABC**), the need to explicitly evaluate $\Pr\{Data|Parameters\}$ is circumvented by ignoring $\Pr\{Data\}$ on the right-hand side and simulating $\Pr\{Data|Parameters\}$ by choosing the parameters according to some assumed prior multidimensional distribution given by $\Pr\{Parameters\}$. Ignoring $\Pr\{Data\}$ allows the relative magnitude of $\Pr\{Parameters|Data\}$ to be estimated for various sets of parameters, without knowing the absolute magnitude, and various sets of parameters can be chosen at random from the prior distribution $\Pr\{Parameters\}$. The computation therefore consists of choosing a set of *Parameters* and simulating a set of *Data* given the chosen *Parameters*. An important feature of approximate Bayesian computation is that one rejects sets of *Parameters* that result in *Data* deviating from the actual observations (or summary statistics of the actual observations) by more than a predetermined magnitude. The outcome of the ABC procedure is a sample of *Parameters* that are approximately distributed as Pr{*Parameters*|*Data*} without explicitly evaluating Pr{*Data*|*Parameters*}. The many considerations, caveats, and limitations of this method are discussed by Sisson et al. (2018).

Another potentially powerful approach to analyzing complex models is **supervised machine learning** (Schrider and Kern 2018). In this approach, the parameters in a demographic model are used as input and simulated data from the model are used as output to train the machine-learning algorithm to find a mapping function that will predict the simulated data as a function of the parameters to any desired degree of accuracy. The trained algorithm is then used to analyze actual data and hopefully reveal which sets of parameters would yield the realized data. The use of simulation to produce training sets means that this approach, like other methods currently in use, is vulnerable to errors in the specification of the model.

In whatever way the analysis of demographic models is carried out, it is important to be cautious in interpreting the estimates because their standard errors tend to be large, and the more complex the model, the larger the standard errors. Even for estimating the demographic history of single population, the standard errors decrease as the logarithm of the number of independently segregating sites analyzed (Terhorst and Song 2015), which is exponentially slower than most estimation problems in statistics.

7.5 Ancient DNA in Studies of Human Populations

Not so long ago, it would have been inconceivable that the origins of human populations could be studied by sequencing the genomes of individuals who lived thousands of years in the past. But in recent years advances in laboratory techniques, sequencing technologies, and computational analysis have made this not only possible but commonplace. At the present time, over a thousand genomes from human remains dating from 45–5 ka have been sequenced and analyzed. (The abbreviation **ka** stands for **kiloannum;** 1 ka equals 1000 years before the present time.) This section contains an overview of the major findings, but to set the stage we need to look more broadly at human origins.

Human Origins

The earliest stages in human history must be inferred from fossils of individuals and the tools and other artifacts associated with them. The story begins in Africa where the genus *Homo* ("human being") evolved a few million years ago. From this lineage evolved *H. erectus* about 2 ma (**ma** stands for millions of years before the present time). This species prospered and multiplied, colonizing large parts of the huge African continent and spreading out into Eurasia where it persisted for more than a million years. Back in Africa the genus *Homo* continued to evolve, splitting about 700 ka into a lineage that eventually gave rise to modern humans and another that eventually gave rise to **Neanderthals**. Both lineages flourished and dispersed out of Africa in several waves. Beginning about 450–430 ka, the Neanderthals spread widely in the Middle East, Europe, and Southwest and Central Asia. An offshoot of the Neanderthal lineage appeared in East Asia about 400 ka and came to occupy a large region ranging from Siberia to Southeast Asia. These were the **Denisovans**. While widespread, these groups had relatively small effective population sizes owing to their population structure consisting of semi-isolated populations of limited size.

In the meantime, the lineage leading to modern humans prospered in Africa and dispersed into the Middle East in several waves beginning as far back as 250 ka. Most of the genome of modern humans can be traced to migration out of Africa that took place about 100 ka (Ingman et al. 2000; Slatkin and Racimo 2016). By about 45 ka this founding population had evolved into **modern humans** with skeletal variations in the same range as found today, but as early as 80 ka the population had split into two groups, one that colonized Western Europe and another that dispersed along the southern coast of Asia to colonize Eurasia and Oceania (Yang and Fu 2018). As these groups moved northwest and southeast they encountered Neanderthal and Denisovan populations, and they overlapped with them for at least 10 thousand years (Slatkin and Racimo 2016).

The Neanderthals became extinct about 30 ka; the Denisovans also went extinct, but the timing is less clear. Both groups seem to have been in decline for a long period prior to extinction, and the old notion that they were driven to extinction by modern humans has been discredited. Indeed, as we shall see, modern humans interbred with both groups at least to a limited extent. Human history more recently than about 45 ka has been illuminated by DNA from the bones of ancient ancestors, but these studies are still in their earliest stages and much more is to be learned.

Technical Challenges of Ancient DNA

DNA in living organisms is constantly being damaged and repaired by variety of specialized enzymes. When repair no longer happens, DNA is subject to fragmentation, crosslinking, deamination, depurination, and other kinds of degradation produced by hydrolysis, oxidation, and reduction. Under ideal conditions DNA can be expected to maintain sufficient integrity for current extraction and sequencing technologies for only about a million years. Most conditions are far from ideal, however.

Studies of DNA from ancient specimens from museums and mummified tissue began in the 1980s but were limited by the need for large amounts of high-quality DNA. The development of the polymerase chain reaction enabled the small amounts of DNA from insects and other organisms preserved in amber to be amplified, but not until about 10 years ago did procedures begin to be developed that allowed **ancient DNA** (often abbreviated **aDNA**) to be studied on a large scale (Reich 2018). Improved laboratory methods included DNA extractions being carried out in scrupulously clean and sterile rooms, with the extracted DNA fragments being immediately joined with synthetic tags of known sequence to identify their origin. Improved computational methods included bioinformatic tools to identify contaminating microbial sequences, possible contamination from other organisms, and characteristic patterns of damage in aDNA. Contamination of aDNA is a recurring issue. The oldest aDNA thus far sequenced is genomic DNA from a 700,000-year-old leg bone of an ancestor of horses, zebras, and donkeys preserved in permafrost in the Canadian Yukon (Orlando et al. 2013). This material was heavily contaminated: 99.7 per cent of the 12 billion sequenced DNA fragments were derived from microbes.

Insights into Human History from Ancient DNA

Most aDNA from human specimens dates from 45–5 ka, and the chief lesson that emerges is that most present-day human populations are admixtures of populations that came from somewhere else. Population **admixture** takes place when previously isolated or semi-isolated subpopulations come together and interbreed. Populations have migrated from place to place throughout human history: in some places leaving as refugees fleeing famine, drought, pestilence, or the ravages of war; in other places entering as raiders, traders, colonizers, or conquerors. The genomes of modern Europeans, for example, are an admixture of at least three earlier populations: Early Western European hunter-gatherers, who about 9–8 ka admixed with Early European farmers from the Near East, who then about 4.5 ka underwent further admixture with Ancient North Eurasians from the Caucasus in central Europe and the Pontic Steppe on the northern shores of the Black Sea (Patterson et al. 2012; Slatkin and Racimo 2016). The population history in Africa is also complex, with evidence for admixture among disparate populations of hunter-gatherers and herders (Skoglund et al. 2017).

The earliest evidence of population admixture comes from aDNA isolated from Neanderthal remains ranging in age from 38 to 45 ka found in a cave in Croatia (Green et al. 2010). With only 1 per cent contamination, the quality of the extracted DNA was remarkable. Detailed analysis revealed that the genomes in present-day non-African populations contain regions derived through introgression from Neanderthal ancestors. In the process of **introgression**, genes from one population are introduced into the gene pool of another by admixture followed by repeated backcrossing to the majority population.

The evidence for Neanderthal introgression comes from the patterns of **lineage sorting** of SNPs, which means the patterns in which polymorphisms resolve into fixation of one allele or the other as time progresses through many generations. Lineage sorting of polymorphic alleles is illustrated in Figure 7.15, in which the broad bands represent the ancestral relations among three populations designated P1, P2, and P3 along with an outgroup O. We assume that the ancestral allele state is *A* and that this allele is fixed in the

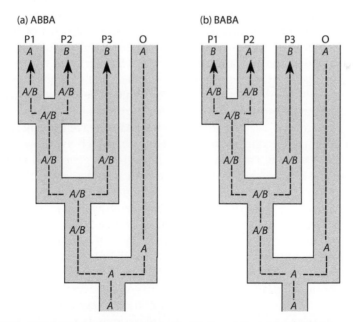

Figure 7.15 ABBA–BABA test for introgression. P1, P2, P, and O represent distinct populations, and *A/B* represents a single-nucleotide polymorphism that becomes fixed for *B* in population P3 and for either *A* or *B* in populations P1 and P2. Introgression is indicated by a significant difference in the number of *ABBA* sites versus the number of *BABA* sites.

outgroup, whereas population P3 is fixed for an alternative allele *B*. In this scenario only four patterns of *A* and *B* are possible between P1, P2, P3, and O, namely *BBBA*, *AABA*, *ABBA*, and *BABA*. Assuming that only one mutation is likely to occur over this span of time, if the *A*-to-*B* mutation took place between the common ancestor and the split between P1/P2 and P3, then the observed pattern would be *BBBA*; and if the *A*-to-*B* mutation took place along the lineage leading to P3, then the observed pattern would be *AABA*. Another possibility is that an *A/B* polymorphism arises and resolves into fixation of either *A* or *B* in various lineages. The pattern *BA* in population P3 and O requires that the polymorphism arise no later than the common ancestor of P1, P2, and P3. If *A/B* resolves into *B* in all three lineages, we get *BBBA*; and if it resolves into *B* in the P3 lineage and *A* in the P1/P2 lineage, we get *AABA*. But if the polymorphism persists into the individual populations P1 and P2, it could resolve into the possibilities *AA*, *BB*, *BA*, or *AB*. The first two yield the overall patterns *AABA* and *BBBA*, which are not informative, but the latter two resolve into *BABA* and *ABBA*, which are highly informative because they are at odds with the ancestral history of the populations. Furthermore, the *BABA* and *ABBA* patterns are expected to be equally likely if they are due only to the resolution of an inherited, selectively neutral *A/B* polymorphism.

This reasoning is the basis of the **ABBA–BABA test** for introgression (Durand et al. 2011; Kulathinal et al. 2009). Introgression of *B* from P3 into P2 is detected by significant excess of *ABBA* over BABA because the introgressed *B* reduces the number of *BABA* by changing *BABA* into *BBBA* and increases the number of *ABBA* by changing *AABA* into *ABBA*.

The *ABBA–BABA* test was applied to the Neanderthal data by letting P1 be any indigenous present-day African population, P2 be any present-day non-African population, P3 be the Neanderthal genome, and O be a chimpanzee outgroup (Green et al. 2010). The result was a significant excess of *ABBA* over *BABA*, which is consistent with introgression from Neanderthals into non-African populations. Current estimates of the fraction of introgressed nucleotides are 1.8–2.4 per cent among European populations and slightly more among Asian populations (Yang and Fu 2018). Similar tests with the Denisovan genome indicate 4–6 per cent introgression into the genomes of Oceanians. Estimates of introgression from *ABBA–BABA* tests based only on limited regions of the genome can be misleading, but alternative estimates are available (Martin et al. 2015).

Some of the introgressed regions from Neanderthals or Denisovans give evidence of having been maintained due to positive selection, which is known as **adaptive introgression**. For example, Tibetans have an introgressed gene from Denisovans that aids adaptation to the low oxygen pressure at the high altitudes where they live (Huerta-Sanchez et al. 2014). Other evidence for adaptive introgression includes genes influencing in skin pigmentation, digestive capability, as well as defense against infectious diseases and especially RNA viruses (Enard and Petrov 2018; Yang and Fu 2018).

Throughout human history, many populations have disappeared having made no detectable contribution to the present gene pool. An example comes from the 45 ka Ust'-Ishim individual recovered in western Siberia (Q. Fu et al. 2014), whose genome represents a population that became a dead end. The Ust'-Ishim genome is also noteworthy in having tracts of Neanderthal introgression 2–4 times longer than present-day populations, which enables dating of the Neanderthal admixture to approximately 60–50 ka.

Other populations that have contributed to the present-day gene pool have disappeared from their original place of origin. An example is the 24 ka Mal'ta individual from south-central Siberia, which represents a population that contributed 14–38 per cent of the genes in present-day Native Americans but has no close affinity with modern East Asians

(Raghavan et al. 2014a). The colonization of North America took place 17–22 ka across a land bridge that then connected Eastern Siberia with the continent. This migration spread to the east and south, populating southern North America and Central and South America. The colonization of northern North America took place later and included two distinct Paleo-Eskimo migrations from Siberia that spread widely across northern North America 4.5–1 ka and then vanished leaving no trace in present-day populations (Slatkin and Racimo 2016). Modern northern North Americans descend from migration of a Siberian people about 1 ka known as the Thule, who became the nearest genetic ancestors of today's Inuit (Raghavan et al. 2014b).

Modern human genomes have become even more admixed because of major population movements within historical times. These include the Muslim conquest across the Middle East, North Africa, and the Iberian Peninsula in the seventh and eighth centuries; the Viking colonization of Scandinavia, the British Isles, and Northern Europe in the eighth–eleventh centuries; the Mongol expansion that conquered a large swath of Central Asia in the thirteenth century; the trans-Atlantic slave trade in the sixteenth–nineteenth centuries; and the European colonization of the Americas starting in 1492, Oceania starting in the sixteenth century, and Africa in the late nineteenth and early twentieth centuries.

7.6 Transposable Elements

Most species of organisms contain DNA sequences called **transposable elements (TEs)** that can move from one location in the genome to another, some via a DNA intermediate and others via an RNA intermediate. Because of their ability to replicate and transpose, TEs can increase in copy number in the genome. Many TEs are also capable of **horizontal transmission** between reproductively isolated species (Capy et al. 1994; Clark et al. 1994; Robertson 1993). In some cases, horizontal transmission is evident from close sequence similarity of TEs in two otherwise very distantly related species. For example, *Drosophila erecta* and the cat flea *Ctenocephalides felis* are very distantly related species, as evidenced by the 40 per cent nucleotide identity found in the coding sequence for their sodium-potassium transmembrane pump, a generally very conserved gene. Yet both species contain a TE called *mariner* that shows 96–99 per cent nucleotide identity (Lohe et al. 1995). The vectors of horizontal transmission of TEs have not been identified, nor is the rate of horizontal transmission well estimated.

Models for the population dynamics of TEs usually incorporate several features (Charlesworth et al. 1994; Charlesworth and Charlesworth 2012; Langley et al. 1983; Sawyer and Hartl 1986):

- A rate of infection, in which genomes previously lacking the TE become infected with it.
- A rate of transposition, which determines how rapidly the copy number increases; the effects of regulation are taken into account by assuming that the rate of transposition is a decreasing function of copy number.
- A mechanism, or combination of mechanisms, for eliminating elements from the population or at least controlling their proliferation. Many transposons have evolved mechanisms for controlling their own transposition so as not to kill their hosts (Charlesworth and Langley 1986). In addition, many organisms have evolved mechanisms for repressing transposition to protect themselves from the deleterious effects

of TE mobilization (Navarro 2017). These include epigenetic silencing associated with heavy methylation (Bennetzen and Wang 2014) as well as silencing by means of small interfering RNAs (Girard and Hannon 2008; van Rij and Berezikov 2009). To suggest that most TE insertions are deleterious is not to deny that some TE insertions have acquired important functions in evolution (Biemont 2010; Chuong et al. 2017). But TEs seem to persist mainly because they are efficient molecular invaders of the genome, not because a few individual insertions have opportunistically been put to use (Barron et al. 2014).

Insertion Sequences and Transposons in Bacteria

Bacteria contain several types of TEs, the simplest of which are **insertion sequences**, typically 1–2 kb in length, containing short nucleotide sequences repeated at each end in inverted orientation and at least one long, open reading frame coding for the transposase protein that catalyzes transposition. The intestinal bacterium *E. coli* contains six well-characterized insertion sequences, the numbers of which have been determined among 71 natural isolates (Sawyer et al. 1987). These distributions have been used to estimate the parameters in population models.

Models of TEs in bacteria are greatly simplified because of asexual reproduction, a low rate of recombination among strains, and a low rate of deletion of insertion sequences. The "state" of a strain may be described in terms of the number of copies i of an element it contains. The parameters that govern the population genetics of a bacteria TE are:

- u, the rate at which uninfected cells become infected (i.e., go from state $i = 0$ to state $i = 1$).
- T, the rate of transposition in infected strains (i.e., the probability that a cell in state i goes to state $i + 1$ in the interval of one generation).
- s, the selection coefficient measuring the decrease in fitness of infected cells, relative to uninfected cells.

The most general models of this type allow T and s to be functions of i, but here we will assume that they are constant. However, assuming a constant value for T does incorporate regulation of transposition, because the rate of transposition *per element* in a strain in state i equals T/i, which is a decreasing function of i.

With these assumptions, the population attains an equilibrium distribution of numbers of elements, in which the probability p_i that a strain contains exactly i elements is given by:

$$p_0 = \alpha \ \text{ and } \ p_i = (1 - \alpha)(1 - \Theta)\Theta^{i-1} \ (i \geq 1) \tag{7.25}$$

where $\alpha = 1 - (u/s)$ and $\Theta = T/(T + s - u)$ (Sawyer and Hartl 1986; Sawyer et al. 1987).

Equation 7.25 may be applied to the concrete case of insertion sequence IS*30* in *E. coli*, in which the distribution of numbers among 71 strains fits a model with $\alpha = 1/2$ and $\Theta = 1/2$ (Sawyer et al. 1987). With these parameters, the expected distribution reduces to the remarkably simple formula $p_i = (1/2)^{i+1}$ for $i \geq 0$. Among 71 natural isolates, the observed and expected number of isolates containing 0, 1, 2, 3, 4, and ≥ 5 elements are shown in Figure 7.16. The fit is obviously very good, even though α and Θ were estimated from the data.

Aside from their own evolutionary dynamics, insertion sequences are important because they can mobilize other sequences in the genome. When two copies of an insertion sequence are on flanking sides of an unrelated sequence, the inverted repeats that are

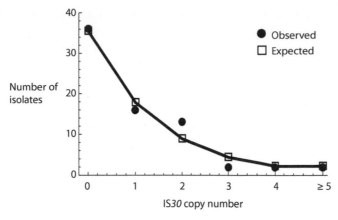

Figure 7.16 Observed and expected copy number of IS*30* elements among 71 isolates of *E. coli*. (Data from Sawyer et al. (1987).)

used in transposition are preferentially those at the extreme ends, creating a composite TE or **transposon** that transposes as a single unit. For example, the transposon *Tn5* consists of a central DNA sequence containing three genes conferring resistance to the antibiotics neomycin, streptomycin, and bleomycin, which are flanked by copies of the insertion sequence IS*50* enabling mobilization of *Tn5* as an intact unit (Reznikoff 2008). Antibiotic-resistance transposons are also important building blocks of infectious bacterial plasmids called **resistance transfer factors (R factors)** that contain multiple antibiotic-resistance genes (Bengtsson-Palme et al. 2018; Partridge et al. 2018).

Transposable Elements in Eukaryotes

Long-term persistence of TEs is possible in asexual organisms owing to transmissible plasmids and other mechanisms that allow them to be transmitted from one cell to the next (Hartl and Sawyer 1988). Such mechanisms are not necessary in sexual organisms. Sexual reproduction itself enables TEs to spread because the sexual process brings different genomes into physical association (Charlesworth 1985; Charlesworth and Charlesworth 2012; Charlesworth and Langley 1989; Charlesworth et al. 1994). As in asexual populations, an indefinite increase in TE copy number is checked in part by regulatory mechanisms that have evolved that reduce the net rate of transposition as copy number increases (Bennetzen and Wang 2014; Girard and Hannon 2008; van Rij and Berezikov 2009). Equally as important are selective effects, which need not be very large to offset even unregulated transposition. If TEs are distributed at random among the genomes in a population, and if the relative fitness of an organism decreases with the number i of copies in its genome, then at equilibrium the average fitness of the population $\bar{w}$, relative to a population lacking TEs, is given by:

$$\bar{w} = e^{-\bar{i}T} \tag{7.26}$$

where $\bar{i}$ is the mean number of copies per individual and T is the transposition rate (Charlesworth 1985).

In *D. melanogaster*, the average number of TEs per genome is about 1500, comprising about 100 different families of TEs averaging about 15 copies per family (Charlesworth and Charlesworth 2012). The average transposition rate is on the order of 10^{-4}

(Charlesworth and Charlesworth 2012). These values would put $\bar{w}$ at about 0.9985, yielding a very rough estimate of the fitness cost per copy of 10^{-6}, which justifies Charlesworth's remark that there is no great difficulty in explaining how natural selection can balance transposition in maintaining multiple families of TEs in a population (Charlesworth 1985).

Host mechanisms controlling TEs sometimes break down and growth in copy number can be explosive. The genomes of flowering plants afford a good example. The range of genome size in angiosperm species ranges from about 60 Mb to about 150 Gb (Bennetzen and Wang 2014). Some of the variation in size is due to polyploidy, but the largest genomes are so large primarily because of massive amplification of a small number of LTR retrotransposons, which are TEs that have long terminal repeats and transpose via an RNA intermediate. Most of these TEs are found as nested repeats (transposons that have inserted into previously existing transposons), with the oldest TEs at the bottom of each nest (Bennetzen and Wang 2014).

Another example of explosive TE amplification is seen in the strawberry poison-dart frog *Oophaga pumilio* (Rogers et al. 2018). Its genome size is 6.76 Gb, of which 70 per cent consists of TEs with very high copy number. TEs within each amplified family are similar in sequence to one another, implying that the amplification took place in the recent past. The amplified TEs encompass those that transpose via a DNA intermediate as well as those that transpose via an RNA intermediate. Some types of TEs are especially prevalent. The LTR retrotransposon *gypsy* accounts for about half of the amplified TEs, and the *mariner* element constitutes about 20 per cent. The *mariner* elements in *O. pumilio* strongly resemble those in fish, suggesting recent horizontal transmission, and the transposase of these elements is highly expressed in the germ line. The authors suggest that the unusual abundance of TEs in this organism results from repeated horizontal transmission and an ongoing proliferation that has so far evaded host mechanisms of control (Rogers et al. 2018).

Problems

7.1 The first 18 amino acids present at the amino terminal end of the human and mouse immune γ-interferon protein constitute a signal peptide that is used in secretion of the molecule.

```
Human: M-K-Y-T-S-Y-I-L-A-F-Q-L-C-I-V-L-G-S
Mouse: M-N-A-T-H-C-I-L-A-L-Q-L-F-L-M-A-V-S
```

a. Calculate the average number of amino acid replacements per site and its 95 per cent confidence interval, correcting for multiple hits.

b. Use these data to estimate the average rate and 95 per cent confidence interval of amino acid replacements in this peptide during the divergence of mice and humans, which based on fossils and other evidence, took place approximately 90 million years ago.

7.1 ANSWER a. The sequences of 18 aligned amino acids differ at 10 amino acid sites, hence d_0 in Equation 7.6 is given by $d_0 = 10/18 = 0.5556$, and therefore the corrected number of replacements per amino acid site is estimated as $d = -(19/20)\ln[1 - (20/19)d_0] = 0.8351$ amino acid replacements per site. The standard deviation is estimated as the square root of $(1/L)d_0(1 - d_0)/[1 - (20/19)d_0]^2$, where

$L = 18$, and this equals 0.2821. Hence, the 95 per cent confidence interval is $\pm 1.96 \times 0.2821 = 0.5529$. **b.** The estimated rate of amino acid replacement per amino acid site per year equals $d/(2 \times 90 \times 10^6) = 4.6397 \times 10^{-9}$. (The divergence time is doubled because an evolutionary branch of 90×10^6 years applies to each species.) The 95 per cent confidence interval in the rate equals $\pm 3.0716 \times 10^{-9}$, which is quite large relative to the mean, but then again it is based on only 18 amino acids.

7.2 The following 60 nucleotides are found in the coding region of the *trpA* genes in strains of the related enteric bacteria *Escherichia coli* strain K12 and *Salmonella typhimurium* strain LT-2. The *trpA* gene encodes one of the subunits of the enzyme tryptophan synthetase used in the synthesis of tryptophan.

```
K12:   V   A   P   I   F   I   C   P   P   N   A   D   D   D   L   L   R   Q   I   A
K12:   GTCGCACCTATCTTCATCTGCCCGCCAAATGCCGATGACGACCTGCTGCGCCAGATAGCC
LT2:   ATCGCGCCGATCTTCATCTGCCCGCCAAATGCGGATGACGATCTTCTGCGCCAGGTCGCA
LT2:   I   A   P   I   F   I   C   P   P   N   A   D   D   D   L   L   R   Q   V   A
```

a. Estimate the amount of nucleotide divergence d_n and amino acid divergence d_a and their standard deviations.

b. Assuming that *Escherichia* and *Salmonella* diverged at around the time of the mammalian radiation about 70 million years ago, estimate the rates of nucleotide substitution and amino acid replacement.

7.2 ANSWER a. The corrected amount of nucleotide divergence is given by Equation 7.5 as $d_n = -(3/4)\ln[1 - (4/3 d_0]$ where $d_0 = 9/60 = 0.1500$ is the proportion of differences, hence $d_n = 0.1674$. The variance of d_n is given by $(0.1500)(0.8500)/[(60)(1 - 4d_0/3)^2] = 0.003320$, which yields a standard deviation of 0.05762. For amino acids, the corrected divergence is given by Equation 7.6 as $d_a = -(19/20)\ln[1 - (20/19)d_0]$ where $d_0 = 2/20 = 0.1000$, hence $d_a = 0.1057$. The standard deviation is given by the square root of $(0.1000)(0.9000)/[(20)(1 - (20/19)(0.1000))^2]$, which equals 0.0750. **b.** In each case estimate λ from $\lambda = d/(2t)$ where t is the divergence time, in this case 70×10^6 years. For nucleotide substitutions the estimated rate is $\lambda = d_n/(2t) = 0.1674/(140 \times 10^6) = 1.1957 \times 10^{-9}$ substitutions per nucleotide site per year. For amino acid replacements the estimated rate is $\lambda = d_a/(2t) = 0.1057/(140 \times 10^6) = 7.5500 \times 10^{-10}$ replacements per amino acid site per year.

7.3 The mitochondrial DNA molecules (mtDNA) of 21 humans of diverse geographic and racial origin were digested with 18 restriction enzymes, 11 of which exhibited one or more fragments in which size polymorphism occurred. All fragment polymorphisms could be explained by single-nucleotide differences, and there was no evidence for insertions, deletions, or other mtDNA rearrangements. Altogether, 868 nucleotide sites were assayed for differences among individuals, and the average number of differences per nucleotide site per individual was estimated at 0.0018. Assuming that mammalian DNA undergoes sequence divergence at the rate of $5-10 \times 10^{-9}$ nucleotide substitutions per site per year, and that the rate is uniform in time, estimate the length of time since all of the 21 contemporary mtDNA molecules last shared a common ancestor.

7.3 ANSWER The problem specifies that the rate of divergence equals $5-10 \times 10^{-9}$ nucleotide substitutions per site per year, and that the average number of differences between mtDNA molecules is $d_0 = 0.0018$. Equation 7.5 yields $d = 0.001802$ (the correction is actually unnecessary because d_0 is so small), and because for any two sequences $2\lambda t = d$, the estimates of time are $t = 0.001802/(10 \times 10^{-9}) \approx 180{,}000$ years ago to $0.001802/(5 \times 10^{-9}) \approx 360,000$ years ago. That is, the present mtDNA genetic diversity in this sample of individuals could have originated from a single mitochondrion present in a female as recently as 9,000–18,000 generations ago, assuming 20 years per generation. One possible explanation of this finding is that the human population might have undergone a bottleneck in population size at about this time. However, Equation 6.15 in Chapter 6 implies that, among new neutral alleles destined to be fixed in a haploid population (appropriate for thinking about mitochondrial DNA), the average time to fixation is $2N$ generations. Assuming that a single mitochondrial genome became fixed in 9,000–18,000 generations, and assuming that these values bracket the average, the corresponding range of estimates of N for mitochondrial DNA (and therefore the N of females) is 4,500–9000, which is not unreasonable for small societies of the sort that have characterized most of human evolution. Moreover, the average size of human populations in regard to mtDNA may not be representative of the average size in regard to nuclear DNA.

7.4 You have the chance to study a diverse assemblage of microorganisms returned from a space probe. Their genetics and physiology is much like that of organisms on Earth, except that (1) their DNA includes an third pair of nucleotides resulting in six possible nucleotide pairs instead of four, and (2) their codons consist of two nucleotides instead of three with a pattern of redundancy in their genetic code in which the 36 possible codons specify 12 amino acids.

   a. To correct for multiple hits, what is the best one-parameter estimate of the proportion of nucleotide differences between two DNA sequences and its variance, given an observed proportion of differences of d_0?

   b. What is the best one-parameter estimate of the proportion of amino acid differences between two protein sequences and its variance?

7.4 ANSWER a. For a pair of nucleotide sequences, the best one-parameter correction is analogous to Equation 7.5, which becomes $d = -\frac{5}{6} \ln\left[1 - \frac{6}{5}d_0\right]$. Its variance is estimated as $Var(d) = \frac{d_0(1-d_0)}{L[1-(6/5)d_0]^2}$ where L is the number of nucleotides in each sequence. **b.** For a pair of amino acid sequences, the best estimate is analogous to Equation 7.6, which becomes $d = -\frac{11}{12} \ln\left[1 - \frac{12}{11}d_0\right]$ with variance $Var(d) = \frac{d_0(1-d_0)}{L[1-(12/11)d_0]^2}$.

7.5 Polymorphism and divergence were studied among synonymous and replacement sites in the *bride of sevenless* (*boss*) gene in samples from four species of *Drosophila*. In these samples, there were 71 fixed and 106 polymorphic synonymous differences as well as 8 fixed and 13 polymorphic replacement differences. Carry out a chi-square test of polymorphism and divergence and interpret the result.

7.5 ANSWER The chi-square statistic equals 0.032 with 1 degree of freedom for which $P = 0.858$. In these data the ratio of amino acid fixations to amino acid polymorphisms is not significantly different from the ratio of synonymous fixations to synonymous polymorphisms, hence no evidence for selection on amino acid sites.

7.6 A 300 bp highly repetitive sequence in the human genome is present in more than 300,000 copies and accounts for about 3 per cent of the total DNA. The sequence is known as *Alu* because of an *Alu*I restriction site it contains. Two randomly chosen *Alu* sequences differ, on the average, at 15–20 per cent of their nucleotide sites. Assuming a rate of sequence evolution of 5 nucleotide substitutions per nucleotide site per 10^9 years (approximately the rate for pseudogenes), estimate the average time of divergence of two randomly chosen *Alu* sequences.

7.6 ANSWER d_0 is given as 0.15–0.20 which, from Equation 7.5 corresponds to $d = 0.16736 - 0.23262$, and $2\lambda t = d$, where λ is the rate of sequence evolution given as $\lambda = 5 \times 10^{-9}$ per site per year. Hence the range of t values implied is $t = 0.16736/(2 \times 5 \times 10^{-9}) = 16.7$ million years to $t = 0.23262/(2 \times 5 \times 10^{-9}) = 23.3$ million years.

7.7 The distribution of copy number of several types of transposable insertion sequences has been studied in 71 natural isolates of *E. coli*, yielding the following data:

	Number of copies					
Type of element	0	1	2	3	4	≥ 5
IS1 No. strains	11	14	8	6	7	25
IS2 No. strains	28	8	12	5	5	13
IS4 No. strains	43	5	5	3	5	10

Equation 7.25 can be fit to these data to estimate α and Θ. Then a chi-square test can be used to assess goodness of fit to the model. Calculate the expected numbers for each of the models below and carry out a chi-square test. (Each test has three degrees of freedom because two parameters were estimated from the data.)

a. IS1 with $\alpha = 1/5$ and $\Theta = 5/6$.
b. IS2 with $\alpha = 2/5$ and $\Theta = 2/3$.
c. IS4 with $\alpha = 2/3$ and $\Theta = 3/4$.

7.7 ANSWER a. For IS1, the distribution in Equation 7.25 is given by $p_0 = 1/5$ and $p_i = (4/5)(1/6)(5/6)^{i-1}$ for $i \geq 1$. Hence the expected proportions are $p_0 = 0.2000$, $p_1 = 0.1333$, $p_2 = 0.0111$, $p_3 = 0.0926$, $p_4 = 0.0772$, and $p_{\geq 5} = 1 - p_0 - p_1 - p_2 - p_3 - p_4 = 0.3858$, yielding the expected number for 71 strains of 14.20, 9.47, 7.89, 6.57, 5.48, and 27.39, respectively. The $X^2 = 3.58$ with three degrees of freedom and $P = 0.311$, so the fit is acceptable. **b.** For IS2, $p_0 = 2/5$ and $p_i = (3/5)(1/3)(2/3)^{i-1}$ for $i \geq 1$. The expected numbers from 0 copies to ≥ 5 copies are 28.40, 14.20, 9.47, 6.31, 4.21, and 8.41, respectively. The $X^2 = 6.31$ and $P = 0.097$, which is also acceptable. **c.** For IS4, the distribution is given by $p_0 = 2/3$ and $p_i = (1/3)(1/4)(3/4)^{i-1}$ for $i \geq 1$. The expected numbers from 0 copies to ≥ 5 copies are 47.33, 5.92, 4.44, 3.33, 2.50, and 7.49, respectively. The $X^2 = 4.00$ and $P = 0.2621$, which is, once again, acceptable.

References

Abbott, S. and Fairbanks, D. J. (2016), 'Experiments on Plant Hybrids by Gregor Mendel', *Genetics*, 204 (2), 407–22.

Akashi, H. (1993), 'Synonymous codon usage in Drosophila melanogaster: natural selection and translational accuracy', *Genetics*, 136 (3), 927–35.

Akashi, H. (1995), 'Inferring weak selection from patterns of polymorphism and divergence at "silent" sites in Drosophila DNA', *Genetics*, 139 (2), 1067–76.

Arenas, M. (2015), 'Trends in substitution models of molecular evolution', *Front Genet*, 6, 319.

Ayala, D. J., Chang, B. S. W., and Hartl, D. L. (1993), 'Molecular evolution of the *Rh3* gene in Drosophila', *Genetica*, 92 (1), 23–32.

Barrandeguy, M. E., Sanabria, D. J., and García, M. V. (2017), 'The revitalization of population genetics: from coalescence theory to phylogeography', *Population Genetics*, http://www.smgebooks.com, https://smjournals.com/ebooks/population-genetics/chapters/PG-17-03.pdf, 1–10.

Barron, M. G., et al. (2014), 'Population genomics of transposable elements in Drosophila', *Annu Rev Genet*, 48, 561–81.

Beaumont, M. A., Zhang, W., and Balding, D. J. (2002), 'Approximate Bayesian computation in population genetics', *Genetics*, 261 (4), 2025–35.

Bengtsson-Palme, J., Kristiansson, E., and Larsson, D. G. J. (2018), 'Environmental factors influencing the development and spread of antibiotic resistance', *FEMS Microbiol Rev*, 42 (1).

Bennetzen, J. L. and Wang, H. (2014), 'The contributions of transposable elements to the structure, function, and evolution of plant genomes', *Annu Rev Plant Biol*, 65, 505–30.

Bernardi, G. (2000), 'Isochores and the evolutionary genomics of vertebrates', *Gene*, 241 (1), 3–17.

Biemont, C. (2010), 'A brief history of the status of transposable elements: from junk DNA to major players in evolution', *Genetics*, 186 (4), 1085–93.

Bustamante, C. D., et al. (2005), 'Natural selection on protein-coding genes in the human genome', *Nature*, 437 (7062), 1153–7.

Cannataro, V. L. and Townsend, J. P. (2018), 'Neutral theory and the somatic evolution of cancer', *Mol Biol Evol*, 35 (6), 1308–15.

Capy, P., et al. (1994), 'Horizontal transmission versus ancient origin: mariner in the witness box', *Genetica*, 93 (1–3), 161–70.

Castle, W. E. (1903), 'The laws of heredity of Galton and Mendel and some laws governing race improvement by selection', *Proc Am Acad Sci*, 39 (8), 233–42.

Charlesworth, B. (1985), 'The population genetics of transposable elements', in T. Ohta and K. Aoki (eds.), *Population Genetics and Molecular Evolution* (Berlin: Springer-Verlag), 213–32.

Charlesworth, B. and Charlesworth, D. (2012), *Elements of Evolutionary Genetics* (Greenwood Village, CO: Roberts and Company).

Charlesworth, B. and Langley, C. H. (1986), 'The evolution of self-regulated transposition of transposable elements', *Genetics*, 112 (2), 359–83.

Charlesworth, B. and Langley, C. H. (1989), 'The population genetics of Drosophila transposable elements', *Annu Rev Genet*, 23, 251–87.

Charlesworth, B., Sniegowski, P., and Stephan, W. (1994), 'The evolutionary dynamics of repetitive DNA in eukaryotes', *Nature*, 371 (6494), 215–20.

Charlesworth, J. and Eyre-Walker, A. (2008), 'The McDonald-Kreitman test and slightly deleterious mutations', *Mol Biol Evol*, 25 (6), 1007–15.

Chuong, E. B., Elde, N. C., and Feschotte, C. (2017), 'Regulatory activities of transposable elements: from conflicts to benefits', *Nat Rev Genet*, 18 (2), 71–86.

Clark, J. B., Maddison, W. P., and Kidwell, M. G. (1994), 'Phylogenetic analysis supports horizontal transfer of P transposable elements', *Mol Biol Evol*, 11 (1), 40–50.

Cogni, R., et al. (2017), 'On the long-term stability of clines in some metabolic genes in Drosophila melanogaster', *Sci Rep*, 7, 42766.

Corbett-Detig, R. B., Hartl, D. L., and Sackton, T. B. (2015), 'Natural selection constrains neutral diversity across a wide range of species', *PLoS Biol*, 13 (4), e1002112.

Cvijovíc, I., Good, B. H., and Desai, M. M. (2018), 'The effect of strong purifying selection on genetic diversity', *Genetics*, 209 (4), 1235–78.

Darwin, C. (1859), *On the Origin of Species* (London: John Murray).

David, J. R., et al. (1986), 'Alcohol tolerance and Adh frequencies in European and African populations of Drosophila melanogaster', *Genet Sel Evol*, 18 (4), 405–16.

DePristo, M. A., Weinreich, D. M., and Hartl, D. L. (2005), 'Missense meanderings in sequence space: a biophysical view of protein evolution', *Nature Rev Genet*, 6 (9), 678–87.

Desai, M. M. and Plotkin, J. B. (2008), 'The polymorphism frequency spectrum of finitely many sites under selection', *Genetics*, 180 (4), 2175–91.

Drake, J. W. (1991), 'Spontaneous mutation', *Ann Rev Genet*, 25, 125–46.

Durand, E. Y., et al. (2011), 'Testing for ancient admixture between closely related populations', *Mol Biol Evol*, 28 (8), 2239–52.

Eanes, W. F., Kirchner, M., and Yoon, J. (1993), 'Evidence for adaptive evolution of the G6pd gene in Drosophila melanogaster and Drosophila simulans', *Proc Natl Acad Sci U S A*, 90, 7475–9.

Enard, D. and Petrov, D. A. (2018), 'Evidence that RNA viruses drove adaptive introgression between Neanderthals and modern humans', *Cell*, 175 (2), 360–71.e13.

Ewens, W. J. (1972), 'The sampling theory of selectively neutral alleles', *Theor Popul Biol*, 3 (1), 87–112.

Eyre-Walker, A. (1996), 'Synonymous codon bias is related to gene length in *Escherichia coli*: selection for translational accuracy', *Mol Biol Evol*, 13 (6), 864–72.

Eyre-Walker, A. and Hurst, L. D. (2001), 'The evolution of isochores', *Nat Rev Genet*, 2 (7), 549–55.

Fay, J. and Wu, C.-I. (2000), 'Hitchhiking under positive Darwinian selection', *Genetics*, 155 (3), 1405–13.

Fay, J., Wyckoff, G. J., and Wu, C.-I. (2001), 'Positive and negative selection on the human genome', *Genetics*, 158 (3), 1227–34.

Ferretti, L., et al. (2017), 'Decomposing the site frequency spectrum: the impact of tree topology on neutrality tests', *Genetics*, 207 (1), 229–40.

Fu, Q., et al. (2014), 'Genome sequence of a 45,000-year-old modern human from western Siberia', *Nature*, 514 (7523), 445–9.

Fu, Y. -X. (1995), 'Statistical properties of segregating sites', *Theor Popul Biol*, 48 (2), 172–97.

Fu, Y. -X. and Li, W. -H. (1993), 'Statistical tests of neutrality of mutations', *Genetics*, 133 (3), 693–709.

Gillespie, J.H. (1989), 'Lineage effects and the index of dispersion of molecular evolution', *Mol Biol Evol*, 6 (6), 636–47.

Gillespie, J.H. and Langley, C. H. (1979), 'Are evolutionary rates really variable?', *J Mol Evol*, 13 (1), 27–34.

Girard, A. and Hannon, G. J. (2008), 'Conserved themes in small-RNA-mediated transposon control', *Trends Cell Biol*, 18 (3), 136–48.

Graur, D. (1985), 'Amino acid composition and the evolutionary rates of protein-coding genes', *J Mol Evol*, 22 (1), 53–62.

Green, R. E., et al. (2010), 'A draft sequence of the Neandertal genome', *Science*, 328 (5979), 710–22.

Griffiths, R. C. and Tavaré, S. (1998), 'The age of a mutation in a general coalescent tree', *Commun Stat Stoch Models*, 14 (1–2), 273–95.

Gutenkunst, R. N., et al. (2009), 'Inferring the joint demographic history of multiple populations from multidimensional SNP frequency data', *PLoS Genet*, 5 (10), e1000695.

Hahn, M. W. (2019), *Molecular Population Genetics* (New York: Oxford University Press).

Harris, H. (1966), 'Enzyme polymorphisms in man', *Proc R Soc Lond B*, 164 (995), 298–310.

Hartl, D. L. and Sawyer, S. A. (1988), 'Why do unrelated insertion sequences occur together in the genome of Escherichia coli?', *Genetics*, 118 (3), 537–41.

Hartl, D. L., Moriyama, E. N., and Sawyer, S. A. (1994), 'Selection intensity for codon bias', *Genetics*, 138 (1), 227–34.

Hey, J. (1999), 'The neutralist, the fly and the selectionist', *Trends Ecol Evol*, 14 (1), 35–8.

Hey, J. and Kliman, R. M. (1993), 'Population genetics and phylogenetics of DNA sequence variation at multiple loci within the Drosophila melanogaster complex', *Mol Biol Evol*, 10 (4), 804–22.

Hudson, R. R., Kreitman, M., and Aguadé, M. (1987), 'A test of neutral molecular evolution based on nucleotide data', *Genetics*, 116 (1), 153–9.

Huerta-Sanchez, E., et al. (2014), 'Altitude adaptation in Tibetans caused by introgression of Denisovan-like DNA', *Nature*, 512 (7513), 194–7.

Ikemura, T. (1985), 'Codon usage and tRNA content in unicellular and multicellular organisms', *Mol Biol Evol*, 2 (1), 13–34.

Ingman, M., et al. (2000), 'Mitochondrial genome variation and the origin of modern humans', *Nature*, 408 (6813), 708–13.

Jacobs, W. M. and Shakhnovich, E. I. (2017), 'Evidence of evolutionary selection for cotranslational folding', *Proc Natl Acad Sci U S A*, 114 (43), 11434–9.

Jouganous, J., et al. (2017), 'Inferring the joint demographic history of multiple populations: beyond the diffusion approximation', *Genetics*, 206 (3), 1549–67.

Jukes, T. H. and Cantor, C. R. (1969), 'Evolution of protein molecules', in H. N. Munro (ed.), *Mammalian Protein Metabolism* (New York: Academic Press), 21–132.

Kimura, M. (1968), 'Evolutionary rate at the molecular level', *Nature*, 217 (5129), 624–6.

Kimura, M. (1980), 'A simple method for estimating evolutionary rates of base substitution through comparative studies of nucleotide sequences', *J Mol Evol*, 16, 111–20.

Kimura, M. (1983), *The Neutral Theory of Molecular Evolution* (Cambridge: Cambridge Universiry Press).

Kliman, R. M. and Hey, J. (1993), 'DNA sequence variation at the period locus within and among species of the Drosophila melanogaster complex', *Genetics*, 133 (2), 375–87.

Kong, A., et al. (2012), 'Rate of de novo mutations and the importance of father's age to disease risk', *Nature*, 488 (7412), 471–5.

Kreitman, M. and Hudson, R. R. (1991), 'Inferring the evolutionary histories of the Adh and Adh-dup loci in Drosophila melanogaster from patterns of polymorphism and divergence', *Genetics*, 127 (3), 565–82.

Kudla, G., et al. (2009), 'Coding-sequence determinants of gene expression in Escherichia coli', *Science*, 324 (5924), 255–8.

Kulathinal, R. J., Stevison, L. S., and Noor, M. A. (2009), 'The genomics of speciation in Drosophila: diversity, divergence, and introgression estimated using low-coverage genome sequencing', *PLoS Genet*, 5 (7), e1000550.

Langley, C. H., Brookfield, J. F. Y., and Kaplan, N. (1983), 'Transposable elements in Mendelian populations. I. A theory', *Genetics*, 104 (3), 457–71.

Langley, C. H., et al. (2012), 'Genomic variation in natural populations of Drosophila melanogaster', *Genetics*, 192 (2), 533–98.

Lewontin, R. C. (1991), 'Electrophoresis in the development of evolutionary genetics: milestone or millstone?', *Genetics*, 128 (4), 657–62.

Lewontin, R. C. and Hubby, J. L. (1966), 'A molecular approach to the study of genic heterozygosity in natural populations II. Amount of variation and degree of heterozygosity in natural populations of Drosophila pseudoobscura', *Genetics*, 54 (2), 595–609.

Li, W.-H. (1977), 'Distribution of nucleotide differences between two randomly chosen cistrons in a finite population', *Genetics*, 85 (2), 331–7.

Li, W.-H. (1997), *Molecular Evolution* (Sunderland, MA: Sinauer).

Li, W.-H. (2011), 'A new test for detecting recent positive selection that is free from the confounding impacts of demography', *Mol Biol Evol*, 28 (1), 365–75.

Lohe, A. R., et al. (1995), 'Horizontal transmission, vertical inactivation, and stochastic loss of *mariner*-like transposable elements', *Mol Biol Evol*, 12 (1), 62–72.

Long, M. and Langley, C. H. (1993), 'Natural selection and origin of jingwei, a chimeric processed functional gene in Drosophila', *Science*, 260 (5104), 91–5.

Martin, S. H., Davey, J. W., and Jiggins, C. D. (2015), 'Evaluating the use of ABBA-BABA statistics to locate introgressed loci', *Mol Biol Evol*, 32 (1), 244–57.

McDonald, J. H. and Kreitman, M. (1991), 'Adaptive protein evolution at the Adh locus in Drosophila', *Nature*, 351 (6328), 652–4.

Mendel, G. (1866), 'Versuche über Pflanzen Hybriden', *Verh Naturforsch Ver in Brünn*, 4, 3–47.

Navarro, C. (2017), 'The mobile world of transposable elements', *Trends Genet*, 33 (11), 771–2.

Navascués, M., Leblois, R., and Burgarella, C. (2017), 'Demographic inference through approximate-Bayesian-computation skylike plots', *PeerJ*, 5, e3530.

Nielsen, R., et al. (2005), 'A scan for positively selected genes in the genomes of humans and chimpanzees', *PLoS Bio* 3, e170.

Nielsen, R., et al. (2009), 'Darwinian and demographic forces affecting human protein coding genes', *Genome Res*, 19 (5), 838–49.

Orlando, L., et al. (2013), 'Recalibrating Equus evolution using the genome sequence of an early Middle Pleistocene horse', *Nature*, 499 (7456), 74–8.

Ortega-Del Vecchyo, D., Marsden, C. D., and Lohmueller, K. E. (2016), 'PReFerSim: fast simulation of demography and selection under the Poisson random field model', *Bioinformatics*, 32 (22), 3516–18.

Parsch, J., Stephan, W., and Tanda, S (1998), 'Long-range base pairing in Drosophila and human RNA sequences', *Mol Biol Evol*, 15 (7), 820–6.

Partridge, S. R., et al. (2018), 'Mobile genetic elements associated with antimicrobial resistance', *Clin Microbiol Rev*, 31 (4), e00088-17.

Patterson, N., et al. (2012), 'Ancient admixture in human history', *Genetics*, 192 (3), 1065–93.

Petrov, D. A., Lozovsky, E. R., and Hartl, D. L. (1996), 'High intrinsic rate of DNA loss in Drosophila', *Nature*, 384 (6607), 346–9.

Petrov, D. A., et al. (2000), 'Evidence for DNA loss as a determinant of genome size', *Science*, 287 (5455), 1060–2.

Plotkin, J. B. and Kudla, G. (2011), 'Synonymous but not the same: the causes and consequences of codon bias', *Nat Rev Genet*, 12 (1), 32–42.

Raghavan, M., et al. (2014a), 'Upper Palaeolithic Siberian genome reveals dual ancestry of Native Americans', *Nature*, 505 (7481), 87–91.

Raghavan, M., et al. (2014b), 'The genetic prehistory of the New World Arctic', *Science*, 345 (6200), 1255832.

Reich, D. (2018), *Who We Are and How We Got Here: Ancient DNA and the New Science of the Human Past* (New York: Pantheon).

Reid, S. D., Selander, R. K., and Whittam, T. S. (1999), 'Sequence diversity of flagellin (*fliC*) alleles in pathogenic Escherichia coli', *J Bacteriol*, 181 (1), 153–60.

Reznikoff, W. S. (2008), 'Transposon Tn5', *Annu Rev Genet*, 42, 269–86.

Robertson, H. M. (1993), 'The mariner transposable element is widespread in insects', *Nature*, 362 (6417), 241–5.

Rogers, R. L., et al. (2018), 'Genomic takeover by transposable elements in the Strawberry poison frog', *Mol Biol Evol*, 35 (12), 2913–27.

Rosen, Z., et al. (2018), 'Geometry of the sample frequency spectrum and the perils of demographic Inference', *Genetics*, 210 (2), 665–82.

Sawyer, S. and Hartl, D. L. (1986), 'Distribution of transposable elements in prokaryotes', *Theor Pop Biol*, 30 (1), 1–17.

Sawyer, S. A. and Hartl, D. L. (1992), 'Population genetics of polymorphism and divergence', *Genetics*, 132 (4), 1161–76.

Sawyer, S. A., et al. (2007), 'Prevalence of positive selection among nearly neutral amino acid replacements in Drosophila', *Proc Natl Acad Sci U S A*, 104 (16), 6504–10.

Sawyer, S. A., et al. (1987), 'Distribution and abundance of insertion sequences among natural isolates of Escherichia coli', *Genetics*, 115 (1), 51–63.

Schrider, D. R. and Kern, A. D. (2018), 'Supervised machine learning for population genetics: a new paradigm', *Trends Genet*, 34 (4), 301–12.

Sethupathy, P. and Hannenhalli, S. (2008), 'A tutorial of the Poisson random field model in population genetics', *Adv Bioinformatics*, 2008, 257864.

Sharp, P. M. and Li, W.-H. (1987), 'The codon adaptation index: a measure of directional codon usage bias and its potential application', *Nucleic Acids Res*, 15 (3), 1281–95.

Shields, D. C., et al. (1988), '"Silent" sites in Drosophila are not neutral: evidence of selection among synonymous codons', *Mol Biol Evol*, 5 (6), 704–16.

Simonsen, K. L., Churchill, G. A., and Aquadro, C. F. (1995), 'Properties of statistical tests of neutrality for DNA polymorphism data', *Genetics*, 141 (1), 413–29.

Sisson, S. A., Fan, Y., and Beaumont, M. (2018), *Handbook of Approximate Bayesian Computation* (Boca Raton, FL: CRC Press).

Skoglund, P., et al. (2017), 'Reconstructing prehistoric African population structure', *Cell*, 171 (1), 59–71 e21.

Slatkin, M. and Racimo, F. (2016), 'Ancient DNA and human history', *Proc Natl Acad Sci U S A*, 113 (23), 6380–7.

Smith, N. G. C. and Eyre-Walker, A. (2002), 'Adaptive protein evolution in Drosophila', *Nature*, 415 (6875), 1022–4.

Solar, R., et al. (2019), 'A service-oriented platform for approximate Bayesian computation in population genetics', *J Comput Biol*, 26 (3), 266–79.

Spielman, S. J. and Wilke, C. O. (2015), 'The relationship between dN/dS and scaled selection coefficients', *Mol Biol Evol*, 32 (4), 1097–108.

Spielman, S. J. and Kosakovsky Pond, S. L. (2018), 'Relative evolutionary rates in proteins are largely insensitive to the substitution model', *Mol Biol Evol*, 35 (9), 2307–17.

Stam, L. F. and Laurie, C. C. (1996), 'Molecular dissection of a major gene effect on a quantitative trait: the level of alcohol dehydrogenase expression in Drosophila melanogaster', *Genetics*, 14 (4), 1559–64.

Stoletzki, N. and Eyre-Walker, A. (2011), 'Estimation of the neutrality index', *Mol Biol Evol*, 28 (1), 63–70.

Tajima, F. (1983), 'Evolutionary relationships of DNA sequences in finite populations', *Genetics*, 105 (2), 437–60.

Templeton, A. R. (1996), 'Contingency tests of neutrality using intra/interspecific gene trees: the rejection of neutrality for the evolution of the mitochondrial cytochrome oxidase II gene in hominoid primates', *Genetics*, 144 (3), 1263–70.

Terhorst, J. and Song, Y. S. (2015), 'Fundamental limits on the accuracy of demographic inference based on the sample frequency spectrum', *Proc Natl Acad Sci U S A*, 112 (25), 7677–82.

Thornlow, B. P., et al. (2018), 'Transfer RNA genes experience exceptionally elevated mutation rates', *Proc Natl Acad Sci U S A*, 115 (36), 8996–9001.

van Delden, W., Boerma, A. C., and Kamping, A. (1978), 'The alcohol dehydrogenase polymorphism in Drosophila melanogaster. I. Selection in different environments', *Experientia*, 31 (4), 418–20.

van Rij, R. P. and Berezikov, E. (2009), 'Small RNAs and the control of transposons and viruses in Drosophila', *Trends Microbiol*, 17 (4), 163–71.

Wang, H., Tzeng, Y. H., and Li, W. H. (2008), 'Improved variance estimators for one- and two-parameter models of nucleotide substitution', *J Theor Biol*, 254 (1), 164–7.

Wang, W., et al. (2000), 'The origin of the Jingwei gene and the complex modular structure of its parental gene, yellow emperor, in Drosophila melanogaster', *Mol Biol Evol*, 17 (9), 1294–301.

Watterson, G. A. (1975), 'On the number of segregating sites in genetical models without recombination', *Theor Pop Biol*, 7 (2), 256–76.

Watterson, G. A. (1985), 'Estimating species divergence times using multilocus data', in T. Ohta and K. Aoki (eds.), *Population Genetics and Molecular Evolution* (Berlin: Springer-Verlag), 1–44.

Whittam, T. S., Reid, S. D., and Selander, R. K. (1998), 'Mutators and long-term molecular evolution of pathogenic Escherichia coli O157:H7', *Emerging Infect Dis*, 4 (4), 615–17.

Wilson, D. J., et al. (2011), 'A population genetics-phylogenetics approach to inferring natural selection in coding sequences', *PLoS Genet*, 7 (12), e1002395.

Wright, F. (1990), 'The "effective number of codons" used in a gene', *Gene*, 87 (1), 23–9.

Wright, S. I. and Charlesworth, B. (2004), 'The HKA test revisited: a maximum-likelihood-ratio test of the standard neutral model', *Genetics*, 168 (2), 1071–6.

Yang, L. and Gault, B. S. (2011), 'Factors that contribute to variation in evolutionary rate among Arabidopsis genes', *Mol Biol Evol*, 28 (8), 2359–69.

Yang, M. A. and Fu, Q. (2018), 'Insights into modern human prehistory using ancient genomes', *Trends Genet*, 34 (3), 184–96.

Yang, Z., et al. (2018), 'Detecting recent positive selection with a single locus test bipartitioning the coalescent tree', *Genetics*, 208 (2), 791–805.

Zeng, K., et al. (2006), 'Statistical tests for detecting positive selection by utilizing high-frequency variants', *Genetics*, 174 (3), 1431–9.

Zhang, J. and Yang, J. R. (2015), 'Determinants of the rate of protein sequence evolution', *Nat Rev Genet*, 16 (7), 409–20.

Zharkikh, A. (1994), 'Estimation of evolutionary distances between nucleotide sequences', *J Mol Evol*, 39 (3), 315–29.

CHAPTER 8

Population Genetics of Complex Traits

No study of population genetics is sufficient without consideration of traits influenced both by multiple genes and environmental factors that act together. These traits are often called **complex traits**. Most readily observed variation in phenotype that you encounter in everyday life consists of differences in complex traits such as height, weight, hair color, eye color, and skin color. Most congenital abnormalities including heart defects, neural tube defects, and cleft lip or palate are complex traits. And, with the exception of cancer, the most frequent adult disorders including heart disease, diabetes, and bipolar disorder are complex traits. On the whole, complex traits account for far more disorders than do simple Mendelian traits. The most common simple Mendelian traits affect approximately one per thousand individuals or less, whereas the most common complex traits affect approximately one per hundred individuals or more.

The **genetic architecture** of a complex trait includes all of the genetic and environmental factors that affect the trait, along with the magnitudes of their individual effects and the magnitudes of any interactions among the factors. The genetic architecture of a complex trait depends not only on the trait but also on the particular population. The genetic architecture is affected by the genotype frequencies, the distributions of environmental factors, and biological properties such as age and sex. The approaches described in this chapter allow predictions of the expected phenotype of an individual based on the phenotypes of the individual's parents and other relatives. Being able to make such predictions is the foundation for the improvement of plant and animal populations by selecting superior individuals from each generation to be parents of the next generation. Large-scale genomic sequencing has enabled the genes affecting many complex traits to be identified and located along a genetic map. Environmental factors affecting complex traits are less easily sorted out, but major lifestyle factors affecting the risk of complex traits such as heart disease and diabetes have been identified from public health surveys and follow-up studies. But let's start from the beginning by discussing various types of complex traits.

8.1 Phenotypic Variation in Complex Traits

In human populations, the most common medical disorders with a genetic component are complex traits, including heart disease, asthma, peptic ulcer, schizophrenia, and many congenital abnormalities. In domesticated animals and plants, most traits of commercial interest are complex traits, such as egg production, milk production, and yield

A Primer of Population Genetics and Genomics. Fourth Edition. Daniel L. Hartl, Oxford University Press (2020). © Daniel L. Hartl.
DOI: 10.1093/oso/9780198862291.003.0008

of grain. Even as modern methods of genetic engineering are applied to animal and plant improvement, complex traits continue to be important because most desirable traits result from interactions of multiple genes and environmental factors.

Three Types of Complex Traits

Complex traits are often called **quantitative traits** to distinguish them from traits that appear in discrete categories, like round versus wrinkled pea seeds. Three types of quantitative traits may be distinguished:

- **Metric traits**, which are measured on a continuous, uninterrupted scale, such as height or weight.
- **Categorical** or **meristic traits**, which are measured by counting and include such traits as litter size or number of bristles. When the number of possible phenotypes is large, there is little distinction between a categorical trait and a metric trait.
- **Threshold traits**, which are discrete in that they are either present or absent in any one individual, but which are complex in that the underlying risk or liability toward the trait is determined by multiple genetic and environmental factors.

Because complex traits are affected by multiple genetic factors, they are also often called **polygenic traits**. Typically, each of the multiple genes underlying a complex trait has the feature that the mean difference in phenotype between the alternative genotypes is relatively small in comparison with the total variance in phenotype in the population. These genes are sometimes called **polygenes**, but more commonly they are referred to as **quantitative trait loci** or **QTLs**. When a QTL affects the level of gene expression, it is known as an **expression QTL** or **eQTL**.

Phenotypic Variation

Complex traits can differ in phenotype from one individual to the next. For quantitative and categorical traits, the phenotype of a particular individual is called its **phenotypic value**. An example of phenotypic variation for a categorical trait is illustrated in Figure 8.1. The trait is the number of bristles on an abdominal segment in *Drosophila*. For bristle number in males, the phenotypic values typically range from 13 to 25 with a modal number of about 19. The smooth curve is a normal distribution approximating the histogram. Two parameters are of great interest for any quantitative trait:

- The **mean** or average, usually denoted μ, which is defined as the expected phenotypic value across all individuals in the population, $E(x)$.
- The **variance**, usually denoted σ^2, which is defined as the expected value of the squared deviation of each phenotypic value from the population mean, $E(x - \mu)^2$. The variance is a measure of dispersion of the phenotypic values around the mean. The more clustered the values, the smaller the variance; the more scattered, the bigger the variance. An important related quantity is the **standard deviation**, which is the square root of the variance and symbolized as σ.

The true mean μ and variance σ^2 in a population are rarely known for certain but must be estimated from the phenotypic values observed in a random sample of individuals. The best estimates of the mean and variance are usually denoted as $\bar{x}$ and s^2, and from a random sample of n individuals these estimates are given by:

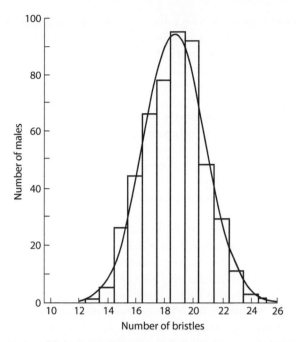

Figure 8.1 Number of bristles on the fifth abdominal sternite in 530 males of a strain of *D. melanogaster*. (Data courtesy of Trudy Mackay.)

$$\bar{x} = \frac{\sum\limits_{i=1}^{n} x_i}{n} \quad \text{and} \quad s^2 = \frac{\sum\limits_{i=1}^{n} (x_i - \bar{x})^2}{n-1} \tag{8.1}$$

where x_i is the phenotypic value of the ith individual in the sample. The need to use $n-1$ in the denominator of s^2 is a correction for sampling error. In practice, it is a reminder that the variance in a population cannot be estimated from the phenotypic value of a single individual. When n is reasonably large, the correction is minor.

For the bristle-number data in Figure 8.1, $\bar{x} = 18.74$ bristles, $s^2 = 4.306$ bristles-squared, and $s = \sqrt{(4.306)} = 2.075$ bristles. This example shows why the standard deviation is often preferred over the variance as a measure of dispersion: It is because the standard deviation has the same units of measurement as the mean.

Properties of the Normal Distribution

The bell-shaped curve in Figure 8.1 is a **normal distribution**, often denoted $N(\mu, \sigma)$, which has the functional form:

$$f(x) = \frac{1}{\sqrt{2\pi}\sigma} e^{-\frac{(x-\mu)^2}{2\sigma^2}} \qquad (-\infty < x < \infty) \tag{8.2}$$

where, in the case of Figure 8.1, we use $\bar{x} = 18.74$ as an estimate of μ and $s^2 = 4.306$ as an estimate of σ^2. The values of $\pi = 3.14159$ and $e = 2.71828$ are constants. Note that the theoretical range of x is minus infinity to plus infinity, but for these values of the mean and variance the expected proportion of negative phenotypic values is negligible.

It is often convenient to measure x as a deviation from the population mean in units of standard deviation, that is, as $(x - \mu)/\sigma$. This measure transforms the normal distribution in Equation 8.2 into a **standard normal distribution** that has a mean of 0 and a variance of 1. In a normal distribution, the proportion of individuals in the population that have phenotypic values that deviate from the mean by less than one standard deviation equals 68.3 per cent. This proportion is indicated by the darkest shading in Figure 8.2. Since the normal distribution is symmetrical, half the deviations larger than σ are positive and half are negative, which means that 15.8 per cent of the population has $(x - \mu)/\sigma > 1$ and 15.8 per cent of the population has $(x - \mu)/\sigma < 1$. The proportion of a normal distribution that has values $-2 \leq (x - \mu)/\sigma \leq +2$ and $-3 \leq (x - \mu)/\sigma \leq +3$ are 95.5 and 99.7 per cent, respectively (Figure 8.2).

As an example, we can use data for pupa weight in the flour beetle *Tribolium castaneum*, which is normally distributed with μ and σ estimated from Equation 8.1 as 2246.9 mg and 176.86 mg, respectively (Enfield 1980). From the values for one, two, and three standard deviations in Figure 8.2 we can assert that ≈ 68 per cent of the pupae are in the range of weight $\mu \pm \sigma$ (2070.0–2423.8 mg), ≈ 95 per cent are in the range $\mu \pm 2\sigma$ (1893.2–2600.6 mg), and ≈ 99.7 per cent are in the range $\mu \pm \sigma$ (1716.3–2777.5 mg).

The normal distribution occupies such a prominent place in the study of complex traits because of a statistical theorem called the *central limit theorem*, which states, roughly speaking, that the summation of many random, independent quantities conforms to the normal distribution. Since complex phenotypes are determined by multiple genetic and environmental factors, a normal distribution is to be expected if the factors are independent and their effects approximately additive. Francis Galton, a cousin of Charles Darwin, spent a good deal of his professional life investigating applications of the normal distribution to phenotypic variation. He was sufficiently impressed by the wide applicability of the normal distribution to write (Galton 1889):

I know of scarcely anything so apt to impress the imagination as the wonderful form of cosmic order expressed by the "law of frequency of error" [the normal distribution]. Whenever a large sample of chaotic elements is taken in hand and marshaled in the order of their magnitude, this unexpected and most beautiful form of regularity proves to have been latent all along. The law

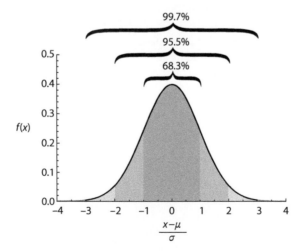

Figure 8.2 Proportions of a normal distribution with values that deviate from the mean by less than one, two, or three standard deviations.

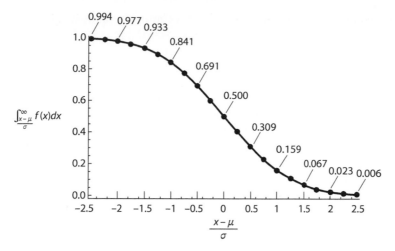

Figure 8.3 For a normal distribution, the curve depicts expected proportion of individuals with a phenotypic value that exceeds some specified deviation from the mean in units of standard deviation.

would have been personified by the Greeks if they had known of it. It reigns with serenity and complete self-effacement amidst the wildest confusion. The larger the mob and the greater the apparent anarchy, the more perfect is its sway. It is the supreme law of unreason.

In some applications to complex traits one needs to know what proportion of the population exceeds some specified threshold. This proportion is given by the integral of Equation 8.2 from the threshold value to infinity. Figure 8.3 shows representative values of these proportions for the standard normal distribution.

Although many quantitative and categorical traits are distributed normally or approximately so, some are not. For example, data in proportions (p) are often not normally distributed, but they may become nearly so when expressed as $x = \arcsin \sqrt{(p)}$. Another such normalizing transformation useful in other cases is to take the natural logarithm of x.

8.2 Genes and Environment

A major assumption of quantitative genetics is that genetic and environmental effects on complex traits are additive. This assumption cannot be verified explicitly because the individual factors are not directly observable. But it allows the phenotypic value (P) of any individual to be written as the sum of the genetic effects (G) and the environmental effects (E) on the trait, which is to say:

$$P = G + E \tag{8.3}$$

where for simplicity P, G, and E are all assumed to be measured as deviations from their respective means. As noted, G and E cannot be estimated, which means that one cannot specify the relative importance of genotype ("nature") or environment ("nurture") for any given trait. A moment's reflection will reveal that the question of whether nature or nurture is more important is meaningless. For example, if you were told that genes account for half of adult height, you might well ask whether that would be the top half or the bottom half, or maybe the left half or the right half. The statement makes no sense because each cell in each bone and sinew that accounts for adult height is affected both by genes and environment, and their effects are inseparable.

Genotypic Variance and Environmental Variance

Whereas Equation 8.3 does not and cannot say anything about the relative importance of genotype and environment with respect to the *mean* of a trait, it does have an important implication regarding the *variance* of the trait among individuals. In particular, Equation 8.3 implies that the variance in phenotypic value (the **phenotypic variance**, σ_p^2) can be partitioned into one component due to variation among the genetic factors, which is called the **genotypic variance**, σ_g^2, and another component due to variation among the environmental factors, which is called the **environmental variance**, σ_e^2. The basis of this partitioning is the definition of the variance in terms of the expected value of the squared deviations:

$$\sigma_p^2 = Var(P) = Var(G+E) = \sigma_g^2 + \sigma_e^2 \tag{8.4}$$

Strictly speaking, the right-hand side of Equation 8.4 should also include a covariance term, but its value is 0 if the genetic and environmental factors are independent. Later we'll examine some reasons why the genetic and environmental factors might not independent and how lack of independence effects the interpretation of Equation 8.4.

The biological meaning of Equation 8.4 is shown for the alleles of one gene in Figure 8.4. The solid curves represent the phenotypic distributions of the three genotypes AA, AA', and $A'A'$ in a randomly mating population with their means denoted G_1, G_2, and G_3. The dashed curve represents the phenotypic distribution in the entire population. The phenotypic variance σ_p^2 is the variance of the dashed distribution. The genotypic variance σ_g^2 is the variance among the G's, which is given by $\sigma_g^2 = p^2 G_1^2 + 2pq G_2^2 + q^2 G_3^2$ where p is the allele frequency of A. Although the G's are not generally known, σ_g^2 must equal zero in a genetically uniform population. Hence the observed variance of a genetically uniform population provides an estimate of σ_e^2, whereas the observed variance of a randomly bred population provides an estimate of $\sigma_g^2 + \sigma_e^2$. An estimate of σ_g^2 can be obtained by

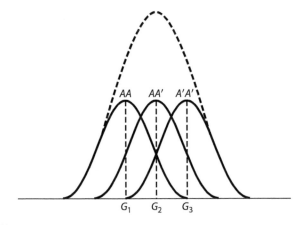

Figure 8.4 Overall phenotypic distribution (dashed curve) of a quantitative trait in a hypothetical randomly mating population, showing the underlying phenotypic distributions for the three genotypes of one gene.

Table 8.1 Estimation of the genotypic variance $\left(\sigma_g^2\right)$ and environmental variance (σ_e^2) of thorax length in *Drosophila melanogaster*[a]

	Populations	
Variance	Random-bred	Uniform
Theoretical	$\sigma_g^2 + \sigma_e^2$	σ_e^2
Observed	0.366	0.186

$\sigma_e^2 = 0.186$

$\sigma_g^2 = \left(\sigma_g^2 + \sigma_e^2\right) - \sigma_e^2 = 0.366 - 0.186 = 0.180$

Source: data from Robertson (1957).
[a] In units of 10^{-2} mm^2.

subtraction, since $\sigma_g^2 = \left(\sigma_g^2 + \sigma_e^2\right) - \sigma_e^2$. An example with thorax length in *Drosophila* is shown in Table 8.1. In this case, genetic variation among flies in the randomly bred population accounts for about $0.180/0.366 = 49.2$ per cent of the phenotypic variance in the population. This method of separating the genotypic and environmental variance has also been used in studies of monozygotic twins (identical twins) in human populations, because MZ twins have identical genotypes but different environmental experiences.

Broad-Sense Heritability

For a complex trait, one measure of the aggregate effect of all genetic factors combined is the **broad-sense heritability** H^2, defined as the ratio of the genotypic variance to the phenotypic variance:

$$H^2 = \frac{\sigma_g^2}{\sigma_p^2} = \frac{\sigma_g^2}{\sigma_g^2 + \sigma_e^2} \tag{8.5}$$

Hence, if $H^2 = 0$, all of the phenotypic variance is attributable to differences in environment, and if $H^2 = 1$, all of the phenotypic variance is attributable to differences in genotype.

Equation 8.4 is important in suggesting how genetic and environmental effects on the variance may be separated. What about effects on the mean phenotypic value? For example, suppose two human populations differ in average height and that the values of σ_g^2 and σ_e^2 are known in both populations. What can be inferred about the genetic versus environmental cause of the difference in the mean? Nothing whatsoever. At one extreme, the two populations could be quite similar genetically but their environments different. At the other extreme, the two populations could differ genetically with the environments being similar. Without additional data, nothing more can be said. The variance components have a very restricted utility and apply only to differences in phenotype within populations, not between populations.

To underline this point, imagine flocks of chickens on two different farms. On one farm the hens average 250 eggs per year and on the other they average 200 eggs per year. The broad-sense heritability in both flocks is 50 per cent, which tells you that half of the variation in egg production within each flock is due to genetic differences among the hens. But it tells you nothing about why the flocks produce different numbers of eggs. One extreme possibility is that the hens represent genetically distinct breeds that differ intrinsically in egg production, for example, Rhode Island Reds versus New Hampshire Reds. The other extreme possibility is that the hens are genetically similar but the feed or water on one farm is inadequate. The difference in egg production lies somewhere between these extremes, but without further investigation the heritability tells you nothing about the cause of the phenotypic difference between the flocks.

Genotype-by-Environment and Other Interactions

Equation 8.4 ignores several possible sources of nonindependence between genotype and environment.

- **Genotype-by-environment interaction (GEI).** GEI occurs when the genotypic and environmental effects are not additive but differ according to which genotype is in which specific environment. An example of GEI in maize is illustrated in Figure 8.5. The two strains are hybrids formed by crossing different pairs of inbred lines, and the index of environmental quality is based on soil fertility, moisture, and other factors. The overall mean yields of A and B, averaged across all environments, are both about 8500 kg/ha. However, A clearly outperforms B in the very stressful environments (negative), whereas B outperforms A in the very favorable environments (positive). A curve showing the phenotype of a genotype across the range of environments is called the **norm of reaction** for the genotype. GEI is indicated when the norms of reaction cross. The important implication of GEI is that the deviations due to genotype are not independent of the deviations due to environment. When GEI occurs but is not explicitly accounted for, the variance in phenotype due to GEI is included in the environmental variance in Equation 8.4.

- **Genotype-by-environment association (GEA).** This is another cause of nonindependence of genotypic and environmental deviations. It occurs when the genotypes in a population are not distributed randomly in all the possible environments. With GEA it is difficult, if not impossible, to separate genetic and environmental causes of variation because there is a systematic association of certain genotypes with certain environments. One example of a deliberate GEA is the practice of many dairy farmers to

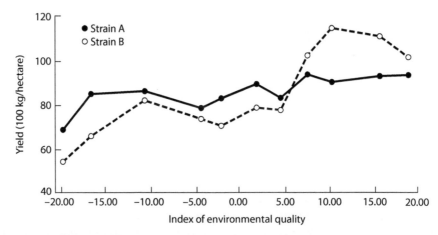

Figure 8.5 Genotype-by-environment interaction in two strains of maize. Each curve is the norm of reaction of the strain. (Data from Russell (1974).)

provide more feed supplements to cows that produce more milk, hence cows that have superior milk-yield genotypes are also provided a superior nutritional environment. Unless explicitly accounted for, the variance in phenotype due to GEA is included in the genetic variance in Equation 8.4.

- **Genotype-by-sex interaction (GSI)**. GSI occurs when the magnitude of a genotypic deviation depends on the sex of the individual. It is a potential problem in genetic analysis because the expression of many complex traits depends in part on developmental, hormonal, or other factors associated with sex (Nuzhdin et al. 1997). An example is seen in human height, in which the female average is about 10 cm smaller than the male average. Yet there is no reason for thinking that height-related genotypes are distributed differently in females and males. In this case, considering a mixed population of females and males, there is a GSI because the majority of genotypes will yield a negative deviation from the overall average if the individual is a female, and a positive deviation if the individual is a male. As with GEI, the variance in phenotype due to GSI is included in the environmental variance in Equation 8.4.

Genetic Effects on Complex Traits

To illustrate the genetic term in Equation 8.4 with actual data, we may use a genetic factor affecting coat coloration in guinea pigs (Table 8.2). The alleles at the locus are c^r and c^d, but for consistency with other symbols in this chapter we will designate them as A and A'. The phenotypic value of each animal is measured as $\arcsin\sqrt{(x)}$, where x is the proportion of black coloration on the animal. The mean phenotypes of AA, AA', and $A'A'$ genotypes are denoted a, d, and $-a$, respectively, which are measured as a deviation from the average of the homozygous genotypes (in this case 61.60). The calculations of a and d are shown in column 3. The symbols a and d represent the effects of the alleles. The quantity $2a$ measures the difference between means of the homozygous genotypes, because $a - (-a) = 2a$, and d/a serves as a measure of dominance:

- $d = a$ means that A is dominant to A'.
- $d = 0$ implies additivity (i.e., the phenotype of AA' is exactly between the phenotypes of AA and $A'A'$).
- $d = -a$ means that A' is dominant to A.

In this example, $a = 7.27$ and $d = -0.93$. The negative sign on d means that the A' allele (c^d) is partially dominant, and $d/a = -0.128$.

To obtain the G_i values needed in Equation 8.4, we need to express the mean phenotype of each genotype as a deviation from the overall population mean. If A and A' have allele frequencies p and q, then in a random mating population the overall mean phenotypic value μ is given by $p^2 a + 2pqd + q^2(-a)$. The deviations are obtained by subtracting μ from the mean of each genotype as shown in column 4. The algebra is somewhat tedious, but each deviation can be expressed in terms of the quantity $a + (q-p)d$, which appears so often in the equations of quantitative genetics that it is assigned the special symbol α and given the special name **average effect**:

$$\alpha = a + (q-p)d \tag{8.6}$$

The average effect can also be written as $q[d - (-a)] + p(a - d)$, which more easily shows its biological meaning. It is the average change in mean phenotype that would result from choosing an A' allele at random (in whatever genotype it happens to be) and changing it into an A allele. In Table 8.2, the values $2q\alpha$, $(q - p)\alpha$, and $-2p\alpha$ are called

Table 8.2 Expressions for allelic effects of one locus affecting coat coloration in guinea pigs

Genotype	Mean phenotype value	Deviation from average of the homozygous genotypes	Deviation from population mean (allele frequencies p and q)
$c^r c^r$ (AA)	68.87	$a = 68.87 - 61.60 = 7.27$	$\begin{aligned} G_1 &= a - \mu \\ &= -2q[a+(q-p)d] - 2q^2d \\ &= 2qa - 2q^2d \end{aligned}$
$c^r c^d$ (AA')	60.67	$d = 60.67 - 61.60 = -0.93$	$\begin{aligned} G_2 &= d - \mu \\ &= (q-p)[a+(q-p)d] + 2pqd \\ &= (q-p)a + 2pqd \end{aligned}$
$c^d c^d$ (A'A')	54.33	$-a = 54.33 - 61.60 = -7.27$	$\begin{aligned} G_3 &= -a - \mu \\ &= -2p[a+(q-p)d] - 2p^2d \\ &= -2pa - 2p^2d \end{aligned}$

$$\text{Population mean } \mu = p^2 a + 2pqd + q^2(-a)$$
$$= (p-q)a + 2pqd$$

Source: data from Wright (1968).

the **breeding values** of the AA, AA', and $A'A'$ genotypes, and $-2q^2d$, $2pqd$, and $-2p^2d$ are the **dominance deviations**.

Components of Genotypic Variation

The genotypic variance σ_g^2 due to the locus in Table 8.2 is given by $\sigma_g^2 = p^2 G_1 + 2pq G_2 + q^2 G_3$, which equals:

$$\sigma_g^2 = p^2 \left[2q\alpha - 2q^2 d\right]^2 + 2pq[(q-p)\alpha + 2pqd]^2 + q^2 \left[-2p\alpha - 2p^2 d\right]^2 \tag{8.7}$$

Here again the algebra is rather tedious, but it helps to consider the coefficients of the α^2, αd, and d^2 terms separately. These are:

$$\left[4p^2 q^2 + 2pq(q-p)^2 + 4p^2 q^2\right]\alpha^2 = 2pq\alpha^2$$
$$\left[-8q^2 q^3 + 8p^2 q^2 (q-p) + 8p^3 q^2\right]\alpha d = 0$$
$$\left[4p^2 q^4 + 8p^3 q^3 + 4p^4 q^2\right]d^2 = (2pqd)^2$$

Hence, we can also write Equation 8.7 as:

$$\sigma_g^2 = 2pq\alpha^2 + (2pqd)^2 \tag{8.8}$$

The first term in Equation 8.8 is called the **additive genetic variance**, symbolized σ_a^2:

$$\sigma_a^2 = 2pq\alpha^2 = 2pq[a+(q-p)d]^2 \tag{8.9}$$

The second term in Equation 8.8 is called the **dominance variance**, symbolized σ_d^2:

$$\sigma_d^2 = (2pqd)^2 \tag{8.10}$$

It is not difficult to show that the additive genetic variance $2pq\alpha^2$ equals the variance in the breeding values of the genotypes, and the dominance variance $(2pqd)^2$ equals the variance of the dominance deviations.

Whereas the additive variance depends both on a, which equals half the difference in phenotype between the homozygous genotypes, and d, which is the dominance effect, the dominance variance depends only on d. Both variance components also depend on allele frequencies, as shown in Figure 8.6. The case of overdominance (Figure 8.6d) shows that the additive genetic variance can equal 0 (in this example at $p = q = 1/2$), even though the genotypic variance is nonzero. It is a worthwhile exercise to show that the additive genetic variance equals $8pq^3a^2$ for A dominant and $8p^3qa^2$ for A recessive.

When multiple loci are considered, $2pq\alpha^2$ in Equation 8.9 is replaced with the sum of such terms, one for each locus, and likewise $(2pqd)^2$ in Equation 8.10 is replaced with a sum of similar terms. Each locus may have different values of p, q, a, and d. There is also a term corresponding to nonadditive interactions between the genotypes at different loci, which is called the **epistatic variance** or **interaction variance**, symbolized σ_i^2. Still more general models allow for a component of variation due to **assortative mating**, in which mating pairs have a positive correlation in phenotype, as occurs in human populations for height. With all these complications taken into account, the genotypic variance can be written as:

$$\sigma_g^2 = \sigma_a^2 + \sigma_d^2 + \sigma_i^2 + \sigma_{am}^2 \tag{8.11}$$

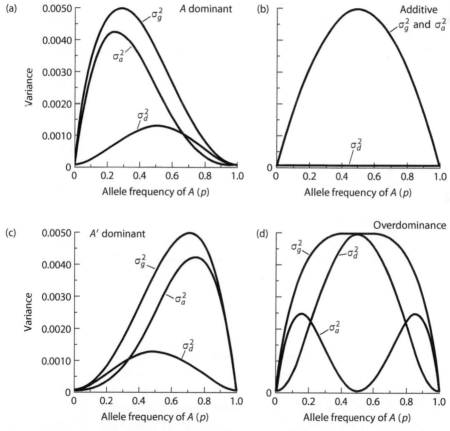

Figure 8.6 Genotypic variance σ_g^2, additive genetic variance σ_a^2, and dominance variance σ_d^2 for a gene with two alleles A and A' at allele frequencies p and q in a randomly mating population. The mean phenotypes of AA, AA', and $A'A'$ are a, d, and $-a$, respectively, with values chosen to maximize σ_g^2 at 0.005. (a) $a = d = 0.07071$. (b) $a = 0.1$, $d = 0$. (c) $-a = d = 0.07071$. (d) $a = 0$, $d = 0.14142$.

where σ^2_{am} is the genotypic variance due to assortative mating. Because of assortative mating for genetic or environmental factors that affect longevity, the σ^2_{am} term contributes about half of the genotypic variance in human lifespan (Ruby et al. 2018).

In principle, one could also partition the environmental variance. For example, with respect to mouth and throat cancer, there would be an environmental variance component due to cigarette smoking (σ^2_s), which by itself increases the risk by a factor of about 2; one due to alcohol consumption (σ^2_d), which by itself increases the risk by a factor of about 1.5; and one due to the synergistic effects of smoking and drinking ($\sigma^2_{s\times d}$), which increases the risk by a factor of about 30. For these two environmental factors we could then write:

$$\sigma^2_e = \sigma^2_s + \sigma^2_d + \sigma^2_{s\times d} \tag{8.12}$$

However, the analysis of environmental causes of variation in complex traits is more difficult than that of genetic causes. While the elementary genetic factors are genes, it is often very difficult to identify critical environmental factors.

Physiological Epistasis Versus Statistical Epistasis

At this point we need to take a closer look at the epistatic variance σ^2_i in Equation 8.11 because it is a continuing source of confusion between how genes interact in physiological networks versus what these physiological interactions contribute to the genotypic variance of a trait in a population (Cheverud and Routman 1995; Sackton and Hartl 2016). The direct or indirect interaction between genes or gene products in cellular or organismal physiology is known as **physiological epistasis** (Cheverud and Routman 1995). Consider the simplest case of two genes, each with a dominant and recessive allele. Denote the phenotypes of the genotypes as $P_{[A-B-]}$, $P_{[A-bb]}$, $P_{[aaB-]}$, and $P_{[aabb]}$, where the subscripts are the genotypes and the dash is a wild card indicating either allele at that position. The baseline phenotype can be taken as $P_{[aabb]}$, and relative to the baseline the effect of the $A-$ genotype can be measured as $P_{[A-bb]} - P_{[aabb]}$ and that of the $B-$ genotype as $P_{[aaB-]} - P_{[aabb]}$. The predicted phenotype of $A- B-$ can be written in terms of the $A-$ and $B-$ effects as:

$$P_{[A-B-]} = P_{[aabb]} + \left(P_{[A-bb]} - P_{[aabb]}\right) + \left(P_{[aaB-]} - P_{[aabb]}\right) + \varepsilon \tag{8.13}$$

where ε is a deviation from additivity that serves as measure of physiological epistasis. Suppose, for example, that the baseline phenotype equals -1, that each of $A-$ and $B-$ have an effect of $+1$, and that $\varepsilon = 0$. Then according to Equation 8.13, the predicted phenotype of $A- B-$ would be $-1 + [1 - (-1)] + [1 - (-1)] = 3$. When $\varepsilon \neq 0$, its value can be found by solving Equation 8.13 for ε, which yields:

$$\varepsilon = P_{[A-B-]} - P_{[A-bb]} - P_{[aaB-]} + P_{[aabb]} \tag{8.14}$$

Consider the example of **complementary epistasis**, in which a particular phenotype results from homozygosity of a recessive allele at either or both of two loci. In this case we can write the phenotypes of $A- B-$, $A- bb$, $aa B-$, and $aa bb$ as 1, -1, -1, and -1, respectively. The baseline equals -1 and the $A-$ and $B-$ genotypes each have an effect of $-1 - (-1) = 0$. The 0 effect makes sense because with complementary epistasis $A-$ and $B-$ have an effect only when they are present together. According to Equation 8.14, the degree of epistasis is large: $\varepsilon = 1 - (-1) - (-1) + (-1) = 2$, which equals the range of the phenotypic values.

Let's now consider the same example of complementary epistasis in regard to its contribution to the components of genotypic variation. The variance component σ_i^2 due to epistasis is called **statistical epistasis** (Cheverud and Routman 1995; Sackton and Hartl 2016). Paradoxically, the magnitude of statistical epistasis does not closely track the magnitude of physiological epistasis. Calculation of the variance components is tedious (Crow and Kimura 1970), but in a randomly mating population in which the allele frequencies of A and B are both 0.5, the variance components with complementary epistasis are $\sigma_a^2 = 0.57$, $\sigma_d^2 = 0.29$, and $\sigma_i^2 = 0.14$, where these have been normalized to sum to 1. In other words, even though all of the phenotypic differences in this example are due to epistasis, the epistatic variance accounts for only 14 per cent of the genotypic variance.

How is it that physiological epistasis can account for all phenotypic differences yet statistical epistasis accounts for only 14 per cent of the genotypic variation? The reason is that, in calculating the genotypic variance, the variance components are fitted hierarchically, one at a time. The procedure is shown graphically in Figure 8.7 for the example of complementary epistasis with equal allele frequencies, where the actual phenotype of each genotype is depicted as a large purple sphere. The first-level fit is with a purely additive model, which is indicated by the tilted plane. The phenotypes predicted by the additive model are shown as blue dots, and it is the variance among these points that constitutes the additive genetic variance. The next-level fit is a model with dominance, and the phenotypes predicted from this model are shown as orange points. The dominance deviations are indicated by the dashed orange lines, and it is the variance among these deviations that accounts for the dominance variance. Finally, the genotypic variance due to epistasis is fit as deviations from the dominance model, and these deviations are shown

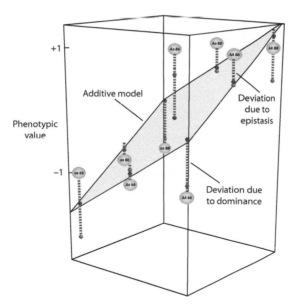

Figure 8.7 Hierarchical fitting of variance components obscures physiological epistasis in population studies. The plane is the least-squares fit to an additive model when there is actually complementary epistasis and the allele frequencies of A and B are 0.5. Phenotypic values are represented as purple spheres. The blue dots are the predicted phenotypes based on a purely additive model, and the orange dots are those predicted with dominance. The orange and purple dashed lines indicate deviations from the additive model due to dominance and epistasis, respectively.

by the dashed purple lines. It is evident from Figure 8.7 that there is little epistasis left to fit once the additive and dominance components have been extracted from the data, and this is why one can have a great deal of physiological epistasis but very little statistical epistasis (Hill et al. 2008; Sackton and Hartl 2016).

The magnitude of statistical epistasis also depends on the allele frequencies. With complementary epistasis, for example, the epistatic variance accounts for only 2 per cent of the genotypic variance when the allele frequencies of A and B are both 0.8, but it accounts for 47 per cent of the genotypic variance when the allele frequencies are both 0.2. Empirical estimates from hundreds of studies in humans, as well as theoretical studies of the types of epistasis and the allele frequencies likely to occur in natural populations, indicate that the additive genetic variance typically accounts for over half and often a much greater fraction of the total genotypic variation (Hill et al. 2008).

Like the overall genotypic variance, the epistatic variance can be resolved into components. For two genes, the components of the epistatic variance can be partitioned into a variance that depends on the value of a for each gene (recall that a equals half the difference in phenotype between the homozygous genotypes), on the value of a for one gene and d for the other, and on the value of d for both genes. These are denoted respectively as $\sigma^2_{a \times a}$ (the additive-by-additive variance), $\sigma^2_{a \times d}$ (additive-by-dominant), $\sigma^2_{d \times a}$ (dominant-by-additive), and $\sigma^2_{d \times d}$ (dominant-by-dominant) (Bulmer 1985; Lynch and Walsh 1998). Again, because of the hierarchical model fitting, the largest component of the epistatic variance is usually the $a \times a$ component. In the case of complementary epistasis when the allele frequencies of A and B are both 0.2, for example, $\sigma^2_{a \times a}$ accounts for 79 per cent of the epistatic variance, $\sigma^2_{a \times d}$ and $\sigma^2_{d \times a}$ each account for 10 per cent, and $\sigma^2_{d \times d}$ for only 1 per cent. Importantly, insofar as the transmission of complex traits from one generation to the next is concerned, the component of the epistatic variance that depends only on the additive-by-additive values plays the same role as the additive genetic variance itself.

8.3 Artificial Selection

Selection cannot change the genotype of a population in which every individual has an identical genotype, such as a highly inbred line. Selection is ineffective in such cases because the only causes of variation are environmental, which are not transmitted from generation to generation. Genetic changes can sometimes occur slowly in traits affected by many genes in populations that are large enough, because then selection can act on the genetic variation contributed by new mutations. But for bristle number in inbred strains of *Drosophila*, new mutations arising in each generation account for only 0.1–1 per cent of the variance in bristle number (Mackay et al. 1994).

Although genetic variation is essential for progress under selection, it is not sufficient. The reason is really quite subtle. Only the additive genetic variance σ^2_a contributes to the response to selection. The genotypic variance σ^2_g is not the key quantity. We will examine this principle by examining **artificial selection**, which refers to the deliberate choice of a select subset of individuals to be used for breeding. The most common type of artificial selection is **directional selection**, in which phenotypically superior individuals are chosen. Artificial selection has been practiced empirically for thousands of years, for example, in the body size of domesticated dogs. But understanding the genetic principles permits the breeder to predict the speed and degree to which a population can be changed through artificial selection in any small number of generations.

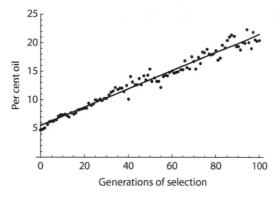

Figure 8.8 Long-term artificial selection for oil content in corn. (Data from https://www.ideals. illinois.edu/handle/2142/3525.)

Figure 8.8 shows the results of 100 generations of artificial selection for increased oil content in corn kernels in an experiment that began in 1896 (Dudley and Lambert 2004; Hill 2005). In each generation, about 100 ears of corn were analyzed for oil content, and the 15 or so with the greatest percentage of oil were selected to produce the next generation. The large increase in oil content illustrates the important point that, after many generations of directional selection, the mean phenotypic value of the selected population can greatly exceed the maximum phenotypic value among individuals in the original population. In the original population, the oil content averaged 4.6 per cent with a standard deviation of 0.42 per cent. By generation 100 the average oil content was 20.4 per cent—an increase of 37 standard deviations over the original mean. In a parallel experiment selecting for increased protein, 100 generations of selection yielded a strain with an increase of 17 standard deviations over the original mean (Dudley and Lambert 2004).

Such a sustained response to long-term selection is not usually observed. The gain in each generation typically slows and eventual reaches a **selection limit** in which the population no longer responds to selection. In some cases a selection limit occurs because all of the polymorphisms affecting the trait in the original population have become fixed. More commonly a selection limit is reached because individuals with more extreme phenotypes have reduced fitness so that artificial selection in one direction is balanced by natural selection acting in the opposite direction. In a typical long-term selection experiment, a total selection response of 3–5 times the original phenotypic standard deviation is not unusual, and in an ideal population of size N it requires about $N/2$ generations for a selected trait to evolve halfway to its selection limit (Falconer 1977).

The sustained response to selection for increased oil and protein content in corn implies first that natural selection is either indifferent to these phenotypes or that the artificial selection remained strong enough to overcome its counteracting effects. The sustained response also implies that the number of genes affecting these traits must be relatively large and that the continued response is due to the contribution of new mutations. The situation in regard to oil content has been analyzed rather carefully (Laurie et al. 2004). At generation 70, individuals from the line selected for high oil content were crossed with those from a line that had been selected for low oil content. The offspring were randomly mated for ten generations to reduce linkage disequilibrium, and then 500 plants were selected for continued self-pollination to render them homozygous. Among these lines,

440 single-nucleotide polymorphisms (SNPs) were studied to detect QTLs affecting oil content.

About 50 QTLs affecting oil content were identified (Laurie et al. 2004). The QTL with the largest effect accounted for only 2 per cent of the difference in phenotype between the high-oil and low-oil lines at generation 70. The average effect was about 0.1 per cent of the difference. This estimate suggests that the total number of gene affecting oil content in these strains must be 100 or more and that the majority went unidentified because their effect size was too small to be statistically significant (Hill 2005; Laurie et al. 2004).

Prediction Equation for Individual Selection

When each individual is selected for breeding based solely on its own phenotypic value, the type of artificial selection is called **individual selection**. This is by no means the only type of artificial selection (Falconer 1960; Falconer and Mackay 1996; Hartl 1980; Walsh and Lynch 2018). For example, instead of selecting on an individual's own phenotype, one could practice **family selection**, in which whole sibships are selected or rejected based on the family mean. Or one could practice **within-family selection**, in which the selected individuals are those that deviate the most from their family mean. Still another mode of artificial selection is **sib selection**, in which individuals are selected according to the phenotype of their siblings. These alternative modes of selection are designed to circumvent or minimize one obstacle or another. For example, family selection is useful when the heritability is low but the environmental effects are independent from one individual to the next, within-family selection is helpful when the heritability is low and the environmental effects are common to members of a family, and sib selection is necessary for traits such as egg or milk production that are expressed in only one sex. We focus on individual selection because it is most similar to natural selection in that each individual contributes to future generations according to its own fitness.

Figure 8.9 illustrates a type of individual selection called **truncation selection**. The trait is seed weight in edible beans, and the experiment is one of the first of its kind (Johannsen 1903). The histogram in Figure 8.9a represents the distribution of seed weight in the parental population. The shaded part of the distribution to the right of the phenotypic value denoted T (650 mg) indicates those seeds selected to germinate and grow for breeding among themselves. The value T is called the **truncation point**. The mean phenotype in the entire population is denoted μ (403.5 mg), and that of the selected parents is denoted μ_S (691.7 mg). When the selected parents were mated at random, their offspring seeds had the phenotypic distribution shown in Figure 8.9b, where the mean phenotype is denoted μ' (609.1 mg). It is typical of truncation selection that the offspring mean μ' is greater than μ but less than μ_S. The reason μ' is greater than μ is that some of the selected parents have favorable genotypes and so transmit favorable genes to their offspring. At the same time, μ' is generally less than μ_S for two reasons:

- Some of the selected parents do not have favorable genotypes. Their exceptional phenotypes result from chance exposure to exceptionally favorable environments.
- Alleles, not genotypes, are transmitted from parents to offspring, and the favorable genotypes in the parents are disrupted by Mendelian segregation and recombination.

The difference in mean phenotype between the selected parents and the entire parental population is called the **selection differential** and designated S:

$$S = \mu_s - \mu \tag{8.15}$$

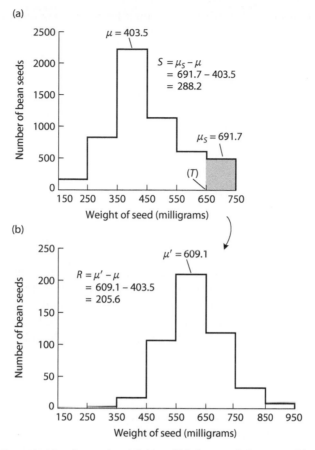

Figure 8.9 Truncation selection for seed weight in edible beans of the genus *Phaseolus*. The truncation point is $T = 650$ mg. (Data from Johannsen (1903).)

The difference in mean phenotype between the progeny generation and the previous generation is called the **response to selection** and designated R:

$$R = \mu' - \mu \tag{8.16}$$

For the data in Figure 8.8, $S = 288.2$ mg and $R = 205.6$ mg.

Any equation that defines the relationship between the selection differential S and the response to selection R is known as a **prediction equation**. Each type of artificial selection has its own prediction equation. For individual truncation selection, the prediction equation is:

$$R = h^2 S \tag{8.17}$$

where h^2 is called the **narrow-sense heritability** of the trait. As with the broad-sense heritability in Equation 8.5, the narrow-sense heritability is a ratio of variances, but not the ratio of genotypic variance to phenotypic variance. Rather, as we shall see, the narrow-sense heritability is the ratio of the additive genetic variance to the phenotypic variance:

$$h^2 = \frac{\sigma_a^2}{\sigma_p^2} \tag{8.18}$$

When statistical epistasis is taken into account, the numerator of Equation 8.18 also includes terms like $\sigma^2_{a \times a}$, $\sigma^2_{a \times a \times a}$, and so forth, but as noted these are usually small relative to σ^2_a.

Before going into the genetic details justifying Equation 8.18, for the moment let's interpret the narrow-sense heritability merely as a description of an observed result. In Figure 8.8, for example, $S = 288.2$ mg and $R = 205.6$; hence, $h^2 = R/S = 205.6$ mg/288.2 mg = 71.3 per cent. When estimated like this from an observed result, h^2 is called the **realized heritability**.

Intensity of Selection

Some special properties of the normal distribution lead to different expressions for $\mu_S - \mu$ and therefore different forms of Equation 8.17. To obtain these alternatives we start with Figure 8.10, which shows a normal distribution of a complex trait in a hypothetical randomly mating population. The mean is denoted μ and the phenotypic variance σ^2_p. In truncation selection, all individuals with phenotypes above the truncation point T are saved for breeding, and the shaded area B of the distribution represents the proportion of the population selected. Values of B are shown in Figure 8.3 for truncation points located various numbers of standard deviations greater than or less than the mean. In Figure 8.10, the height of the normal distribution at the point T is denoted Z, and so $Z = f(T)$ in Equation 8.2. The mean phenotype among the selected individuals is μ_S.

One of the properties of the normal distribution is that:

$$\frac{\mu_S - \mu}{\sigma^2_p} = \frac{Z}{B}$$

(8.19)

hence the selection differential expressed as a deviation from the mean in units of the standard deviation can be written as:

$$\frac{S}{\sigma_p} = \frac{\sigma_p Z}{B} = i$$

(8.20)

where i is known as the **intensity of selection**. The intensity of selection is an informative quantity because it allows comparison of different modes of selection and

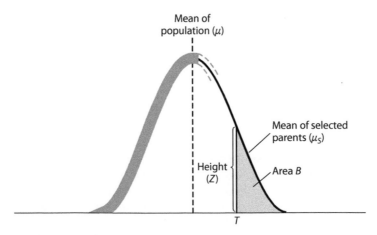

Figure 8.10 Normal distribution of a quantitative trait in a population subjected to truncation selection with truncation point T.

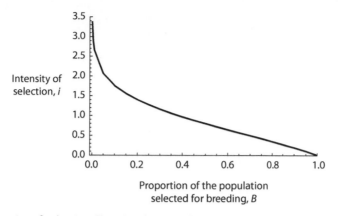

Figure 8.11 Intensity of selection (i) as it relates to the proportion of the population selected for breeding (B).

different choices of the truncation point. For example, suppose a breeder wishes to convert from a breeding program in which $B = 0.10$ to one in which $B = 0.01$. How much greater selection would this change bring about? The answer is shown in Figure 8.11, which depicts the intensity of selection i against the proportion of the population chosen for breeding B. A change from $B = 0.10$ to $B = 0.01$ reduces the proportion of selected individuals by a factor of 10, yet the intensity of selection changes only from 1.76 to 2.66, a factor of 1.5. Considering how much larger a population would be required to sustain selection at $B = 0.01$, the modest increase in selection intensity makes the change hardly worth it. Figure 8.11 also indicates that, in natural populations in which most individuals can breed or are only mildly selected against, the intensity of selection is on the order of 1 or less.

Expressing Equation 8.17 $(R = h^2 S)$ in terms of the intensity of selection in Equation 8.20 leads to an alternative expression for the response to selection:

$$R = i h^2 \sigma_p \tag{8.21}$$

Genetic Basis of the Prediction Equation

To identify the genetic underpinnings of the prediction equation $R = h^2 S$ (Equation 8.17), we must consider how alternative alleles of a gene affect a complex trait, how truncation selection changes the allele frequencies, and how much any change in allele frequency changes the mean phenotypic value.

First we need an equation for the mean phenotypic value of a complex trait in a randomly mating population. For this purpose, imagine a gene that affects the trait with alleles A and A' at respective allele frequencies p and q. Because of random mating, genotypes AA, AA', and $A'A'$ are present in the population with frequencies p^2, $2pq$, and q^2, respectively, but the individual genotypes cannot be identified through their phenotypic values because of the variation in phenotype caused by environmental factors and genetic differences at other loci. If the genotypes could be identified, their distributions of phenotypic value would have slightly different means, as shown in Figure 8.12.

As in Table 8.2, the mean phenotypic values of AA, AA', and $A'A'$ genotypes are denoted a, d, and $-a$, respectively. The mean phenotype of a population in Hardy–Weinberg equilibrium for A and A' is therefore given by:

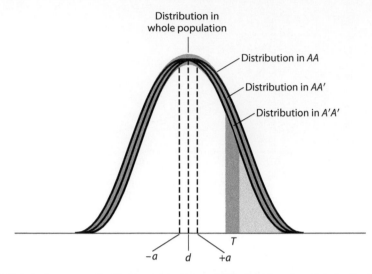

Figure 8.12 This is the same distribution as in Figure 8.10, showing the slightly displaced distributions of phenotypic value among the three genotypes *AA*, *AA'*, and *A'A'* for a gene that contributes to the quantitative trait. The means of the distributions are *a*, *d*, and −*a*, respectively.

$$\mu = p^2 a + 2pqd + q^2(-a) = (p-q)a + 2pqd \tag{8.22}$$

Next we need to calculate the change in allele frequency that takes place in one generation of selection, as that will allow us to calculate the change in the phenotypic mean. Suppose for the moment that we were practicing artificial selection for increased amount of black coat coloration in the guinea pigs in Table 8.2. Selection for black coat coloration in a population containing both the c^r (*A*) and c^d (*A'*) alleles would be successful in increasing the allele frequency of *A*. The magnitude of change in allele frequency with natural selection was derived in Chapter 5 (Equation 5.7), which remains valid for artificial selection if we agree to interpret the "fitness" of an individual as the probability that the individual is included among the group selected as parents of the next generation. With this interpretation of fitness, differences in fitness (reproductive success) of the *AA*, *AA'*, and *A'A'* genotypes correspond to the differences in area to the right of the truncation point in Figure 8.12, because only those individuals in the shaded area are selected to reproduce.

The differences in area are easy to calculate if you shift or slide each curve horizontally until the means coincide. Shift the *A'A'* curve *a* units to the right, and shift the *AA'* and *AA* curves *d* and *a* units to the left. This brings the distributions into coincidence, but it slides the truncation points slightly out of register, as shown in Figure 8.13. The difference in "fitness" between *AA* and *AA'*, denoted $w_{AA} - w_{AA'}$, is equal to the small area indicated in Figure 8.13, as is the difference in fitness between *AA'* and *A'A'*, denoted $w_{AA'} - w_{A'A'}$. The areas corresponding to $w_{AA} - w_{AA'}$ and $w_{AA'} - w_{A'A'}$ are approximately rectangles, and the area of a rectangle is the product of the base and the height.

Therefore, since *Z* represents the height of the normal distribution at the point *T*, we can make the following approximations:

$$w_{AA} - w_{AA'} = Z[(T+a) - (T+d)] = Z(a-d)$$
$$w_{AA'} - w_{A'A'} = Z[(T+d) - (T-a)] = Z(a+d) \tag{8.23}$$

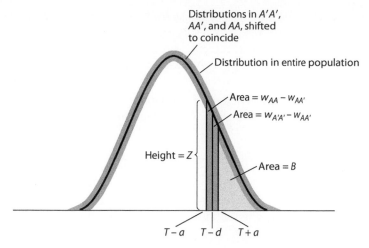

Figure 8.13 This is the same distribution as in Figure 8.12 but with the distributions of *AA*, *AA′*, and *A′A′* shifted to coincide. Shifting the distributions slides the truncation points slightly out of register, so the truncation points become *T−a*, *T−d*, and *T + a*, respectively.

The average fitness $\bar{w}$ of the entire population simply equals B, because B is the proportion of the population selected for breeding. From Equation 5.7 we know that the change in frequency of the allele A in one generation of selection equals:

$$\Delta p = pq \frac{[p(w_{AA} - w_{AA'}) + q(w_{AA'} - w_{A'A'})]}{\bar{w}} \qquad (8.24)$$

Substituting from Equation 8.23 and using $\bar{w} = B$ leads to:

$$\Delta p = \frac{pq[pZ(a-d) + qZ(a+d)]}{B} \qquad (8.25)$$

or, since $p + q = 1$:

$$\Delta p = \frac{Z}{B}pq[a + (q-p)d] \qquad (8.26)$$

An equation corresponding to 8.26 could be obtained for any gene affecting the trait, but the values of p, q, a, and d would differ for each gene.

Change in Mean Phenotype from One Generation of Selection

Equation 8.26 provides an expression for Δp that can be used to calculate the mean phenotypic value in the next generation. Among the progeny of selected parents the allele frequencies of A and A' are $p + \Delta p$ and $q - \Delta p$, respectively. With random mating, the mean phenotype in the next generation is given by Equation 8.22 as:

$$\mu' = (p + \Delta p)^2 a + 2(p + \Delta p)(q - \Delta p)d + (q - \Delta p)^2(-a)$$

When the right-hand side of this expression is multiplied out and terms in $(\Delta p)^2$ are ignored because Δp is usually small, then μ' is found to be approximately:

$$\mu' = \mu + 2[a + (q-p)d]\Delta p \qquad (8.27)$$

Now move μ to the left-hand side, substitute Δp from Equation 8.26, and replace Z/B with the left-hand side of Equation 8.19. The result is:

$$\mu' - \mu = (\mu_s - \mu)\frac{2pq[a + (q-p)d]^2}{\sigma_p^2} \tag{8.28}$$

But $\mu_s - \mu = S$ is the selection differential (Equation 8.15), $\mu' - \mu = R$ is the response to selection (Equation 8.16), and $h^2 = R/S$ (Equation 8.17). Hence, Equation 8.28 implies that:

$$h^2 = \frac{2pq[a + (q-p)d]^2}{\sigma_p^2} = \frac{2pq\alpha^2}{\sigma_p^2} = \frac{\sigma_a^2}{\sigma_p^2} \tag{8.29}$$

This is the equation that we wanted, because it defines the narrow-sense heritability in terms of its variance components, and it justifies Equation 8.18. We have thereby shown that:

- The narrow-sense heritability h^2 equals the ratio of the additive genetic variance to the phenotypic variance (Equation 8.29).
- The narrow-sense heritability h^2 is the heritability used in the prediction equation for individual selection (Equations 8.17 and 8.21).

Application of Equation 8.29 can be illustrated using the genetic difference for coat color in Table 8.2, where $a = 7.27$ and $d = -0.93$. Consider a randomly mating population in which $p = q = 1/2$. With equal allele frequencies the term in d in Equation 8.29 disappears, and then $\sigma_p^2 = 2(1/2)(1/2)a^2 = 26.43$. This is close to its maximum possible value for $a = 7.27$ and $d = -0.93$. The maximum occurs when $p = 0.562$, in which case $\sigma_a^2 = 26.852$. The narrow-sense heritability due to this locus depends on the phenotypic variance σ_p^2, but h^2 is minimized at $p = 0$ or $p = 1$ (for which $h^2 = 0$) and maximized at $p = 0.562$.

Equation 8.29 is valid only when a single gene affects the trait, but for multiple genes the right-hand side is replaced by a summation of such terms, one for each gene. The equivalent equation for multiple genes is therefore:

$$h^2 = \frac{\sum_i 2p_iq_i[a_i + (q_i - p_i)d_i]^2}{\sigma_p^2} = \frac{\sum_i 2p_iq_i\alpha_i^2}{\sigma_p^2} = \frac{\sigma_a^2}{\sigma_p^2} \tag{8.30}$$

where p_i, q_i, a_i, and d_i are the values of p, q, a, and d for the ith gene, and the summation is over all genes affecting the complex trait. The summation in the numerator is the additive genetic variance of the trait due to all loci. When statistical epistasis is significant, the numerator also includes all of the additive-by-additive terms of the epistatic variance.

Effect of Selection on a Constituent Locus of a Complex Trait

How large is the conventional selection coefficient affecting any one gene underlying a complex trait? This is a question that we've already answered but not explicitly. Equation 8.26 asserts that $\Delta p = (Z/B)pq[a + (q-p)d]$. Substituting $i/\sigma_p = Z/B$ from Equation 8.20 yields:

$$\Delta p = \frac{ipq[a + (q-p)d]}{\sigma_p} \tag{8.31}$$

An alternative expression for Δp in terms of the conventional relative fitnesses w_{AA}, $w_{AA'}$ and $w_{A'A'}$ is given by Equation 8.24. If the alleles of the gene are additive and A is favored, then we can write $w_{AA} = 1$, $w_{AA'} = 1 - s/2$, and $w_{A'A'} = 1 - s$ where $s > 0$. If we also assume that s is small enough that $\overline{w} = 1$, then Equation 8.24 implies that:

$$\Delta p = \frac{pqs}{2} \tag{8.32}$$

Setting the right-hand sides of Equations 8.31 and 8.32 equal to each other with $d = 0$ because the alleles are additive yields:

$$s = \frac{2ia}{\sigma_p} \tag{8.33}$$

In other words, the selection coefficient for a favorable allele that influences a complex trait increases with the intensity of selection and the effect of the allele relative to the magnitude of the phenotypic standard deviation. Alleles make a small contribution to the phenotypic standard deviation either when their effect size is small or when they are rare, and in either case selection for (or against) the allele is weak. The relation in Equation 8.33 also holds when the favored allele is dominant and when the favored allele is recessive (Milkman 1978).

Genomic Selection

In traditional plant and animal breeding, the assessment of an individual's breeding value is based on the individual's own phenotype. In much of agriculture, the time and cost of phenotyping sets a limit on how many individuals can be evaluated, not only in the obvious cases of livestock breeds or forest trees but also in many other agricultural plants and animals (Rincent et al. 2018).

One possible solution is to identify the genes underlying a desirable phenotype and to use molecular methods to identify the genotypes associated with superior performance. In most cases, however, many genes affect the traits of interest, and the proportion of the genotypic variance explained by any one gene is small and its effect difficult to estimate accurately (Goddard 2009).

Another approach is to identify many thousands of SNPs or other genetic markers throughout the genome. In this case, all of the genotypic variance can be explained by the SNPs without knowing the causal variants, and regression or artificial intelligence can be used to create a SNP-based model that best explains the breeding values of organisms that have already been phenotyped. The model can then be used to predict the breeding value of individuals that have not been phenotyped, and individuals with the highest predicted breeding values can be selected for breeding (Meuwissen et al. 2001). In this way, more individuals can be evaluated quickly and cheaply, and with a suitable SNP density, breeding value can be predicted with an accuracy of up to 85 per cent (Meuwissen et al. 2001).

Selection of organisms based on breeding values predicted from molecular data is known as **genomic selection**, and it has become extremely popular since its inception (Meuwissen et al. 2001, 2013). The accuracy of prediction decreases with the number of selection cycles because the allele frequencies in the population change due to selection, hence the model must be modified and updated from time to time. The advantages of genomic selection are enhanced when prediction is based on whole-genome sequences, because in this case the causal mutations are actually included in the data.

Correlated Response to Selection

Selection that changes the phenotypic value of one trait often results in a change in the phenotypic value of another trait, which is known as a **correlated response** to selection. Correlated responses result from two conditions:

- **Linkage disequilibrium**, discussed in Chapter 2, when the alleles affecting the selected trait are nonrandomly associated along the chromosome with alleles affecting the nonselected trait.
- **Pleiotropy**, when alleles affecting the magnitude of the selected trait are components of a genetic network that also affects the magnitude of the nonselected trait.

Whereas the extent of linkage disequilibrium depends on linkage relationships, allele frequencies, population structure, and population history, the extent of pleiotropy is intrinsic to the development and physiology of the traits themselves. And whereas linkage disequilibrium can be broken up by recombination, pleiotropy persists.

The magnitude of any correlated response due to pleiotropy depends on the genetic correlation between the traits (Conner and Hartl 2004; Falconer and Mackay 1996; Lynch and Walsh 1998). Just as the variance of one trait can be expressed in terms of the additive variance plus other terms, the covariance between two traits can be resolved into an additive covariance and other terms. Let's call the two traits T1 and T2, and express the genetic effects of the AA, AA', and $A'A'$ genotypes in terms of a and d values corresponding to each trait as in Table 8.2. Then a_{T1} and a_{T2} are the a values of the two traits, d_{T1} and d_{T2} are the dominance deviations, and α_{T1} and α_{T2} are the average effects as defined in Equation 8.6. The genetic covariance between T1 and T2 is given by the average of the product of the deviations of each value of T1 and T2 from its respective mean, and arguments analogous to those leading to Equation 8.9 indicate that the additive genetic covariance between T1 and T2 is given by:

$$\mathrm{cov}_a\,(\mathrm{T1},\mathrm{T2}) = 2pq\alpha_{T1}\alpha_{T2} = 2pq\left[a_{T1} + (q-p)\,d_{T1}\right]\left[a_{T2} + (q-p)\,d_{T2}\right] \tag{8.34}$$

The genetic correlation between T1 and T2 is therefore:

$$r_a = \frac{\mathrm{cov}_a\,(\mathrm{T1},\mathrm{T2})}{\sigma_{aT1}\sigma_{aT2}} \tag{8.35}$$

where σ_{aT1} and σ_{aT2} are the additive standard deviations of the two traits, in each case equal to the square root of the additive genetic variance (Equation 8.9).

When individual selection is practiced in trait T1, the expected correlated response in trait T2 (CR_{T2}) is given by:

$$CR_{T2} = ih_{T1}h_{T2}r_a\sigma_{pT2} \tag{8.36}$$

Equation 8.36 is the correlated-response analog of Equation 8.21 for the direct response. In Equation 8.36, h_{T1} and h_{T2} are the square roots of the narrow-sense heritabilities of T1 and T2, and σ_{pT2} is the phenotypic standard deviation of the trait T2. Equation 8.36 can be derived in a similar fashion as Equation 8.21, starting with the change in allele frequency expected from direct selection (Equation 8.26). The product $h_{T1}h_{T2}r_a$ is called the **coheritability** because it plays the same role in Equation 8.36 as h^2 plays in Equation 8.21.

The genetic correlation can be estimated in several ways (Conner and Hartl 2004; Falconer and Mackay 1996; Lynch and Walsh 1998), but the simplest conceptually is a double-selection experiment in which selection is carried out on each trait separately, with the same intensity of selection, and the direct and correlated responses are recorded. The genetic correlation is then estimated as:

$$r_a^2 = \frac{CR_{T2}}{R_{T1}} \frac{CR_{T1}}{R_{T2}}$$

(8.37)

where R indicates the direct response of each trait. Equation 8.37 follows immediately from Equations 8.36 and 8.21 (Falconer and Mackay 1996; Walsh and Lynch 2018). An important caveat is that, whatever the method, estimates of the genetic correlation tend to have large sampling variances.

Genetic correlations are important in at least three contexts:

- **Indirect selection**, in which desired changes in one trait are obtained by means of selection on another trait. This approach is useful when the trait of interest has a low heritability, or is difficult to measure, and a second trait has a positive genetic correlation and a high heritability. In such cases greater progress in improving the trait of interest may come about by selecting on the correlated trait.
- **GEI** occurs when the phenotypic value of a genotype depends on the environment. For two environments, if the genetic correlation between the environments is known, then the mean phenotype of a population in one environment may be predicted as a correlated response to selection in the other environment according to Equation 8.36.
- **Selection limits** often occur when a selected trait has a negative genetic correlation with a component of fitness, because in this case the gain expected from artificial selection is offset by natural selection owing to the correlated response that reduces fitness. In mice, for example, there is a negative genetic correlation between body weight and litter size, and selection for increased body weight eventually plateaus because, among the selected parents, the largest females have the smallest litters.

8.4 Resemblance Between Relatives

Estimation of the additive genetic variance might at first seem to be very difficult, but in fact it is quite straightforward. The reason is that the theoretical covariance between certain types of relatives is a simple multiple of σ_a^2. The relation between a parent and its offspring affords a good example.

Parent–Offspring Covariance

The parent–offspring covariance is often used to estimate the narrow-sense heritability because the covariance equals the additive genetic variance, which when divided by the phenotypic variance yields h^2. The covariance can be calculated either from a single parent (usually chosen to be the father, to avoid possible nongenetic maternal effects on the offspring) or from the average of the parents (called the midparent). The result is the same.

The calculations will be illustrated using father–offspring pairs in a random mating population, as set out in Table 8.3. The first two columns give each parental genotype with its frequency and the mean phenotypic deviation. The next two columns show the gametes produced by the parent and the frequencies of offspring genotypes with random mating. The two columns at the right give the frequencies of parent–offspring pairs and the mean phenotypic deviation of the offspring. The covariance is calculated as the product of the parent–offspring deviations, each weighed by the frequency of the

Table 8.3 Frequencies and phenotype deviations of parent-offspring pairs with random mating

Parental genotype (frequency)	Parental deviation	Gametes	Uniting gamete (frequency)	Offspring genotype (frequency)	Offspring deviation
AA (p^2)	$2qa - 2q^2d$	A	A (p)	AA (p^3)	$2qa - 2q^2d$
			A' (q)	AA' (p^2q)	$(q-p)a + 2pqd$
AA' $(2pq)$	$(q-p)a + 2pqd$	A	A (p)	AA (p^2q)	$2qa - 2q^2d$
			A' (q)	AA' (pq^2)	$(q-p)a + 2pqd$
		A'	A (p)	AA' (p^2q)	$(q-p)a + 2pqd$
			A' (q)	$A'A'$ (pq^2)	$-2pa - 2p^2d$
$A'A'$ (q^2)	$-2pa - 2p^2d$	A'	A (p)	AA' (pq^2)	$(q-p)a + 2pqd$
			A' (q)	$A'A'$ (q^3)	$-2pa - 2p2d$

parent–offspring pair. The first term, for example, is $p^3(2q\alpha - 2q^2d)^2$. In the sum of these products, the coefficient of the term α^2 is:

$$4p^3q^2 + 4p^2q^2(q-p) + pq^2(q-p)^2 + p^2q(q-p)^2 - 4p^2q^2(q-p) + 4p^2q^3 = pq$$

As an exercise, you may verify that the coefficients of the αd and the d^2 terms are both 0. Therefore, letting $Cov(PO)$ be the covariance between a single parent and its offspring:

$$Cov(PO) = pq\alpha^2 = \frac{\sigma_a^2}{2} \tag{8.38}$$

Alternatively we can write the correlation coefficient between a single parent and its offspring r_{PO} as a function of the phenotypic variance σ_p^2:

$$r_{PO} = \frac{Cov(PO)}{\sigma_p^2} = \frac{(1/2)\sigma_a^2}{\sigma_p^2} = \frac{h^2}{2} \tag{8.39}$$

where h^2 is the narrow-sense heritability. Either Equation 8.38 or 8.39 may be used as a basis for estimating the additive genetic variance.

Covariance Between Relatives

Theoretical covariances for several common relationships are shown in Table 8.4. The additive genetic variance can be estimated directly from covariance between parent and offspring, midparent (the average of the parents) and offspring, half siblings, uncle–nephew (or aunt–niece), or first cousins. What these degrees of relationship have in common is that the relatives can share at most one allele at any locus. The covariances between the other degrees of relationship in Table 8.4 include a contribution due to dominance because the relatives can share two alleles at any locus. The theoretical covariance between monozygotic twins is equal to the covariance of an individual with itself, or σ_g^2 (Equation 8.8).

The values for the pairs of relatives in Table 8.4 illustrate the use of a simple equation that allows the covariance to be deduced from any degree of relationship. For any pair of relatives, neither of which is itself inbred, define r_1 as the probability that the pair of relatives share exactly one allele that is identical by descent, and define r_2 as the

Table 8.4 Theoretical covariance in phenotype between relatives[a]

Degree of relationship	Covariance
Offspring and one parent	$\sigma_a^2/2$
Offspring and average of parents (midparent)	$\sigma_a^2/2$
Half siblings	$\sigma_a^2/4$
Full siblings	$(\sigma_a^2/2) + (\sigma_d^2/4)$
Monozygotic twins	$\sigma_a^2 + \sigma_d^2$
Nephew and uncle	$\sigma_a^2/4$
First cousins[b]	$\sigma_a^2/8$
Double first cousins[b]	$(\sigma_a^2/4) + (\sigma_d^2/16)$

[a]Variance terms due to interaction between loci (epistasis) are ignored.
[b]First cousins are the offspring of matings between siblings and unrelated individuals; double first cousins are the offspring of matings between siblings from two different families.

probability that the relatives share both alleles. Then the covariance between the pair of relatives is given by:

$$Cov = \left(\frac{r_1}{2} + r_2\right)\sigma_a^2 + r_2\sigma_d^2 \tag{8.40}$$

(Gillespie 2004). For siblings, for example, $r_1 = 1/2$ and $r_2 = 1/4$; for double first cousins $r_1 = 6/16$ and $r_2 = 1/16$; and for monozygotic twins, $r_1 = 0$ and $r_2 = 1$.

Heritability Estimates from Covariance

Figure 8.14 shows estimates of narrow-sense heritabilities of diverse quantitative traits as estimated from the covariance between relatives. The data are presented merely to show the values of heritability that breeders typically consider. The heritabilities pertain to one population at one particular time. The same trait in a different population or in a different environment might well have a different heritability. Generally speaking, traits that are closely related to fitness (such as calving interval in cattle or eggs per hen in poultry) tend to have rather low heritabilities. For comparison, Figure 8.14 also shows estimated broad-sense heritabilities of a number of quantitative traits in human populations. Broad-sense heritabilities vary widely for different traits, as they do in other species. Note the relatively low heritability of fertility, a trait that is obviously closely related to fitness.

Heritability Estimates from Regression

Figure 8.15 shows a plot of parental phenotype along the x-axis and offspring phenotype along the y-axis. The straight line is the **regression line** of offspring on parent, determined by finding the slope and intercept that minimizes the sum of the squares of the vertical deviation of the points from the line.

Parent–offspring regression is a convenient method for estimating the narrow-sense heritability because the slope of the line b_{OP}, called the **regression coefficient**, can be shown to satisfy:

$$b_{OP} = \frac{Cov(PO)}{\sigma_p^2} = \frac{(1/2)\sigma_a^2}{\sigma_p^2} = \frac{h^2}{2} \tag{8.41}$$

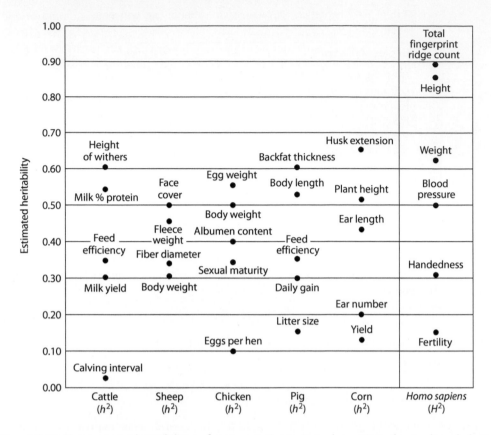

Figure 8.14 Narrow-sense heritabilities of representative traits in domesticated animals and plants, and broad-sense heritabilities of various traits in human beings. (Data from Robinson et al. (1949), Pirchner (1969), and Smith (1975).)

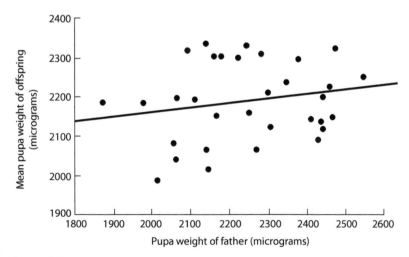

Figure 8.15 Regression of mean pupa weight of male offspring on father's pupa weight in the flour beetle *Tribolium castaneum*. Each point is the mean of about eight male offspring. (Data courtesy of F. D. Enfield.)

where $Cov(PO)$ is the covariance of one parent and its offspring and σ_p^2 is the phenotypic variance among the parents (as well as that among the offspring). Hence, h^2 can be estimated as $2b_{OP}$. It makes no difference whether the offspring of each parent are considered individually or are pooled and their mean used instead. Figure 8.15 uses the mean pupa weight of each set of progeny. The slope of the regression line is $b_{OP} = 0.11$, and thus the estimate of the narrow-sense heritability of pupa weight in this population is $h^2 = 2b_{OP} = 0.22$. The regression coefficient b_{OM} of offspring on midparent (the average of the parents) yields an even simpler estimate of h^2 because $b_{OM} = h^2$. The data points in Figure 8.15 represent 32 sibships and are quite scattered. Because of this sort of scatter, heritability estimates tend to be rather imprecise unless based on data from several hundred families.

8.5 Complex Traits with Discrete Expression

Discrete traits are expressed in an all-or-none fashion, but they may nevertheless have complex inheritance in the sense that pedigrees show no obvious pattern of Mendelian transmission. Even a major gene can escape detection if there is **incomplete penetrance**, which means that the affected phenotype is not always expressed, or if the affected phenotype can also result from other genes or from environmental causes. A trait that can result from two or more different genotypes is said to show **genetic heterogeneity**. When pedigrees of unrelated affected individuals are pooled, as they usually are in human genetics, then genetic heterogeneity makes the pattern of inheritance very complex because two or more major genes are treated as if they were one.

Threshold Traits: Genes as Risk Factors

Some discrete traits are truly multifactorial. There are no single genes that are critical risk factors in themselves. These traits are determined by multiple genetic and environmental factors that act collectively to determine the risk of the trait being expressed. They are **threshold traits**, mentioned earlier in this chapter. An example is pyloric stenosis, an obstruction of the opening at the lower end of the stomach. Pyloric stenosis is a threshold trait because, while each individual is either affected or not affected, the risk of being affected is transmitted in pedigrees as if it were a metric trait. Affected parents do not necessarily transmit the condition to their offspring, but they do transmit genes that increase the risk. The next section shows how the underlying risk toward a threshold trait such a pyloric stenosis can be analyzed as a metric trait.

Heritability of Liability

The basic idea behind the quantitative genetics of threshold traits is illustrated in Figure 8.16. The normal curve in panel (a) represents the (unobservable) distribution of a hypothetical **liability** (or risk) toward the threshold trait, measured on a scale such that the mean value is $\mu = 0$ and the standard deviation is $\sigma = 1$. This curve represents the parental generation, and any parent with a liability above the threshold T_P actually manifests the trait. The proportion of affected individuals in the parental generation is denoted B_P, and the mean liability among affected parents is denoted μ_P.

Among the offspring of matings in which one parent is affected, the distribution of liability is given in panel (b). It has the same variance as in (a), but the mean is shifted

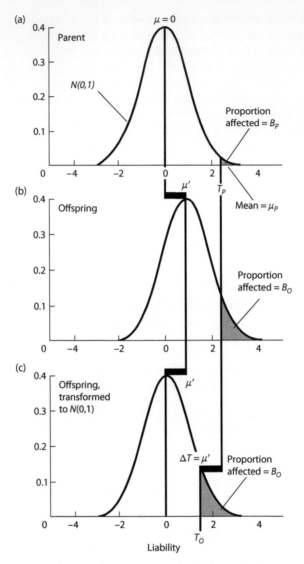

Figure 8.16 (a) Distribution of liability for a threshold trait, assumed to be normal $N(\mu,\sigma)$ with mean $\mu = 0$ and standard deviation $\sigma = 1$. The shaded area denotes parents with liability above a critical threshold (T_P) needed for expression of the trait. (b) Distribution of liability among offspring with one affected parent. The mean is displaced to μ'. (c) Shifting the offspring distribution to the left to coincide with $N(0,1)$ displaces the truncation point to $T_O = T_P - \mu'$.

to the right, to the position μ'. We can estimate the realized heritability of liability using Equations 8.15–8.17 as:

$$h^2 = \frac{R}{S} = \frac{\mu' - \mu}{\mu_S - \mu} = \frac{\mu'}{\mu_S} = \frac{\mu'}{(\mu_P/2)} = \frac{2\mu'}{\mu_P} \tag{8.42}$$

The third equality follows from the fact that liability is measured on a scale in which $\mu = 0$. The next follows from the fact that only one parent is affected, hence the mean of the parents, μ_S, when one parent is affected (mean μ_P) and the other is not (mean $\mu = 0$) is given by $\mu_S = (\mu_P + 0)/2 = \mu_P/2$.

Both μ' and μ_P can be estimated from the proportion of affected individuals in the parental generation (B_P) and among the offspring when one parent is affected (B_O). For pyloric stenosis, the incidence in fathers and their sons $B_P = 0.005$ and $B_O = 0.05$ (Carter 1961). B_P and T_P must satisfy the equation:

$$B_P = \int_{T_P}^{\infty} N(0,1)\,dx \tag{8.43}$$

where $N(0, 1)$ is given in Equation 8.2 with $\mu = 0$ and $\sigma = 1$. Given any value of B_P the corresponding value of T_P can be obtained by numerical integration. When $B_P = 0.005$, as for pyloric stenosis, $T_P = 2.58$.

The way to find μ' is illustrated in Figure 8.16c, where the offspring distribution has been shifted μ' units to the left to coincide with the parental distribution. To maintain the same proportion of affected offspring, B_O, the threshold must also be shifted μ' units to the left, hence the displacement of the threshold is $\Delta T = T_P - T_O = \mu'$. However, the transformed distribution in panel (c) is $N(0,1)$ and therefore T_O satisfies:

$$B_O = \int_{T_O}^{\infty} N(0,1)\,dx \tag{8.44}$$

For $B_O = 0.05$ as in pyloric stenosis, $T_O = 1.64$. Therefore μ' in the numerator of Equation 8.42 is $\mu' = T_P - T_O = 2.58 - 1.64 = 0.94$.

The denominator in Equation 8.42 is the mean liability among the affected parents, which is given by:

$$\mu_P = \frac{\int_{T_P}^{\infty} x N(0,1)\,dx}{\int_{T_P}^{\infty} N(0,1)\,dx}$$

This again requires evaluation using numerical integration, but in the case when $T_P = 2.58$, then $\mu_P = 0.01430/0.00494 = 2.89$. Putting all this together and substituting into Equation 8.42, we obtain the realized heritability of liability toward pyloric stenosis among males as:

$$h^2 = \frac{2\mu'}{\mu_P} = \frac{2(T_P - T_O)}{\mu_P} = \frac{2(2.58 - 1.64)}{2.89} = 65\% \tag{8.45}$$

Although Figure 8.16 depicts the underlying rationale for estimating the heritability of liability, the exact implementation is a little tedious. However, over a broad range of population incidences, the following approximation is sufficiently accurate for most purposes:

$$\log_{10}(B_R) = \left(\frac{-0.274h^2}{1 + 0.742h^2} \right) + \left(\frac{1 - 0.579h^2}{1 + 0.371h^2} \right) \log_{10}(B_G) \tag{8.46}$$

Equation 8.46 generalizes the symbolism in Figure 8.16 in that B_G now represents the incidence of the trait in the general population (corresponds to B_P in Figure 8.16) and B_R represents the risk of the trait in first-degree relatives of affected individuals (corresponds to B_O in Figure 8.16). **First-degree relatives** share half their genes, and the most important first-degree relationships are parent–offspring and full siblings. Equation 8.46 is a satisfactory approximation over the broad range of values of B_G from 0.00001 to 0.20. Therefore, within this range, given any values of B_G and h^2 (or B_R and h^2), the corresponding value of B_R (or B_G) can easily be approximated from Equation 8.46. In the example of pyloric stenosis, $B_G = 0.005$ and $h^2 = 0.65$. Using Equation 8.46, $\log_{10}(B_R)$ is estimated as -1.276, hence $B_R \approx 0.05$. Equation 8.46 can also be adjusted for use with relatives other than those of the first degree. For the relationship between uncles/aunts and their nephews/nieces, replace h^2 by $h^2/2$, and for first cousins, replace h^2 by $h^2/4$.

Applications to Human Disease

Equation 8.46 implies that, for a fixed value of h^2, the relation between $\log_{10}(B_R)$ and $\log_{10}(B_G)$ should be approximately linear. This relation is shown in Figure 8.17 along with the relation expected for simple Mendelian dominant and simple Mendelian recessive inheritance. (The risk plotted for recessive inheritance refers to the full sibling of an affected individual, not the offspring of an affected individual.) Also plotted are observed data for many of the most common clinically relevant conditions. Note that the threshold traits, as a group, tend to be much more common than the simple Mendelian traits.

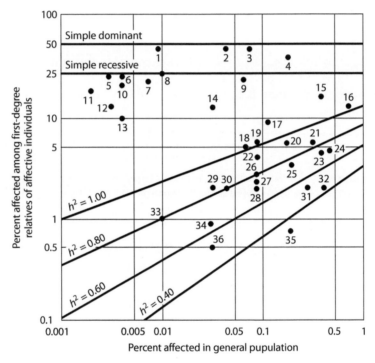

Figure 8.17 Risks of occurrence and recurrence of common abnormalities. The diagonal lines are the theoretical risks in first-degree relatives for threshold traits with the narrow-sense heritabilities indicated. The horizontal line for simple dominant indicates the risk in offspring or siblings of affected individuals, that for simple recessive indicates the risk in siblings. The traits are: (1) achondroplasia, (2) target-cell anemia, (3) periodic paralysis, (4) otosclerotic deafness, (5) retinoblastoma, (6) hemophilia, (7) albinism, (8) retinitis pigmentosum, (9) cystic fibrosis, (10) phenylketonuria, (11) osteogenesis imperfecta, (12) microphthalmos, (13) Hirschsprung disease, (14) deafmutism, (15) bipolar depressive disorder, (16) mental deficiency, (17) schizophrenia, (18) congenital dislocated hip, (19) multiple sclerosis, (20) strabismus, (21) pyloric stenosis, (22) anencephaly, (23) diabetes, (24) rheumatic fever, (25) spina bifida aperta, (26) clubfoot, (27) patent ductus, (28) cleft lip, (29) celiac disease, (30) cleft palate, (31) congenital heart disease, (32) epilepsy, (33) situs inversus viscerum, (34) exomphalos, (35) hydrocephaly, and (36) psoriasis. (Data from Newcombe (1964)).

Problems

8.1 A genetically heterogeneous population of wheat has a variance in the number of days to maturation of 40, whereas two inbred populations derived from it have a variance in the number of days to maturation of 10.

a. What is the genotypic variance, σ_g^2, the environmental variance, σ_e^2, and the broad-sense heritability, H^2, of days to maturation in this population?

b. If the inbred lines were crossed, what would be the predicted variance in days to maturation of the F_1 generation?

8.1 ANSWER a. The variance in the genetically heterogeneous population equals $\sigma_g^2 + \sigma_e^2 = 40$, whereas that of the inbred lines is $\sigma_e^2 = 10$. The genotypic variance σ_g^2 is estimated as $\left(\sigma_g^2 + \sigma_e^2\right) - \sigma_e^2 = 40 - 10 = 30$, and the environmental variance is estimated as $\sigma_e^2 = 10$. The broad-sense heritability H^2 is therefore estimated as $H^2 = 30/40 = 75$ per cent. **b.** The F_1 from a cross of inbred lines is also genetically homogeneous, hence its variance is $\sigma_e^2 = 10$.

8.2 In a cross between two cultivated inbred varieties of tobacco, the variance in leaf number per plant in the F_1 generation is 1.46 and in the F_2 generation it is 5.97. What are the genotypic and environmental variances? What is the broad-sense heritability in leaf number?

8.2 ANSWER The estimated $\sigma_e^2 = 1.46$ and the estimated $\sigma_g^2 = 5.97 - 1.46 = 4.51$. The estimated $H^2 = 4.51/5.97 = 75.5$ per cent.

8.3 Below are data on the number n of abdominal bristles in samples from two consecutive generations G_1 and G_2 of an experiment in directional selection for increased bristle number in *Drosophila*. In the G_1 generation, flies with 22 or more bristles (enclosed in brackets) were mated together at random to form the G_2 generation. Estimate the realized heritability of the number of abdominal bristle in this experiment.

n	G_1	G_2	n	G_1	G_2	n	G_1	G_2
15	0	2	20	20	13	25	[1]	3
16	21	4	21	12	14	26	0	2
17	5	7	22	[13]	12	27	0	0
18	18	16	23	[3]	6	28	0	2
19	17	17	24	[5]	3			

8.3 ANSWER The relevant means are $\mu = 19.304$ of the G_1 generation, $\mu_S = 22.727$ of the selected parents, and $\mu' = 20.149$ of the G_2 generation. Therefore, $S = 22.727 - 19.304 = 3.4229$ is the selection differential and $R = 20.149 - 19.304 = 0.8442$ is the response, yielding a realized heritability of $h^2 = R/S = 0.2466$.

8.4 Use Equations 8.15–8.17 to show that, if h^2 remains constant from generation to generation, the total response after n generations of selection is given by:

$$\mu_n - \mu_0 = \left(S_0 + S_1 + \cdots + S_{n-1}\right) h^2$$

where μ_0 and μ_n are the population means in generations 0 and n, respectively, and S_i is the selection differential applied in generation i. (The sum in this equation is called the *cumulative selection differential*.)

8.4 ANSWER This principle can be demonstrated by the method of successive substitutions using $\mu' - \mu = Sh^2$. Because we are dealing with a population across multiple generations, it is best to use generational subscripts rather than μ' and μ. The initial generation has mean μ_0 and the selection differential in this generation is S_0; hence $\mu_1 - \mu_0 = S_0h^2$. By the same logic $\mu_2 - \mu_1 = S_1h^2$. Adding these two equations yields $\mu_2 - \mu_0 = (S_0 + S_1)h^2$. Note that the μ_1 terms have cancelled. Now $\mu_3 - \mu_2 = S_2h^2$, and adding this to the previous equation yields $\mu_3 - \mu_0 = (S_0 + S_1 + S_2)h^2$, where now the μ_2 terms have cancelled. Continuing in this manner yields the desired formula.

8.5 The regression coefficient of offspring on the average phenotypic value of the parents (often called the *midparent* value) can also be used to estimate the narrow-sense heritability. If b is the regression coefficient of offspring on midparent, then $h^2 = b$. (Note that there is no factor of 1/2 in this case because both parents are involved.) The following data show shell breadth in 119 sibships of the snail *Arianta arbustorum*. For computational convenience, the data have been grouped into six categories.

Number of sibships	Midparent value (mm)	Offspring mean (mm)
22	16.25	17.73
31	18.75	19.15
48	21.25	20.73
11	23.75	22.84
4	26.25	23.75
3	28.75	25.42

Estimate the narrow-sense heritability of shell breadth from these data.

8.5 ANSWER The values needed are the variance of midparent values, which equals 8.1806 mm^2, and the covariance between midparent values and offspring means, which equals 5.1831 mm^2. The narrow-sense heritability h^2 is estimated as the regression coefficient $b = 5.1831/8.1806 = 0.634$, or 63.4 per cent. (There is some loss of accuracy from grouping the data into categories.)

8.6 In a population in Hardy–Weinberg equilibrium:

a. Show that the narrow-sense heritability of a trait completely determined by a single autosomal recessive allele with frequency q equals $2q/(1+q)$, which implies that $h^2 \approx 0$ for $q \approx 0$. (Hint: if $\sigma_e^2 = 0$, then $\sigma_g^2 = \sigma_a^2 + \sigma_d^2$.)

b. Show that the heritability of a trait completely determined by a single autosomal dominant allele with frequency q equals $2(1-q)/(2-q)$, which implies that $h^2 \approx 1$ for $q \approx 0$.

8.6 ANSWER a. For a recessive allele, $d = a$, and therefore $\alpha = a + (q-p)d = 2qa$. Therefore $\sigma_a^2 = 2pq\alpha^2 = 8pq^3a^2$ and $\sigma_d^2 = (2pqd)^2 = 4p^2q^2a^2$. Then set $h^2 = \sigma_a^2/(\sigma_a^2 + \sigma_d^2)$, which reduces to the expression $2q/(1+q)$. **b.** In this case, $d = -a$. Then $\sigma_a^2 = 8p^3qa^2$ and $\sigma_d^2 = (2pqd)^2 = 4p^2q^2a^2$. Then set $h^2 = \sigma_a^2/(\sigma_a^2 + \sigma_d^2)$, which reduces to $2(1-q)/(2-q)$.

8.7. The "trait heritability" of a complex threshold trait is the conditional probability that an offspring is affected, given that one parent is affected, or $\Pr\{O|P\}$. The trait heritability is different, and usually smaller, than the heritability of liability. To express the trait heritability in terms of offspring-on-parent regression, code the phenotypic values such that "affected" = 1 and "not affected" = 0, which yields the following table:

Frequency in population	Parent phenotype	Offspring phenotype
p_1	1	1
p_2	1	0
p_3	0	1
p_4	0	0

where $p_1 + p_2 + p_3 + p_4 = 1$. Note that $p_1 + p_2 = p_1 + p_3 = b$, the overall frequency of affected persons in the population, and $p_1 = \Pr\{O|P\}\Pr\{P\} = \Pr\{O|P\}b$. Show that the covariance between parent and offspring equals $p_1 - b^2$ and that the variance among parents equals $b - b^2$. The regression coefficient of offspring on parent equals the ratio of these two quantities. Assuming $b^2 << b$, show that the ratio $(p_1 - b^2)/(b - b^2)$ is approximately equal to $\Pr\{O|P\}$, which defines the trait heritability.

8.7 ANSWER The mean parental phenotype equals $p_1 + p_2$, and so the deviations of the parents in the parent–offspring pairs are, from top to bottom, $1 - (p_1 + p_2)$, $1 - (p_1 + p_2)$, $- (p_1 + p_2)$, and $- (p_1 + p_2)$. Note that $p_1 + p_2 = b$, the proportion of affected individuals in the population. In terms of b, the deviations can be written as $1 - b, 1 - b, -b$, and $- b$, respectively, and so the variance among the parents equals $p_1(1 - b)^2 + p_2(1 - b)^2 + p_3(-b)^2 + p_4(-b)^2 = b(1 - b)^2 + (1 - b)b^2 = b - b^2$. The offspring mean equals $p_1 + p_3$, which also equals b, and the deviations, from top to bottom, are $1 - b, -b, 1 - b$, and $- b$. The parent–offspring covariance is therefore equal to $p_1 (1 - b)(1 - b) + p_2 (1 - b)(-b) + p_3 (-b)(1 - b) + p_4 (-b)(-b) = p_1 - b(2p_1 + p_2 + p_3) + b^2 = p_1 - b^2$ (because $2p_1 + p_2 + p_3 = 2b$). The regression coefficient is therefore $(p_1 - b^2)/(b - b^2)$, which if $b^2 << b$ becomes p_1/b. In terms of conditional probabilities, $p_1 = \Pr\{O|P\}\Pr\{P\} = \Pr\{O|P\}b$, and therefore the regression coefficient p_1/b in terms of conditional probabilities equals $\Pr\{O|P\}$, which is by definition the "trait heritability." This is a very different concept than the heritability of liability.

8.8 Renal stone disease is a threshold trait that occurs at a frequency of 0.4 per cent in the general population and approximately 2.5 per cent among the offspring of affected individuals. Solve Equation 8.46 to estimate the narrow-sense heritability of liability. Then use Equation 8.46 again to estimate the frequency of the trait in:

a. Brothers or sisters of affected individuals.
b. Nephews or nieces of affected individuals.
c. First cousins of affected individuals.

8.8 ANSWER The $\log_{10}(B_R) = \log_{10}(0.025) = -1.60206$ and $\log_{10}(B_G) = \log_{10}(0.004) = -2.39794$. Equation 8.46, after some rearrangement, yields a quadratic formula in the variable h^2, which has two solutions, $h^2 = -1.273$, which is biologically impossible, and $h^2 = 0.457$, which is the estimate of narrow-sense heritability of liability. **a.** For brothers or sisters of affected individuals, use Equation 8.36 with $h^2 = 0.457$; hence $\log_{10}(B_R) = -1.601$ or $B_R = 0.025$, or about 1 in 40. **b.** For nephews or nieces of affected individuals, use Equation 8.36 with $h^2 = 0.457/2$; hence $\log_{10}(B_R) = -1.972$ or $B_R = 0.0107$, or about 1 in 94. **c.** For cousins of affected individuals, use Equation 8.36 with $h^2 = 0.457/4$; hence $\log_{10}(B_R) = -2.177$ or $B_R = 0.0066$, or about 1 in 150.

References

Bulmer, M. G. (1985), *The Mathematical Theory of Quantitative Genetics* (Oxford: Clarendon).

Carter, C. O. (1961), 'The inheritance of congenital pyloric stenosis', *Br Med Bull*, 17, 251–4.

Cheverud, J.M. and Routman, E.J. (1995), 'Epistasis and its contribution to genetic variance components', *Genetics*, 139 (3), 1455–61.

Conner, J. K. and Hartl, D. L. (2004), *A Primer of Ecological Genetics* (Sunderland, MA: Sinauer Associates).

Crow, J. F. and Kimura, M. (1970), *Introduction to Population Genetics* (New York: Harper & Row).

Dudley, J. W. and Lambert, R. J. (2004), '100 generations of selection for oil and protein in corn', in J. Janick (ed.), *Plant Breeding Reviews* (24 Part 1; Hoboken, NJ: John Wiley & Sons), 79–110.

Enfield, F. D. (1980), 'Long term effects of selection: the limits to response', in A. Robertson (ed.), *Selection Experiments in Laboratory and Domestic Animals* (Slough: Commonwealth Agricultural Bureaux), 69–86.

Falconer, D. S. (1960), *Introduction to Quantitative Genetics* (London: Longman).

Falconer, D. S. (1977), 'Some results of the Edinburgh selection experiments with mice', in E. Pollak, O. Kempthorne, and T. J. Jr. Bailey (eds.), *International Conference on Quantitative Genetics* (Ames, IA: Iowa State University Press), 101–15.

Falconer, D. S. and Mackay, T. F. C. (1996), *Introduction to Quantitative Genetics* (2 edn.; London: Longman).

Galton, F. (1889), *Natural Inheritance* (London: Macmillan).

Gillespie, J.H. (2004), *Population Genetics: A Concise Guide* (2 edn.; Baltimore, MD: Johns Hopkins University Press).

Goddard, M. (2009), 'Genomic selection: prediction of accuracy and maximisation of long term response', *Genetica*, 136 (2), 245–57.

Hartl, D. L. (1980), *Principles of Population Genetics* (1st edn.; Sunderland, MA: Sinauer Associates).

Hill, W. G. (2005), 'A century of corn selection', *Science*, 307 (5710), 683–4.

Hill, W. G., Goddard, M. E., and Visscher, P. M. (2008), 'Data and theory point to mainly additive genetic variance for complex traits', *PLoS Genet*, 4 (2), e1000008.

Johannsen, W. (1903), *Über Erblichkeit in Populationen und in reinen Linien* (Jena, Germany: Gustav Fisher).

Laurie, C. C., et al. (2004), 'The genetic architecture of response to long-term artificial selection for oil concentration in the maize kernel', *Genetics*, 168 (4), 2141–55.

Lynch, M. and Walsh, B. (1998), *Genetics and Analysis of Quantitative Traits* (Sunderland, MA: Sinauer Associates).

Mackay, T. F. C., et al. (1994), 'Polygenic mutation in Drosophila melanogaster: estimates from response to selection in inbred lines', *Genetics*, 136 (3), 937–51.

Meuwissen, T., Hayes, B., and Goddard, M. (2013), 'Accelerating improvement of livestock with genomic selection', *Annu Rev Anim Biosci*, 1, 221–37.

Meuwissen, T. H. E., Hayes, B. J., and Goddard, M. E. (2001), 'Prediction of total genetic value using genome-wide dense marker maps', *Genetics*, 157 (4), 1819–29.

Milkman, R. (1978), 'Selection differentials and selection coefficients', *Genetics*, 88 (4), 391–403.

Newcombe, H. B. (1964), [Panel discussions] in M. Fishbein (ed.), *'Papers and Discussions of the Second International Conference on Congenital Malformations'* (New York: The International Medical Conference).

Nuzhdin, S. V., et al. (1997), 'Sex-specific quantitative trait loci affecting longevity in Drosophila melanogaster', *Proc Natl Acad Sci U S A*, 94 (18), 9734–9.

Pirchner, F. (1969), *Population Genetics in Animal Breeding* (San Francisco, CA: W. H. Freeman).

Rincent, R., et al. (2018), 'Phenomic selection is a low-cost and high-throughput method based on indirect predictions: proof of concept on wheat and poplar', *G3 (Bethesda)*, 8 (12), 3961–72.

Robinson, H. F., Comstock, R. E., and Harvey, P. H. (1949), 'Estimates of heritability and degree of dominance in corn', *Agron J*, 41, 353–9.

Ruby, J. G., et al. (2018), 'Estimates of the heritability of human longevity are substantially inflated due to assortative mating', *Genetics*, 210 (3), 1109–24.

Russell, W. A. (1974), 'Comparative performance for maize hybrids representing different eras of maize breeding', *Annu Corn Sorghum Res Conf*, 29, 81–101.

Sackton, T. B. and Hartl, D. L. (2016), 'Genotypic context and epistasis in individuals and populations', *Cell* (2), 166, 269–87.

Smith, C. (1975), 'Quantitative inheritance', in G. Fraser and O. Mayo (eds.), *Textbook of Human Genetics* (Oxford: Blackwell), 382–441.

Walsh, B. and Lynch, M. (2018), *Evolution and Selection of Quantitative Traits* (New York: Oxford University Press).

CHAPTER 9

Complex Traits in Natural Populations

The concepts relevant to directional selection in agricultural breeding also provide a basis for understanding natural selection in nature, where fitness is the preeminent complex trait. Applications to natural populations raise several special considerations (Barton and Keightley 2002; Moore and Kukuk 2002; A. Wilson and Poissant 2016):

- One issue is whether the genomes of natural populations contain sufficient additive genetic variance in fitness, either from standing variation or from new mutations, to make selection effective.
- Another is that estimating overall fitness entails analysis of all aspects of survival and reproduction of individual genotypes, which is very difficult in natural populations and subject to large sampling errors.
- The alternative is to study components of fitness such as germination time, flowering time, and seed set in plants or developmental time, age at reproductive maturity, and clutch or litter size in animals.
- Studying the components of fitness also requires estimating any positive or negative genetic correlations among them that result either from linkage disequilibrium or from pleiotropy of the underlying genes.
- Beyond these complexities are theoretical issues including the mutation rates of genes affecting fitness, the distribution of fitness effects of new mutations, epistatic effects of genes affecting fitness, and many others.

These issues are more specialized and technical than can be dealt with comprehensively in this book, but they are vitally important in respect to such issues as whether natural populations have sufficient standing or mutational genetic variation to deal successfully with the expected effects of climate change (Anderson et al. 2014; Kopp and Matuszewski 2014).

9.1 Genetic Variation and Phenotypic Evolution

To what extent is variation in a complex trait manifested as variation in fitness? The answer obviously depends on the trait, and Table 9.1 summarizes the findings for several traits in *Drosophila* (Houle 1998). The column labeled *fitness sensitivity* lists for each trait the percentage change in average fitness $\Delta \overline{w}$ resulting from a 1 per cent change in the mean $\Delta \overline{x}$. Variation in some traits including developmental time, viability, and early fecundity

A Primer of Population Genetics and Genomics. Fourth Edition. Daniel L. Hartl, Oxford University Press (2020). © Daniel L. Hartl.
DOI: 10.1093/oso/9780198862291.003.0009

Table 9.1 Genetic variation and fitness sensitivity of traits in *Drosophila melanogaster*

Trait	Fitness sensitivity $\frac{\Delta \overline{w}}{1\% \Delta \overline{x}}$	Mutation $100 \frac{\sqrt{\sigma_m^2}}{\overline{x}}$	Additive effects $100 \frac{\sqrt{\sigma_a^2}}{\overline{x}}$	Additive/mutation $\frac{\sigma_a^2}{\sigma_m^2}$
Abdominal bristles	0.03	0.24	6.11	646.01
Sternopleural bristles	0.04	0.39	7.38	367.97
Developmental time	1.00	0.43	2.47	33.72
Viability	1.00	1.57	10.40	43.75
Early fecundity	0.96	1.22	8.81	52.12
Late fecundity	0.04	2.56	28.79	126.23
Longevity	0.00	1.35	9.06	45.22

have substantial effects on variation in fitness, whereas the effects of variation in bristle number, late fecundity, and longevity are small.

Mutational Variance and Standing Variance

The symbol σ_m^2 in Table 9.1 is the **mutational variance**, defined as the increment of variance generated in each generation by new mutations assuming additive effects and selective neutrality. The additive genetic variance is denoted σ_a^2, and this is the **standing variance** due to polymorphic alleles transmitted from one generation to the next. Each of these sources of variation is compared as the **coefficient of variation**, defined as the standard deviation expressed as a percentage of the mean ($100\sigma/\overline{x}$), which enables comparison of traits measured on different scales. The quantity σ_a^2/σ_m^2 is the standing additive variance expressed as a multiple of the mutational variance.

For a complex trait affected by mutations that are additive in their effects and selectively neutral, the additive genetic variance is expected to change in each generation according to:

$$\sigma_a^2(t+1) = \left(1 - \frac{1}{2N_e}\right)\sigma_a^2(t) + \sigma_m^2 \tag{9.1}$$

where t stands for time in generations and N_e is the inbreeding effective population size (Clayton and Robertson 1955; Lynch and Hill 1986). Hence, at equilibrium for neutral mutations:

$$\sigma_a^2(t+1) = \sigma_a^2(t) = 2N_e\sigma_m^2 \tag{9.2}$$

For *Drosophila*, N_e is of the order of 10^6 and σ_m^2 is of the order of 10^{-3}–$10^{-4} \sigma_a^2$, hence an equilibrium ratio of standing variance to mutational variance (σ_a^2/σ_m^2) in the range 100–1000 might be expected from mutation accumulation alone, especially considering that Equation 9.2 is an overestimate when the assumption of selective neutrality is relaxed.

As might be expected intuitively, for the traits in Table 9.1 that are strongly affected by new mutations and that have a high fitness sensitivity (developmental time, viability, and early fecundity), σ_a^2/σ_m^2 is reduced by an order of magnitude. What about late fecundity and longevity? They have low fitness sensitivity yet a relatively small ratio of σ_a^2/σ_m^2. These traits are probably strongly affected by pleiotropic effects of genes affecting developmental time and early fecundity, and therefore the low fitness sensitivities are probably misleading because the genes are subjected to selection at an earlier stage in life (Houle 1998).

More generally, the extent to which the equilibrium genetic variance is reduced relative to that expected from neutrality depends on the rate of mutation and the distribution of selective effects. If each gene affecting fitness mutates independently and has the same distribution of selective effects, then a key parameter can be written as $4N_e\sigma_m$, where $N_e\sigma_m$ is the product of the effective population size and the standard deviation of the distribution of mutational effects. If $4N_e\sigma_m \leq 1$, most mutations have very small effects, fitness increases slowly as successive favorable mutations are fixed, and the equilibrium genetic variance is not much smaller than that in Equation 9.2. If $4N_e\sigma_m$ is much larger than 1, then some mutations have very large effects and those that are favorable are fixed rapidly; subsequent mutations are mainly deleterious and the equilibrium genetic variance is much smaller than expected with neutrality (Tachida 1991; Turelli 1984). The latter scenario with $4N_e\sigma_m >> 1$ is known as the **house-of-cards** model (Kingman 1978), where the term invokes the image of adaptation as a delicately balanced house of cards in which a single deleterious mutation can cause the whole edifice to collapse.

The house of cards model is one of fifteen mutational models characterized by Kryazhimskiy et al. (2009) that differ according to the expected rates of change in fitness and number of mutations accumulated over time. It is the simplest model that appears to fit gene-substitution data from long-term evolution experiments (Kawecki et al. 2012; Kryazhimskiy et al. 2009) as well as estimates of genetic variation in levels of gene expression (Hodgins-Davis et al. 2015).

Phenotypic Evolution Under Directional Selection

Natural selection will act jointly on a set of traits related to fitness, and the change in the mean of the traits will change according to the type of selection and the magnitude of the additive genetic variances and covariances among them. A prediction equation that does the job of both Equations 8.17 and 8.36 is:

$$\Delta\bar{\mathbf{z}} = \mathbf{G}\boldsymbol{\beta} \qquad (9.3)$$

in which $\Delta\bar{\mathbf{z}}$ is an $n \times 1$ vector of the change in mean of each of n traits in one generation of natural selection, $\mathbf{G}$ is an $n \times n$ matrix of additive genetic variances (along the diagonal) and additive genetic covariances (off the diagonal) among the traits, and $\boldsymbol{\beta}$ is an $n \times 1$ vector called the **selection gradient** (Lande 1979; Lande and Arnold 1983; Turelli 1988). Each entry in $\boldsymbol{\beta}$ is analogous to S/σ_p^2, the selection differential (Equation 8.15) divided by the phenotypic variance. Technically, $\boldsymbol{\beta}$ is called the **linear selection gradient** (also called the **directional selection gradient**) because its entries correspond to the partial linear regression coefficients of each trait on fitness; they equal the change in average fitness due to a small change in the mean of each trait while holding the means of all the other traits constant (Lande 1979).

For three traits we can write the components of Equation 9.3 explicitly as in Conner and Hartl (2004):

$$\Delta\mathbf{z} = \begin{pmatrix} \Delta\bar{z}_1 \\ \Delta\bar{z}_2 \\ \Delta\bar{z}_3 \end{pmatrix} = \mathbf{G}\boldsymbol{\beta} = \begin{pmatrix} \sigma_{a11}^2 & \sigma_{a12} & \sigma_{a13} \\ \sigma_{a21} & \sigma_{a22}^2 & \sigma_{a23} \\ \sigma_{a31} & \sigma_{a32} & \sigma_{a33}^2 \end{pmatrix} \begin{pmatrix} \beta_1 \\ \beta_2 \\ \beta_3 \end{pmatrix}$$

which leads to:

$$\Delta\bar{z}_1 = \beta_1\sigma_{a11}^2 + \beta_2\sigma_{a12} + \beta_3\sigma_{a13}$$
$$\Delta\bar{z}_2 = \beta_1\sigma_{a21} + \beta_2\sigma_{a22}^2 + \beta_3\sigma_{a23}$$
$$\Delta\bar{z}_3 = \beta_1\sigma_{a31} + \beta_2\sigma_{a32} + \beta_3\sigma_{a33}^2$$

in which σ^2_{aii} is the additive genetic variance of trait i and σ_{aij} is the additive genetic covariance between traits i and j.

Analysis of selection gradients can be illustrated with data from Bumpus (1899), which were obtained after an uncommonly severe winter storm of snow, rain, and sleet had knocked a large number of house sparrows (*Passer domesticus*) to the ground around Providence, Rhode Island. Birds were collected from several localities and brought to Brown University where Hermon C. Bumpus classified each by sex and measured each of seven morphological characters. Seventy-two birds ultimately revived overnight, but sixty-four perished. Later analyses of these data revealed that the greatest difference between survivors and nonsurvivors was in total body length of males (Lande and Arnold 1983), with the average length of survivors being smaller than that of nonsurvivors by about one phenotypic standard deviation (Figure 9.1). Although the sample is biased in including only birds that fell to the ground, the standardized value of the component in the selection gradient favoring smaller birds was 0.52 (Lande and Arnold 1983).

The **standardized linear selection gradient**, in which each component is multiplied by the phenotypic standard deviation of the trait, is one way to make the components in the selection gradient comparable (Hereford et al. 2004; Lande and Arnold 1983). Each component in the standardized selection gradient equals the change in average fitness expected from a change of one phenotypic standard deviation in the mean of the trait; each standardized component happens also to equal the intensity of selection i in Equation 8.20. A value of $i = 0.52$ in truncation selection corresponds to a proportion of the population saved for breeding of about $B = 0.70$ (Figure 8.11).

The standardized selection gradient has been estimated in a wide variety of traits in many natural populations of plants and animals (Conner 2001), and well over 1000

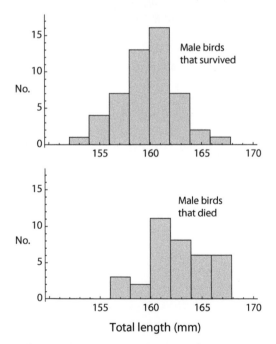

Figure 9.1 Distributions of total body length among knocked-down birds that survived versus those that perished. (Data from https://www.fieldmuseum.org/blog/hermon-bumpus-and-house-sparrows.)

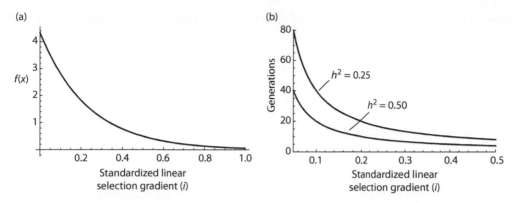

Figure 9.2 Selection intensities in natural populations. (a) Distribution of selection intensities among 753 estimates summarized in Kingsolver et al. (2001). (b) Expected number of generations required for direct selection to change the mean by one phenotypic standard deviation.

studies have been published and reviewed (Kingsolver et al. 2001; Matsumura et al. 2012; Siepielski et al. 2009). Among the studies examined by Kingsolver et al. (2001), the sample sizes tend to be small (median sample size 134) and most studies are unreplicated either spatially or temporally. Across all the studies, the absolute values of the standardized components of the selection gradients were exponentially distributed with a median of 0.16. Such a distribution is shown in Figure 9.2a. Only about 10 per cent of the values exceed 0.5, and many of the larger values are suspect because they come from the smaller studies. The median of $i = 0.16$ corresponds to truncation selection in which $B = 92$ per cent of the population is saved for breeding.

Among 71 traits in human populations measured in nearly 375,000 British individuals, 44 traits yielded a statistically significant standardized linear selection gradient after correcting for multiple tests; these traits included height, systolic blood pressure, pulse rate, and female age at first birth (Sanjak et al. 2018). The median absolute value of the standardized selection differential is only about $i = 0.034$ (Sanjak et al. 2018), which is much weaker selection than observed in other species. An intensity of selection so weak is likely to bring about significant changes in the phenotypic distribution of human traits only on a time scale of hundreds of years.

Figure 9.2b shows another way to interpret the strength of selection in natural populations. It indicates the expected number of generations required for a change of one phenotypic standard deviation (σ_p) in a trait, assuming direct selection on the trait itself with no contributions from correlated response. For the median $i = 0.16$ with a narrow-sense heritability of $h^2 = 0.25$, a change of σ_p requires 25 generations, and with $h^2 = 0.50$ it requires 12.5 generations (Equation 8.2). Compared these values with those for oil content in corn (Figure 8.8), in which a change of σ_p took place every 2.7 generations.

Phenotypic Evolution Under Stabilizing Selection

Many traits are subject to **stabilizing selection**, in which selection favors individuals with phenotypes near the population mean. Deviation from the mean in either direction is deleterious. In *Drosophila*, for example, stabilizing selection for optimal levels of gene expression is important in limiting divergence in expression levels among species (Bedford and Hartl 2009). One consequence is that differences in the average level of gene expression among species are not proportional to their divergence times, as would be expected were expression levels determined by a balance between neutral mutation and random genetic drift. On the other hand, while stabilizing selection plays a major role in shaping gene expression, the selection intensity is rather weak. A deviation in gene expression of ± 1.15 phenotypic standard deviations from the optimum results in a selection coefficient of $s \approx 1/N_e$, and in *Drosophila* N_e is of the order of 10^6 (Bedford and Hartl 2009). A new mutation affecting gene expression changes the level by an average of about one phenotypic standard deviation (Hodgins-Davis et al. 2015); hence, the mutational effects are rather large as expected in the house-of-cards model of mutation-selection

A classic example of stabilizing selection on morphological trait is birth weight in human babies born prior to the advent of neonatal intensive care (Karn and Penrose 1951). The data show a sharp decrease in survival of both males and females as a function of the deviation of birth weight from the mean (Figure 9.3).

Stabilizing selection can be analyzed in the framework of selection gradients. Whereas in directional selection the entries in the selection gradient correspond to the partial linear regression coefficients, in stabilizing selection they correspond to the partial quadratic regression coefficients. The birth-weight data have been analyzed from this perspective, with the result that the change in average fitness expected from a change of one phenotypic standard deviation from the optimal birth weight is 0.06 (Morrissey and Sakrejda 2013).

In the study of 71 traits in humans cited earlier (Sanjak et al. 2018), 26 traits yielded a statistically significant standardized quadratic selection gradient after correcting for multiple tests, including weight, girth, body-fat percentage, and basal metabolic rate. The

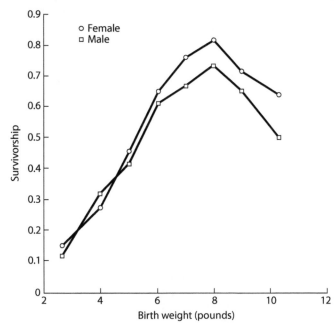

Figure 9.3 Stabilizing selection for birth weight in human babies prior to modern neonatal intensive care. (Data from Karn and Penrose (1951).)

median standardized selection differential was 0.027, which is again much smaller than observed in other species. It nevertheless indicates pervasive but very weak stabilizing selection for these traits in the British population. It should be noted that there is no evidence for stabilizing selection for birth weight in contemporary data (Sanjak et al. 2018), evidently because of the availability of neonatal intensive care when needed in the study population.

9.2 Searching for the Genes Affecting Complex Traits

Once molecular methods for reliable, high-throughput genotyping became available, researchers began to apply these methods to the identification of genes underlying complex traits. The rationale for such studies was compelling. Over and above the intrinsic value of understanding the genetic architecture of complex traits in terms of genetic networks and biological processes were many immediate payoffs. For example, identifying the genes for complex diseases in human populations would help pinpoint protein targets for new drugs to improve treatment; identifying genes for drug resistance in pathogens would help in surveillance and containment of resistance; and identifying genes for agriculturally important traits would aid in plant and animal breeding programs. The following sections provide an overview of several approaches to analyzing the genetics of complex traits as well as a summary of the main findings.

Quantitative Trait Loci

Any gene that is correlated with a quantitative trait is called a **quantitative trait locus (QTL)**. QTLs that affect the level of gene expression are **expression QTLs (eQTLs)**. The opportunity to identify the chromosomal locations of QTLs and eQTLs, if not the causal genes themselves, became possible owing to the prevalence of molecular markers and the relatively high density of such markers in the genetic maps of some species. In this section, we consider some of the approaches for enumerating, genetic mapping, and identifying QTLs. A **causal gene** is a gene responsible for an observed phenotypic difference. A QTL is not necessarily the causal gene that actually affects the trait, but is a correlated surrogate for the causal gene.

Various methods can be used for genetic mapping of QTLs and eQTLs for a quantitative trait, but for an overview of the logic we consider genetic linkage between a single QTL and a causal gene because the theory is straightforward and also informative in regard to fundamental concepts of quantitative genetics.

Figure 9.4 shows the F_2 progeny expected from a cross of F_1 organisms of genotype $QA/Q'A'$, where A and A' are alleles of a causal gene affecting a quantitative trait and Q and Q' are alleles of a putative QTL molecular marker for which heterozygotes can be identified. As usual, we denote the phenotypic contributions of AA, AA', and $A'A'$ genotypes as a, d, and $-a$. The frequency of recombination between the QTL and the causal gene is denoted r with $0 \leq r \leq 0.5$. From the progeny genotypes in Figure 9.4, the following conditional probabilities may be deduced. Each symbol denotes the probability that an individual has a particular causal genotype given that the individual is known to have a particular QTL genotype. For example, $\Pr\{AA|QQ\}$ is the probability than an individual has genotype AA given that it has genotype QQ. The conditional probabilities are as follows:

Gametes and their frequencies	QA $\dfrac{1-r}{2}$	$Q'A'$ $\dfrac{1-r}{2}$	QA' $\dfrac{r}{2}$	$Q'A$ $\dfrac{r}{2}$
QA $\dfrac{1-r}{2}$	$QQ\,AA$ $\left(\dfrac{1-r}{2}\right)^2$	$QQ'\,AA'$ $\left(\dfrac{1-r}{2}\right)^2$	$QQ\,AA'$ $\left(\dfrac{1-r}{2}\right)\left(\dfrac{r}{2}\right)$	$QQ'\,AA$ $\left(\dfrac{1-r}{2}\right)\left(\dfrac{r}{2}\right)$
$Q'A'$ $\dfrac{1-r}{2}$	$QQ'\,AA'$ $\left(\dfrac{1-r}{2}\right)^2$	$Q'Q'\,A'A'$ $\left(\dfrac{1-r}{2}\right)^2$	$QQ'\,A'A'$ $\left(\dfrac{1-r}{2}\right)\left(\dfrac{r}{2}\right)$	$Q'Q'\,AA'$ $\left(\dfrac{1-r}{2}\right)\left(\dfrac{r}{2}\right)$
QA' $\dfrac{r}{2}$	$QQ\,AA'$ $\left(\dfrac{1-r}{2}\right)\left(\dfrac{r}{2}\right)$	$QQ'\,A'A'$ $\left(\dfrac{1-r}{2}\right)\left(\dfrac{r}{2}\right)$	$QQ\,A'A'$ $\left(\dfrac{r}{2}\right)^2$	$QQ'\,AA'$ $\left(\dfrac{r}{2}\right)^2$
$Q'A$ $\dfrac{r}{2}$	$QQ'\,AA$ $\left(\dfrac{1-r}{2}\right)\left(\dfrac{r}{2}\right)$	$Q'Q'\,AA'$ $\left(\dfrac{1-r}{2}\right)\left(\dfrac{r}{2}\right)$	$QQ'\,AA'$ $\left(\dfrac{r}{2}\right)^2$	$Q'Q'\,AA$ $\left(\dfrac{r}{2}\right)^2$

Figure 9.4 Progeny genotypes expected in the F_2 generation when a causal variant (alleles A and A') is genetically linked to a QTL marker (alleles Q and Q') with a frequency of recombination r between the genes.

$$\Pr\{AA|QQ\} = (1-r)^2 \qquad \Pr\{AA'|QQ\} = 2r(1-r) \qquad \Pr\{A'A'|QQ\} = r^2$$
$$\Pr\{AA|QQ'\} = r(1-r) \qquad \Pr\{AA'|QQ'\} = r^2 + (1-r)^2 \qquad \Pr\{A'A'|QQ'\} = r(1-r)$$
$$\Pr\{AA|Q'Q'\} = r^2 \qquad \Pr\{AA'|Q'Q'\} = 2r(1-r) \qquad \Pr\{A'A'|Q'Q'\} = (1-r)^2$$

From these conditional probabilities we can calculate the expected phenotype of each QTL genotype based on the phenotypic values of the causal AA, AA', and $A'A'$ genotypes. These expected values are:

$$E(QQ) = \Pr\{AA|QQ\}\,a + \Pr\{AA'|QQ\}\,d + \Pr\{A'A'|QQ\}\,(-a) = (1-2r)\,a + 2r(1-r)\,d$$
$$E(QQ') = \Pr\{AA|QQ'\}\,a + \Pr\{AA'|QQ'\}\,d + \Pr\{A'A'|QQ'\}\,(-a) = \left[r^2 + (1-r)^2\right]d$$
$$E(Q'Q') = \Pr\{AA|Q'Q'\}\,a + \Pr\{AA'|Q'Q'\}\,d + \Pr\{A'A'|Q'Q'\}\,(-a) = -(1-2r)\,a + 2r(1-r)\,d$$
$$\text{(9.4)}$$

Because the genotype frequencies of QQ, QQ', and $Q'Q'$ in the F_2 generation are 1/4, 1/2, and 1/4, the mean phenotypic value of the F_2 generation can be verified to be $(1/4)E(QQ) + (1/2)E(QQ') + (1/4)E(Q'Q') = d/2$.

It should be intuitively clear that the QTL genotype would have no association with the metric trait if it is unlinked to the causal gene. There will also be no statistically significant association if the phenotypic effect of the causal gene is too small. To see how linkage and causal-gene effects contribute jointly to any association, it is convenient to examine the regression coefficient of phenotype on the QTL genotype. Generalizing the earlier discussion of the regression coefficient of offspring on parent in Equation 8.41, the regression coefficient of a dependent variable Y on an independent variable X equals the covariance between X and Y divided by the variance of X. One type of regression is of the mean phenotype (P) against the number of Q alleles in the QTL genotype. The necessary genotypes, frequencies, and deviations are set out in Table 9.2. The covariance between phenotype and the number of Q alleles equals the product of columns 1, 2, and 4, whereas the variance in the number of Q alleles equals the product of column 1 and the squares of column 4. The regression coefficient of phenotype (P) on number of Q alleles is therefore:

Table 9.2 Frequencies of QTL genotypes in the F_2 generation and phenotypic deviations due to a linked causal locus

Genotype (frequency)	Mean phenotype (as deviation from population mean)	No-Q alleles	No. Q alleles as deviation	Hetero-zygosity	Heterozygosity as deviation
QQ (1/4)	$(1-2r)a + 2r(1-r)d - d/2$	2	1	0	−1/2
QQ′ (1/2)	$\left[r^2 + (1-r)^2\right]d - d/2$	1	0	1	1/2
Q′Q′ (1/4)	$-(1-2r)a + 2r(1-r)d - d/2$	0	−1	0	−1/2
Population mean	0	1	0	1/2	0

$$b_{PQ} = \frac{Cov(P,Q)}{Var(Q)} = \frac{(1/2)(1-2r)a}{(1/2)} = (1-2r)a \qquad (9.5)$$

The association between phenotype and number of Q alleles will therefore be significant if $(1-2r)a$ is large enough, and this regression allows the estimation of a if r is known. Similarly we may regress phenotype (P) against whether or not the QTL genotype is heterozygous. The relevant values are in the last two columns in Table 9.2. In this case, the covariance between phenotype and heterozygosity equals the product of columns 1, 2, and 6, whereas the variance in heterozygosity equals the product of column 1 and the squares of column 6. The regression coefficient of phenotype (P) on QTL heterozygosity is therefore:

$$b_{PH} = \frac{Cov(P,H)}{Var(H)} = \frac{(1/4)(1-2r)^2 d}{(1/4)} = (1-2r)^2 d \qquad (9.6)$$

This regression of phenotype on heterozygosity depends only on the frequency of recombination and the dominance parameter d, and so d can be estimated if r is known.

Equations 9.5 and 9.6 are of special interest in the case $r = 0$, which means biologically that the QTL and the causal gene are one and the same, or at least are inseparable by recombination. When $r = 0$, the regression coefficient in Equation 9.5 estimates a, whereas the regression coefficient in Equation 9.6 estimates d. This special case brings us full circle to classical quantitative genetics, because a and d were originally defined through conceptual regressions of phenotype on number of favorable QTL alleles or heterozygosity (Falconer 1960; Falconer and Mackay 1996; Fisher 1918), long before the beginnings of modern molecular genetics.

For any QTL and causal gene that are associated, the values of a, d, and r could be estimated from the mean phenotypes of QQ, QQ', and $Q'Q'$ using Equation 9.4. This approach is somewhat precarious because there are three means from which to estimate three parameters and no opportunity for an independent test of goodness of fit. A more powerful strategy is to have each causal gene flanked by QTLs, which is called **interval mapping** (Lander and Botstein 1989). For the flanking markers there are nine genotypes that can be used to estimate a, d, and the frequencies of recombination r_1 and r_2 between the causal gene and the flanking markers. In a backcross, to take a concrete example, if there is a significant difference in average phenotype between the nonrecombinant offspring, it tells you that there is a causal gene between the markers; and the average phenotypes of the single recombinants tells you approximately where in the interval between the markers the causal gene is located.

QTL mapping is most powerful in analyzing the progeny of backcrosses, F_2 populations, or recombinant inbred lines (RILs, discussed in Chapter 3). These experimental designs are also limiting in that they can only detect causal variants that are segregating in the backcross or F_2 populations or in the RILs. The amount of recombination the occurs in the creation of the RILs can also limit the mapping resolution of QTLs; however, this can be mitigated by allowing multiple generations of intercrossing before establishing the RILs, as in the *Drosophila* Synthetic Population Resource (King et al. 2012).

QTL mapping has been used for a wide variety of traits in experimental organisms (Mackay 2004), including body weight, growth rate, obesity, atherosclerosis, and cancer susceptibility in the mouse, as well as hypertension, hyperactivity, and arthritis in the rat (Fisler and Warden 1997). It has also been widely used in pigs, poultry, cattle, fish, and in many crop plants (Collard et al. 2005). Nevertheless, once a QTL has been assigned to a chromosomal region, identifying and isolating the causal gene or genes remains a difficult problem. The reason is that the likely position of a causal gene in QTL mapping typically has a rather broad range, so additional data, and often a greater density of molecular markers in the relevant region, are usually necessary to obtain greater precision in locating the causal gene. Even then identifying the causal gene may be problematical, especially in a large genome. In the human genome, $1 \text{ cM} \approx 10^6$ base pairs, which is a lot of DNA to characterize given the rather wide confidence limits (5–10 cM) typically accompanying the location of a causal gene (Grossman et al. 2010).

QTL mapping in natural populations poses additional challenges because it is necessary to estimate the frequencies of the QTL alleles, some of which may be outside the allele frequency range ($p \approx 0.20$ to ≈ 0.80) for which phenotypic values can be accurately estimated. One approach is **composite interval mapping**, in which multiple regression is carried out on all the marker loci simultaneously (Jansen and Stam 1994; Zeng 1994). The analysis is carried out stepwise, identifying first the strongest effect and subtracting this out of the data, then identifying the next strongest effect, and so forth. Choosing appropriate tests for statistical significance is problematical because typically hundreds or thousands of comparisons are made (Austin et al. 2014; Ioannidis 2005; Sham and Purcell 2014; Storey and Tibshirani 2003; Wilson 2019), and in practice it is often preferred to permute the phenotypes randomly among all the genotypes so as to obtain an empirical significance level. More detail on QTL mapping and its applications can be found in Broman (2001), Collard et al. (2005), Doerge (2002), and Mackay (2009).

Candidate Genes

Another approach to QTL identification is to identify **candidate genes**, which are genes chosen on educated guesswork as possibly causal based from their known functions. The causal role of a candidate gene in a quantitative trait is an hypothesis that can be tested independently of genome-wide screens for QTLs, thereby increasing the statistical power of any test of association otherwise eroded by multiple tests.

A good example of the successful use of candidate genes is that of the identification of a polymorphism affecting level of anxiety as assessed by the neuroticism score on a standard personality test (Gelernter 2014; Lesch et al. 1996; Roiser et al. 2005). One class of medications for anxiety and depression, including the widely prescribed Prozac®, selectively inhibits neuronal uptake of the neurotransmitter serotonin (5-hydroxytryptamine). Serotonin uptake is a normal process for helping to terminate neuronal stimulation caused by serotonin release. Hence, the serotonin uptake transporter gene is an obvious candidate gene for traits related to anxiety or depression. The gene is *SLC6A4* in chromosome 17, and as it happens there is polymorphism in the number of copies of a tandem repeat located about 1 kb upstream of the *SLC6A4* transcription start site. The *long* form of the copy-number polymorphism has an allele frequency of about 0.57, whereas the *short* allele frequency of about 0.43. In cells grown in culture, cells of genotype *long/long* have approximately 50 per cent more mRNA for the transporter, and approximately 35 per cent more membrane-bound transporter protein, than cells with *short/long* or *short/short* genotypes (Lesch et al. 1996).

An association between the *SLC6A4* copy-number polymorphism and anxiety was observed in a study of 505 people genotyped for copy number and classified for personality traits from their responses in a questionnaire (Lesch et al. 1996). Significant associations were found between the transporter genotype and the overall neuroticism score, and the highest correlations were with the anxiety-related traits "tension" and "harm avoidance." A comparison between the genotypes with respect to neuroticism score is shown in Figure 9.5. Note that the *short* allele is dominant to *long* with respect to both gene expression and personality score.

Although there is a significant association, it is in the opposite direction than one might have expected from the activity of selective serotonin uptake inhibitors. The effectiveness of these inhibitors suggests that *short/long* and *short/short* genotypes should have reduced anxiety because they have reduced serotonin transporter, but the observed result is the other way around. There is also a great deal of overlap in the distributions: 51 per cent of the homozygous and heterozygous *short* genotypes have a neuroticism score below the average for *long* homozygotes, and 42 per cent of the for *long* homozygotes score above the average for homozygous and heterozygous *short* genotypes.

Subsequent studies have also indicated that the *long* versus *short* polymorphism in *SLC6A4* affected response to the methamphetamine abbreviated as MDMA, often known as Ecstasy or Molly in the social scene. Regular Ecstasy users of genotype *short/long* and *short/short* are more prone to depression and show abnormalities in motor responses to unexpected go/no-go decisions (Roiser et al. 2005). Other behavioral correlations of this *SLC6A4* polymorphism are discussed in Gelernter (2014).

The *SLC6A4* polymorphism accounts for only about 8 per cent of the phenotypic standard deviation in neuroticism score. This example demonstrates that studies of candidate genes can be highly informative when the phenotypic distributions of the

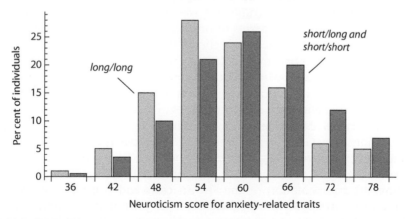

Figure 9.5 Distribution of neuroticism score for anxiety-related traits among genotypes for the serotonin uptake transporter. (Data from Lesch et al. (1996).)

genotypes overlap to such an extent that QTL mapping in a genetically heterogeneous population would lack sufficient statistical power of detection. In the *SLC6A4* example, complete genome sequencing would likely also fail to detect the causal polymorphism, because estimating the number of copies of a tandem repeat in conventional sequencing continues to be problematical.

Genome-Wide Association Studies

A **genome-wide association study (GWAS)** is analogous to a QTL mapping study except that a GWAS uses single-nucleotide polymorphisms (SNPs) as the genetic markers. (A genome-wide association study that uses transcript abundance rather than SNPs is called a *TWAS*.) The number of SNPs in a GWAS is typically much greater than the number of genetic markers in QTL mapping, which in principle increases its power to detect causal variants. SNP genotyping in GWAS was originally based on nucleic acid hybridization with synthetic oligonucleotides affixed in dense arrays to glass slides, but this approach has been largely superseded by the advent of high-throughput, low-cost genome sequencing. As in QTL mapping, GWAS can be carried out in the progeny of backcrosses, F_2 or other populations of known origin, or RILs. Studying such populations has the same limitation as QTL mapping in that the phenotypic associations detectable are limited to those present in the parental crosses or in the inbred strains used to produce the RILs.

Where GWAS can excel is in the study of natural populations including the human population, and these kinds of studies have been its main focus. Within the first 10 years of its inception, more than 3000 GWAS studies of human populations had been carried out to study the genetic risk factors for over 1800 complex traits and diseases (MacArthur et al. 2017; Visscher et al. 2017). Most GWAS are **case–control studies** that compare individuals with a disease or phenotype of interest (the *cases*) with other individuals lacking the disease or phenotype (the *controls*), matched as closely as possible for age, sex, ethnicity, and so forth.

The results of GWAS are often presented in the form of a **Manhattan plot** (Figure 9.6), named for its superficial resemblance to the Manhattan skyline bedecked with towering skyscrapers. The *x*-axis in a Manhattan plot indicates the location of each SNP across the genome, and the *y*-axis plots the negative logarithm of the *P*-value for each SNP. Because the smallest *P*-values indicate the strongest associations between a SNP and a phenotype of interest, the SNPs with the most significant *P*-values rise above those with no association. The plot in Figure 9.6 shows SNPs across six of the 14 chromosomes in the malaria parasite *Plasmodium falciparum* in isolates from Senegal (Park et al. 2012), and the high peaks indicate the locations of genes for resistance to the antimalarial drugs pyrimethamine, mefloquine, and chloroquine. The horizontal line denotes the cutoff for genome-wide statistical significance, corrected for multiple tests. The peaks are exceptionally sharp in *P. falciparum* because recombination among parasites in Africa is sufficient to restrict linkage disequilibrium between SNPs to relatively short distances.

On first consideration, GWAS would seem to be a panacea, the solution to all problems. A little reflection will reveal why this is not the case (Korte and Farlow 2013). First, SNPs at low frequency in the population are uninformative unless their effects are very large because the representation of the SNP in cases and controls is subject to large sampling variance. Second, SNPs with small effects on phenotype will remain undetected except in very large samples. Third, some complex traits are genetically heterogeneous, hence SNPs that are associated with only a subgroup of cases will be more difficult to detect. Fourth, the sample of individuals being studied may consist of a mixture of

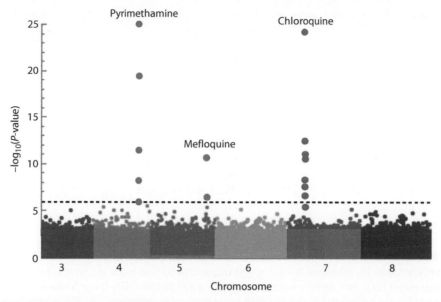

Figure 9.6 Manhattan plot of a GWAS showing SNP locations across six chromosomes of *P. falciparum* and the locations of three genes for drug resistance. (Data from Park et al. (2012).)

different strata of the population (e.g., subsets of individuals with very different genetic ancestries), which violates the assumption of random sampling from a homogeneous population. Fifth, linkage disequilibrium will confound the location of causal SNPs with those associated merely by chance. And sixth, the large number of statistical tests for association presents challenges in balancing false-positive outcomes (type I error, Chapter 2) against false-negative outcomes (type II error) (Ioannidis 2005; Sham and Purcell 2014; P. Zeng et al. 2015).

Despite these and other limitations, GWAS have identified more than 10,000 genetic factors associated with hundreds of quantitative traits and complex diseases in humans (Robinson et al. 2014; Visscher et al. 2017). Some of the main findings are summarized in the next section.

Number of Genes and Magnitude of Effects

Early on in the GWAS era many researchers had hoped that complex traits might be influenced by relatively few genetic factors with large effects that together would account for a major part of the genetic variation in the trait. For most complex traits in most natural populations this hope has been dashed. A good example from humans is the familiar type of pattern baldness (androgenetic alopecia) affecting about 80 per cent of European males that begins in men at the age of about 30 with a receding hairline and hair thinning on the crown and then proceeds gradually to a horseshoe-shaped hair pattern often coiffed in the classic comb-over. In the early twentieth century, when many complex traits were shoehorned into Mendelian patterns, the myth arose that the trait was due to a simple Mendelian mutation that was dominant in males and recessive in females. This notion is not only oversimplified but wildly oversimplified. A GWAS of about 200,000 European males has identified about 650 SNPs scattered throughout the genome that affect the age of onset, rapidity of progression, and pattern of male baldness (Yap et al. 2018).

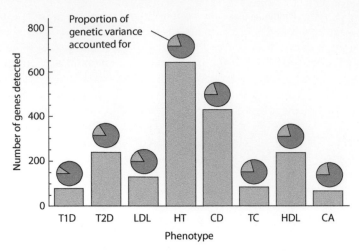

Figure 9.7 Number of genes affecting complex traits identified in GWAS (bar graph) and proportion of the genetic variance explained by these genes (pie charts). The traits are type 1 diabetes (T1D), type 2 diabetes (T2D), low-density lipoprotein (LDL), adult height (HT), Crohn's disease (CD), total cholesterol (TC), high-density lipoprotein (HDL), and cancer susceptibility (CA). (Data from Park et al. (2011).)

Nor is male pattern baldness an exception. Figure 9.7 shows, in the bar graph, the number of genes significantly associated with various complex traits and, in the inset pie charts, the proportion of the genetic variance in each trait explained by these genes (Park et al. 2011). The number of significantly associated genes ranges from about 70 (type I diabetes and susceptibility to cancer) to about 650 (height), whereas the proportion of the genetic variation explained by these genes ranges from only about 10 per cent (type I diabetes) to about 22 per cent (susceptibility to cancer). Note that there is no clear relation between the number of identified genes and the proportion of the genetic variance explained.

The large fraction of the genetic variation in complex traits unexplained by the SNPs in GWAS is known as the **missing heritability**. All sorts of explanations for missing heritability have been put forth (Sackton and Hartl 2016), many of them postulating epistasis. As discussed in Chapter 8, the concept of epistasis as applied to complex traits is tricky. On the one hand there is *physiological epistasis*, which certainly occurs widely because of the structure of genetic networks and biological pathways; on the other hand there is *statistical epistasis*, which is the amount of genetic variation explained by gene interaction over and above that already accounted for by additive and dominance effects (Sackton and Hartl 2016). As Figure 8.7 illustrates, a pair of genes that are connected by substantial physiological epistasis may nevertheless exhibit little or no statistical epistasis.

The most likely current explanation for missing heritability is that GWAS lack the statistical power to detect a large number of causal genes whose effects are simply too small to detect given the sample size (Sackton and Hartl 2016). Adult human height is a classic case showing that an increase in sample size in GWAS results in a greater number of significantly associated SNPs and a higher proportion of the genetic variance explained (Marouli et al. 2017; Wood et al. 2014). Including SNPs that are associated with adult height but lack genome-wide statistical significance accounts for most of the missing heritability (Yang et al. 2011).

Modern GWAS have largely validated a model for complex traits proposed about hundred years ago by Fisher (1918). In this model, called the **infinitesimal model**, Fisher assumed that a complex trait is determined by a very large number of genetic factors, each with a small effect, and he showed how such a model could account quantitatively for the observed correlations between relatives for complex traits. A more specific model, known as the **omnigene model**, imagines a modest number of "core genes," which act directly on the trait and have relatively large effects, coupled with a much larger number of "peripheral genes," which act indirectly on the trait and have relatively small but cumulatively important effects (Boyle et al. 2017; Liu et al. 2019). An alternative view adheres to the traditional infinitesimal model in disputing any clearly definable distinction between core genes and peripheral genes in regard to complex traits (Wray et al. 2018). In this view there is no single coherent core pathway for a complex trait. It supposes that any given complex genetic disorder can result from many different combinations of risk factors that produce the same phenotype, the genetic heterogeneity resulting from the massive amount of genetic variation present in most outbreeding natural populations. And if there is no coherent core pathway for a complex trait, then the omnigene model is merely another term for the same genetic architecture implied by the infinitesimal model (Wray et al. 2018). Empirical support for the infinitesimal model comes from the distribution of effects of individual QTLs in contributing to the overall heritability of complex traits (Hartl et al. 2020).

Just because many genes affect a complex trait does not mean that prediction of phenotype is impossible. For example, in one study of adult height in about 500,000 genomes in the UK Biobank data repository, researchers used machine learning to identify a set of 20,000 SNPs with maximum predictive power for the trait (Lello et al. 2018). The predicted heights of individuals based on these SNPs have a correlation of about 0.65 with their actual heights, and the SNPs account for about 40 per cent of total variance in height (Lello et al. 2018). The remaining variance in height is due largely to environmental factors.

Genetic and Environmental Risk Factors in Complex Traits

The genetic architecture of complex traits is especially important in the interpretation of genetic risk factors assayed in **direct-to-consumer (DTC)** or **over-the-counter (OTC)** genetic tests. On one side are a relatively small number of genes with large effects. These include genes such as *BRCA1* and *BRCA2*, which are major risk factors for breast and ovarian cancer (although they account for only 5–10 per cent of all breast or ovarian cancers). On the other side are a much larger number of genes that have only modest effects on overall risk. How are consumers to decide whether the risk factors in their genomes are in the first category or the second? How are consumers to decide what course of action, if any, should be taken based on DTC or OTC tests?

And the environment is important, too, especially for complex traits. Biology is not destiny. One study of genetic and lifestyle risks in coronary artery disease is revealing (Khera et al. 2016). In this study, about 50,000 individuals were tested for 50 SNPs known from GWAS to be significantly associated with coronary artery disease. The individuals were also classified for each of four healthy lifestyle characteristics related to risk: not smoking, not being obese, physical activity at least once per week, and a diet relatively rich in fruits, nuts, vegetables, whole grains, fish, and dairy products. According to their number of genetic risk factors, each individual was assigned a low, medium, or high genetic risk; and according to their lifestyle choices, each was assigned a low, medium, or

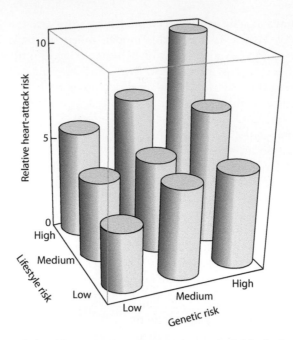

Figure 9.8 Risk of cumulative 10-year coronary events among individuals classified by genetic risk (low, medium, and high) as well as by lifestyle risk (low, medium, and high). (Data from Khera et al. (2016).)

high lifestyle risk. Each individual was also monitored for 10 years to identify significant coronary events.

The results are shown in Figure 9.8 (Khera et al. 2016). As might be expected, the fewest coronary events were observed among people with the lowest genetic and lifestyle risks, and the most coronary events observed among those with the highest genetic and lifestyle risks. The major finding is that a low-risk, healthy lifestyle can completely offset a high-risk genotype, and a high-risk, unhealthy lifestyle can completely offset a low-risk genotype. The take-home lesion is that at least for coronary artery disease, while you cannot choose your genetics you can choose your lifestyle, and your lifestyle choices can really matter.

9.3 Complex Traits in Evolutionary Adaptation

When the genomics revolution began in the 1980s, it didn't take much imagination to realize that two fields of science destined to be profoundly transformed would be medicine and evolutionary biology. In regard to medicine, this insight has been amply proved by progress in increased understanding of the genetic basis of complex traits as well as by applications in tracing genetic ancestry and the development of DTC and OTC genetic testing services. Progress has been equally deep and rapid in evolutionary biology, so much so that in the following brief synopsis we can focus only on a few vignettes to illustrate the main approaches and findings.

Evolutionary Pathways of Drug Resistance

The GWAS plot in Figure 9.6 shows a strong signal on chromosome 4 in the malaria parasite *P. falciparum* for resistance to the drug pyrimethamine. Analysis at a finer scale indicates that the resistance gene encodes for the enzyme dihydrofolate reductase, an essential enzyme that is the target of pyrimethamine. DNA sequencing of resistant and sensitive alleles indicates that resistance is associated with four amino acid replacements, namely, asparagine at position 51 to isoleucine (N51I), cysteine at position 59 to arginine (C59R), serine at positon 108 to asparagine (S108N), and isoleucine at position 164 to leucine (I164L). The level of resistance was estimated in laboratory studies of all sixteen possible combinations of these amino acid replacements, and computer simulations of mutation and selection were carried out to determine the possible pathways by which the fourfold mutant enzyme could arise under continuing drug pressure (Lozovsky et al. 2009). In principle, the number of possible pathways equals 4! = 24; however, fourteen of these are inaccessible because they require a decrease in resistance in spite of continuing drug pressure, and another seven are very uncommon, occurring on average in fewer than 2 per cent of all simulations. The three most common pathways of resistance are shown in Figure 9.9, and they account for 88 per cent of all outcomes.

The finding that the actual number of mutational paths to fitter proteins is far less than the theoretical number is typical, as the requirement for incremental improvement is violated in many (usually most) of the theoretical pathways (Weinreich et al. 2006). Moreover, pathways of resistance typically begin with the fixation of mutations having the largest effect, followed by those with progressively smaller effects.

The pathways of resistance in Figure 9.9 are reminiscent of Maynard Smith's (1970) analogy for evolution in sequence space. He compared the succession of amino acid changes to those in the change-one-letter game, in which the challenge is to take an ordinary English word and to create a sequence of meaningful words by changing one letter at a time. His example was WORD → WORE → GORE → GONE → GENE. As teaching tool, the concept of "fitness" can be incorporated into this analogy by using, as a proxy for fitness, the number of instances that the word was used in any given year as tabulated in, for example, the Google Ngram Viewer (Ogbunugafor and Hartl 2016). With this addendum, a word change is acceptable if and only is it is used more often (i.e., has a greater "fitness") than its predecessor.

The evolution of pyrimethamine resistance illustrates strong selection acting on a single gene, which is typical for antibiotic or insecticide resistance. Strong selection can also affect groups of interconnected genes, changing whole pathways of development or behavior, which is well illustrated in genomic changes that take place under domestication.

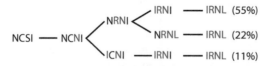

Figure 9.9 The diagram shows the most likely evolutionary pathways for the acquisition of pyrimethamine resistance in the malaria parasite. The amino acids shown in red are those associated with resistance (Data from Lozovsky et al. (2009).)

Genomic Changes Under Domestication

For the great majority of domesticated plants the ancestor is readily identified because of its similarity. Wild tomatoes are morphologically similar to cultivated varieties but have smaller fruit, for example, and wild potatoes are morphologically similar to cultivated varieties but have smaller tubers. Domesticated corn (maize) is different. Maize (*Zea mays* subspecies *mays*) was domesticated by Native Americans in southwestern Mexico beginning about 9000 years ago (Piperno et al. 2009). No native plants are morphologically similar, which led to the hypothesis that the ancestor of domesticated maize had gone extinct. But in 1939, geneticist George Beadle proposed that the ancestor of cultivated maize was teosinte (*Zea mays* subspecies *parviglumis*) (Beadle 1939). This hypothesis at first seemed preposterous because the organisms are morphologically so different in plant and inflorescence morphology (Figure 9.10a), and for years the hypothesis was hotly contested. Genome sequencing of teosinte and maize now leaves no doubt that Beadle's hypothesis is correct.

Detailed studies of teosinte and maize have identified multiple genes responsible for the morphological differences between the subspecies; however, a handful of genes of large effect are among the main players (Doebley 2004; Studer et al. 2017). Especially important is the network of transcription factors shown in Figure 9.10b (Studer et al. 2017). The arrows represent positive regulation (stimulation) of the downstream gene, and the T-bars represent negative regulation (inhibition) of the downstream gene. The gene symbols in red show independent evidence that genetic changes have occurred under domestication.

During domestication, the network in Figure 9.10b that results in such great changes in growth and inflorescence morphology was hijacked from its original function. The central players are the transcription factors *tb1(teosinte branched1)* and *tga1* (*teosinte glume architecture1*). Reduced production of *tb1* results in less branching, fewer inflorescences, and greater elongation of the stem. A reduction of *tb1* also results in less stimulation of *tga1* and hence less inhibition (greater production) of *zag1*, *zag2*, and the other targets of *tga1*. The result is that the stony, silica-laden maternal tissue that surrounds the kernel of teosinte becomes the soft, edible tissue surrounding the kernel of maize.

Being that the network in Figure 9.10b was hijacked from its original function, what was the original function? It was the heart of the *shade-avoidance response*. When a flowering plant is growing in the shade of other plants, the ratio of red light (wavelength 620–

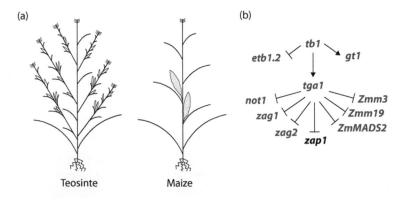

Figure 9.10 (a) Differences in growth habit and inflorescence morphology between teosinte and maize. (b) Network of transcription factors responsible for the main differences in growth habit and inflorescence morphology. (Data in (b) from Studer et al. (2017).)

700 nm) to far-red light (710–850 nm) it receives is increased owing to pigments in the canopy that absorb more of the far-red light. The protein phytochrome B is sensitive to the ratio or red to far-red, and when this ratio is high, phytochromome B inhibits *tb1*, which sets off the shade-avoidance response of reduced branching and inflorescences as well as faster growth of the main stem. The shade-avoidance response was hijacked under domestication when the production of *tb1* became reduced irrespective of the ratio of red to far-red light.

Local Selection Versus Gene Flow

In the evolution of drug resistance and genetic changes under domestication, selection for favored genotypes is sufficiently strong to swamp any effects of gene flow (migration) from drug-sensitive or ancestral populations. Drug-sensitive alleles in migrants from untreated populations are eliminated by drug treatment, and under domestication ancestral types that migrate into the population express undesirable traits and are culled. In regard to natural populations, it is therefore reasonable to ask how much migration is needed to prevent the fixation of alleles that are beneficial in one or more subpopulations but deleterious in others.

An excellent example of selection balanced against migration is afforded by the landscape genetics of cryptic coloration in deer mice (*Peromyscus maniculatus*) in the light-colored Sandhills of north-central Nebraska (Pfeifer et al. 2018). Mice inhabiting the Sandhills have lighter coloration than those in the surrounding dark soil, undoubtedly because the cryptic coloration offers protection from coyotes, foxes, snakes, owls, hawks, and other predators. The lighter coloration is largely due to alleles of the *agouti* gene, which encodes a signaling protein responsible for the distribution of pigment in mammals and whether the pigment cells produce the yellow or red pheomelanin versus the brown or black eumelanin. Genotyping hundreds of individuals at nearly a million SNPs indicated little genetic differentiation among subpopulations across most of the genome (Pfeifer et al. 2018). F_{ST} values among subpopulations ranged from 0.008 to 0.065 depending on distance, which according to Equation 4.43 correspond to a number of migrants per generation (*Nm*) ranging from $Nm \approx 30$ to $Nm \approx 4$, respectively, which are well within the range conventionally regarded as resulting in little divergence (Figure 4.14). Yet the Sandhill subpopulations maintain high frequencies of multiple alleles of the *agouti* gene that result in light body color, which distinguishes these subpopulations from those inhabiting regions with darker soil. In this case, the selection for cryptic coloration is clearly strong enough to overcome levels of migration sufficient to homogenize the genetic background.

9.4 Complex Traits in Speciation

Genomics has also transformed our understanding of species and speciation, but before examining how we need briefly to consider what we mean by the word "species." The classical definition of a biological species is due to Mayr (1942): "Species are groups of actually or potentially interbreeding natural populations, which are reproductively isolated from other such groups." As a working definition this has problems. One is that reproductive isolation is a quantitative trait that comes in degrees ranging from complete lethality or sterility of hybrids to mild impairment of their fitness. The other is that genome sequencing has revealed many cases of gene exchange in nature between what naturalists regard as perfectly good species. An alternative definition is that a species is a recognized taxon (Mallet 2006), but this suggests that species are not real biological entities unless they are assigned a scientific name. Better still is the definition of a **species** as a group of genotypes that, across the genome as a whole, are more similar to each other than to other such groups (Mallet 2006). This definition allows for particular genes or regions of the genome to be similar between species so long as their genetic backgrounds remain largely distinct. This definition also has the advantage of working for asexual organisms or those that undergo limited recombination.

As for speciation, there is a huge and oftentimes controversial literature dating back to Darwin and his predecessors that cannot be adequately summarized in a few paragraphs. It will suffice for our purposes to note that speciation usually requires an ecological opportunity consisting of favorable environmental or biotic circumstances. With this as background, the following examples examine how genomics has afforded new insights into the processes of speciation.

Reinforcement of Mating Barriers

Barriers to gene flow between species are usually classified as prezygotic (affecting mating) or postzygotic (affecting hybrid survival or fecundity). For distinct populations living in the same habitat (sympatry), reproductive barriers that arise can increase in frequency by a type of natural selection known as **reinforcement**. That reinforcement can occur even while gene flow continues between the populations is nicely demonstrated in the native Texas wildflower *Phlox drummondii* in a region where its range overlaps with the related species *P. cuspidata* (Roda et al. 2017). The zone of overlap is a large, roughly triangular region with vertices at Austin, Houston, and Waco, Texas. The flower petals of *P. drummondii* are normally a light blue and resemble those of *P. cuspidata*; however, in the region of overlap the flower color of *P. drummondii* is a deep red owing to the action of two genes in the biochemical pathway for anthocyanin synthesis (Hopkins et al. 2014). Hybrid *Phlox* in the region in question show high but not always complete sterility (Suni and Hopkins 2018), and the deep red flower color substantially reduces hybridization owing to behavior of the major butterfly pollinator, which on visiting a plant with red flowers tends subsequently to visit other plants with red flowers (Briggs et al. 2018). Nevertheless, the ABBA–BABA test (Chapter 7) and other evidence from transcribed sequences indicate that substantial gene flow occurs between the species, with a bias in the direction from the species with blue flowers to that with red flowers (Roda et al. 2017).

Reproducibility of Phenotypic and Genetic Changes in Speciation

At their maximum in the last glacial period, ice sheets covered much of Canada and the northern parts of the United States, the British Isles, Germany, Poland, and Russia. As the glaciers retreated about 12,000 years ago, numerous isolated freshwater lakes were formed entrapping saltwater fish. The freshwater populations began adapting to their new conditions, providing today's scientists a natural evolutionary experiment with many independent replications.

Among the species studied extensively from the standpoint of genetics and genomics is the threespine stickleback *Gasterosteus aculeatus*, in which about twenty QTL studies have been carried out comparing marine and freshwater species, mainly for morphological traits including reduction in body armor and pelvic spines (Peichel and Marques 2017). The body armor consists of about thirty bony plates on each side, and the pelvic spines consist of thorn-like projections jutting out beneath and behind the gills. The body armor and pelvic spines protect against marine predators, which do not survive in freshwater lakes. They are energetically expensive to produce, however, and require minerals that are less abundant in freshwater than in saltwater. For these and other reasons the body armor and pelvic spines in freshwater species are absent or much reduced.

As might be expected from the generally large number of genes affecting complex traits in other organisms, the QTLs for body armor and pelvic spines number in the hundreds (Peichel and Marques 2017). The distribution of gene effects is similar, too, with a small number of genes with large effects and a much larger number with small effects. Among those few with large effects are *ectodysplasia* (*Eda*), in which allelic variation accounts for about 70 per cent of variation in armor plate. *Eda* was discovered by F_{ST} scanning across the genome looking for outliers (Makinen et al. 2008). Another major gene is *pituitary transcription factor 1* (*Pitx1*), which in many populations accounts for most of the reduction in pelvic spines owing to deletion of a pelvic enhancer (Chan et al. 2010).

The QTLs implicated in body armor and pelvic spine reduction can differ from one freshwater lake to the next. In comparing any two populations, only about half of the QTLs are shared (Peichel and Marques 2017). This again attests to the large number of genes involved. Not even *Eda* and *Pitx1* invariably contribute to body armor and pelvic spine reduction. Therefore, while the phenotypic changes in freshwater populations are largely reproducible, the underlying genetic changes are less so.

Accumulation of Genetic Incompatibilities

For over a century *Drosophila* has been a prime organism for speciation research (Barbash 2010; Coyne and Orr 2004). Its advantages include ease of laboratory culture, short generation time, and a cornucopia of genetic and genomic resources. Its biggest limitation is that the species most amenable to genetic analysis are also those whose natural history and ecology are least understood. The species are also long established. Even the most closely related species have diverged for hundreds of thousands or millions of years, quite unlike the situation with *Phlox* or sticklebacks.

Owing to these and other limitations, some commentators hold the view that the emphasis on *Drosophila* has led to a lopsided view of speciation because it overemphasizes genes affecting hybrid sterility or viability at the expense of those affecting mate choice or assortative mating (Mallet 2006). And, indeed, even the most closely related species in the *D. melanogaster* subgroup differ in large numbers of genes affecting hybrid fertility or survival. For example, roughly one hundred genes affect hybrid male sterility between *D. simulans* and *D. mauritiana* (Tao et al. 2003b), which diverged about 250,000 years ago (Garrigan et al. 2012). About half of them are recessive X-linked genes that interact negatively with autosomal dominant genes from the other species, thereby largely accounting for Haldane's rule that hybrid incompatibility first arises in the heterogametic sex (Tao et al. 2003a). The autosomal incompatibilities are partially dominant ($h \approx 0.30$) and act synergistically, with an average of about three required to produce complete hybrid male sterility (Tao and Hartl 2003). These are minimum estimates, as the experimental method excludes recessive–recessive interactions, which outnumber recessive–dominant interactions by about eightfold (Presgraves 2003). In spite of complete hybrid male sterility, up to 5 per cent of genes have sequences consistent with recent gene flow.

For the more distantly related species pair *D. simulans*–*D. melanogaster*, hundreds of genes are estimated to affect hybrid viability (Davis and Wu 1996). Those with the largest effects have been called *speciation genes*; however, they almost certainly arose after the main separation took place. A handful of genes including *Lhr* (*lethal hybrid rescue*) and *Hmr* (*hybrid male rescue*) have also been discovered that allow a few hybrid offspring to survive (Barbash 2010). These genes encode proteins that associate with chromatin, but their precise functions remain elusive (Blum et al. 2017). Echoing the view that inferences based on *Drosophila* may be lopsided, the relevance of these findings to the process of speciation remains unclear, as it remains uncertain which of these incompatibilities if any are operative in the initial stages of reinforcement and which arose in the following hundreds of thousands of years.

Problems

9.1 Use Equation 9.1 to derive an expression for $\sigma_a^2(t)$ in terms of $\sigma_a^2(0)$, and calculate the time required for $\sigma_a^2(t) - 2N_e\sigma_m^2$ to decrease by half from its original value.

9.1 ANSWER Let $\sigma_a^2(t) = \sigma_t^2$. Equation 9.1 implies that:

$$\sigma_t^2 - 2N_e\sigma_m^2 = \left(1 - \frac{1}{2N_e}\right)\left(\sigma_{t-1}^2 - 2N_e\sigma_m^2\right) = \cdots = \left(1 - \frac{1}{2N_e}\right)^t \left(\sigma_0^2 - 2N_e\sigma_m^2\right)$$
$$\approx e^{-t/(2N_e)}\left(\sigma_0^2 - 2N_e\sigma_m^2\right).$$

Setting $\sigma_t^2 - 2N_e\sigma_m^2 = (1/2)\left(\sigma_0^2 - 2N_e\sigma_m^2\right)$ implies that $-t/(2N_e) = \ln(1/2)$ or $t = 1.386N_e$ generations.

9.2 Across multiple traits in human populations the median standardized selection differential is about $i = 0.034$. How many generations would be required to change such a trait by one phenotypic standard deviation assuming $h^2 = 0.25$ and $h^2 = 0.50$?

9.2 ANSWER Equation 8.21 says that $R = ih^2\sigma_p$ per generation, which for t generations is a total change of $ih^2\sigma_p t$. Assuming $R = \sigma_p$ implies that $t = 1/(ih^2)$. For $i = 0.034$ and $h^2 = 0.25$, $t = 117.6$ generations; and for $h^2 = 0.50$, $t = 58.8$ generations. At 25 years per generation, these values represent 2941 years and 1471 years, respectively.

9.3 Consider the cross $Q\ A/Q'\ A' \times Q\ A/Q\ A$, where Q and Q' are alleles of a QTL and A and A' are alleles of a causal gene in which the phenotypic values of AA, AA', and $A'A'$ are a, d, and $-a$, respectively. The frequency of recombination between the Q and A genes is r.

a. Derive an equation for the regression coefficient of phenotype against QTL genotype. (Hint: assign genotypic values of 1 and 0 to QQ and QQ'.)

b. Interpret this equation for $d = a$, $d = 0$, and $d = -a$.

9.3 ANSWER a. Let G be the genotypic value at the Q locus and P be the phenotypic value of the causal genotype. The $Q\ A/Q'A'$ parent produces the gametes $Q\ A$, QA', $Q'A$, and $Q'A'$ with frequencies $(1-r)/2$, $r/2$, $r/2$, and $(1-r)/2$, respectively. The $Q\ A/Q\ A$ parent produces only $Q\ A$ gametes, hence the genotypes of the progeny are $Q\ A/Q\ A$, $Q\ A'/Q\ A$, $Q'A/Q\ A$, and $Q'A'/Q\ A$ with frequencies $(1-r)/2$, $r/2$, $r/2$, and $(1-r)/2$, respectively. The genotypic values at the Q locus are $G = 1$, 0, 1, and 0, respectively, and the phenotypic values of the QTL genotypes are $P = a$, a, d, and d, respectively. Then the means of G and G^2 are both 1/2, and therefore $Var(G) = 1/4$. The mean of P equals $(a+d)/2$ and the mean of $G \times P$ equals $(1/2)(a - ar + dr)$, hence the covariance of $G \times P$ equals $Cov(GP) = (1/4)(a-d)(1-2r)$. The regression coefficient of phenotype on the QTL genotype is therefore $b = Cov(GP)/Var(G) = (a-d)(1-2r)$. **b.** For $d = a$ (that is,

A is dominant), $b = 0$. In this case the cross is not informative because all genotypes of the A gene have the same phenotype. For $d = 0$, $b = a(1 - 2r)$. And for $d = -a$ (i.e., A is recessive), $b = 2a(1 - 2r)$, which means the recessive case is maximally informative.

9.4 You are interested in whether a particular single-nucleotide polymorphism (SNP) is possibly associated with a complex trait. Let's call it SNP* if it truly is associated, and SNP$^-$ if it is not. To pursue this issue, you test the null hypothesis H_0 that the SNP is SNP$^-$; in other words, the hypothesis H_0 is that the SNP is *not* associated with the trait. Your test has a significance level of α and a power of $1 - \beta$ (Chapter 2). Let $p = \Pr\{SNP^*\}$ be the probability that the SNP is truly associated with the trait. In an ideal world, you would reject the null hypothesis if the SNP is SNP* and fail to reject if the SNP is SNP$^-$.

a. Show that $\Pr\{SNP^* \mid P \le \alpha\} = \frac{(1-\beta)p}{(1-\beta)p + \alpha(1-p)}$.
b. What is the value of $\Pr\{SNP^* | P \le \alpha\}$ when $\alpha = p/\{1 - p\}$?
c. Interpret these results in a few short sentences.

9.4 ANSWER a. According to Bayes theorem:

$$\Pr\{SNP^* | P \le \alpha\} = \frac{\Pr\{P \le \alpha | SNP^*\} \Pr\{SNP^*\}}{\Pr\{P \le \alpha | SNP^*\} \Pr\{SNP^*\} + \Pr\{P \le \alpha | SNP^-\} \Pr\{SNP^-\}}.$$

By definition, $1 - \beta = \Pr\{P \le \alpha | SNP^*\}$ and $\alpha = \Pr\{P \le \alpha | SNP^-\}$, and with $\Pr\{SNP^*\} = p$ and $\Pr\{SNP^-\} = 1 - p$, this becomes $(1 - \beta)p/[(1 - \beta)p + \alpha(1 - p)]$. **b.** In this case $\Pr\{SNP^* | P \le \alpha\} = (1 - \beta)/(2 - \beta)$, which even for $1 - \beta = 0.8$ yields 0.44. **c.** The take-home message is that tests of association between SNPs and complex traits are prone to yielding false positives.

9.5 In the ABBA–BABA test for introgression (Chapter 7), the result can be a significant excess of ABBA (ABBA > BABA), a significant excess of BABA (ABBA < BABA), or no significant difference (ABBA ≈ BABA).

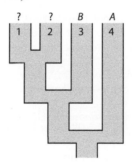

In the diagram shown here, where $1-4$ represent taxa and A and B alleles of a gene:

a. What result is expected with introgression from 3 into 1?
b. What result is expected with introgression from 3 into 2?
c. What result is expected with about equal introgression from 3 into 1 and 3 into 2?

9.5 ANSWER a. With introgression from 3 into 1, some ABBA patterns are changed into BBBA whereas the BABA patterns remain unchanged, and therefore ABBA < BABA. **b.** With introgression from 3 into 2, some BABA patterns are changed into BBBA whereas ABBA patterns remain unchanged, and therefore ABBA > BABA. **c.** With equal introgression from 3 into 1 and 3 into 2, the same number of ABBA patterns are changed to BBBA as BABA patterns are changed to BBBA, and therefore the remaining unchanged patterns have ABBA ≈ BABA.

References

Anderson, J. T., et al. (2014), 'The evolution of quantitative traits in complex environments', *Heredity (Edinb)*, 112 (1), 4–12.

Austin, S. R., Dialsingh, I., and Altman, N. (2014), 'Multiple hypothesis testing: a review', *J Indian Soc Agric Stat*, 68, 303–14.

Barbash, D. A. (2010), 'Ninety years of Drosophila melanogaster hybrids', *Genetics*, 186 (1), 1–8.

Barton, N. H. and Keightley, P. D. (2002), 'Understanding quantitative genetic variation', *Nat Rev Genet*, 3 (1), 11–21.

Beadle, G. W. (1939), 'Teosinte and the origin of maize', *J Hered*, 30, 245–7.

Bedford, T. and Hartl, D. L. (2009), 'Optimization of gene expression by natural selection', *Proc Natl Acad Sci U S A*, 106 (4), 1133–8.

Blum, J. A., et al. (2017), 'The hybrid incompatibility genes Lhr and Hmr are required for sister chromatid detachment during anaphase but not for centromere function', *Genetics*, 207 (4), 1457–72.

Boyle, E. A., Li, Y. I., and Pritchard, J. K. (2017), 'An expanded view of complex traits: from polygenic to omnigenic', *Cell*, 169 (7), 1177–86.

Briggs, H. M., et al. (2018), 'Variation in context-dependent foraging behavior across pollinators', *Ecol Evol*, 8 (16), 7964–73.

Broman, K. W. (2001), 'Review of statistical methods for QTL mapping in experimental crosses', *Lab Anim (NY)*, 30 (7), 44–52.

Bumpus, H. C. (1899), 'The elimination of the unfit as illustrated by the introduced sparrow, Passer domesticus', *Biol Lect Woods Hole Mar Biol Station*, 6, 209–26.

Chan, Y. F., et al. (2010), 'Adaptive evolution of pelvic reduction in sticklebacks by recurrent deletion of a Pitx1 enhancer', *Science*, 327 (5963), 302–5.

Clayton, G. A. and Robertson, A. (1955), 'Mutation and quantitative variation', *Am Nat*, 89, 151–8.

Collard, B. C. Y., et al. (2005), 'An introduction to markers, quantitative trait loci (QTL) mapping, marker-assisted selection for crop improvement: the basic concepts', *Euphytica*, 142, 169–96.

Conner, J. K. (2001), 'How strong is natural selection?', *Trends Ecol Evol*, 16 (5), 215–17.

Conner, J. K. and Hartl, D. L. (2004), *A Primer of Ecological Genetics* (Sunderland, MA: Sinauer Associates).

Coyne, J. A. and Orr, H. A. (2004), *Speciation* (Sunderland, MA: Sinauer Associates).

Davis, A. W. and Wu, C.-I. (1996), 'The broom of the sorcerer's apprentice: the fine structure of a chromosomal region causing reproductive isolation between two sibling species of Drosophila', *Genetics*, 143 (3), 1287–98.

Doebley, J. (2004), 'The genetics of maize evolution', *Annu Rev Genet*, 38, 37–59.

Doerge, R. W. (2002), 'Mapping and analysis of quantitative trait loci in experimental populations', *Nat Rev Genet*, 3 (1), 43–52.

Falconer, D. S. (1960), *Introduction to Quantitative Genetics* (London: Longman).

Falconer, D. S. and Mackay, T. F. C. (1996), *Introduction to Quantitative Genetics* (2 edn.; London: Longman).

Fisher, R. A. (1918), 'The correlation between relatives on the supposition of Mendelian inheritance', *Trans R Soc Edinburgh*, 52, 399–433.

Fisler, J. S. and Warden, C. H. (1997), 'Mapping of mouse obesity genes: a generic approach to a complex trait', *J Nutr*, 127 (9), S1909–16.

Garrigan, D., et al. (2012), 'Genome sequencing reveals complex speciation in the Drosophila simulans clade', *Genome Res*, 22 (8), 1499–511.

Gelernter, J. (2014), 'SLC6A4 polymorphism, population genetics, and psychiatric traits', *Hum Genet*, 133 (4), 459–61.

Grossman, S. R., et al. (2010), 'A composite of multiple signals distinguishes causal variants in regions of positive selection', *Science*, 327 (5967), 883–6.

Hartl, C. L., et al. (2020), 'The region-specific architecture of brain co-expression reveals brain-wide basis of disease susceptibility'. bioRxiv 2020.03.05.965749.

Hereford, J., Hanson, T. F., and Houle, D. (2004), 'Comparing strengths of directional selection: how strong is strong?', *Evolution*, 58 (10), 2133–43.

Hodgins-Davis, A., Rice, D. P., and Townsend, J. P. (2015), 'Gene expression evolves under a house-of-cards model of stabilizing selection', *Mol Biol Evol*, 32 (8), 2130–40.

Hopkins, R., et al. (2014), 'Strong reinforcing selection in a Texas wildflower', *Curr Biol*, 24 (17), 1995–9.

Houle, D. (1998), 'How should we explain variation in in the genetic variance of traits?', *Genetica*, 102/103, 241–53.

Ioannidis, J. P. (2005), 'Why most published research findings are false', *PLoS Med*, 2 (8), e124.

Jansen, R. C. and Stam, P. (1994), 'High resolution of quantitative traits into multiple loci via interval mapping', *Genetics*, 136 (4), 1447–55.

Karn, M. N. and Penrose, L. S. (1951), 'Birth weight and gestation time in relation to maternal age, parity and infant survival', *Ann Eugen*, 16, 147–64.

Kawecki, T. J., et al. (2012), 'Experimental evolution', *Trends Ecol Evol*, 27 (10), 547–60.

Khera, A. V., et al. (2016), 'Genetic risk, adherence to a healthy lifestyle, and coronary disease', *N Engl J Med*, 375 (24), 2349–58.

King, E. G., Macdonald, S. J., and Long, A. D. (2012), 'Properties and power of the Drosophila Synthetic Population Resource for the routine dissection of complex traits', *Genetics*, 191 (3), 935–49.

Kingman, J. F. C. (1978), 'A simple model for the balance between selection and mutation', *J Appl Prob*, 15 (1), 1–12.

Kingsolver, J. G., et al. (2001), 'The strength of phenotypic selection in natural populations', *Am Nat*, 157 (3), 245–61.

Kopp, M. and Matuszewski, S. (2014), 'Rapid evolution of quantitative traits: theoretical perspectives', *Evol Appl*, 7 (1), 169–91.

Korte, A. and Farlow, A. (2013), 'The advantages and limitations of trait analysis with GWAS: a review', *Plant Methods*, 9, 29.

Kryazhimskiy, S., Tkacik, G., and Plotkin, J. B. (2009), 'The dynamics of adaptation on correlated fitness landscapes', *Proc Natl Acad Sci U S A*, 106 (44), 18638–43.

Lande, R. (1979), 'Genetic analysis of multivariate evolution, applied to brain: body size allometry', *Evolution*, 33 (1 Part 2), 402–16.

Lande, R. and Arnold S. J. (1983), 'The measurement of selection on correlated characters', *Evolution*, 37 (6), 1210–26.

Lander, E. S. and Botstein, D. (1989), 'Mapping Mendelian factors underlying quantitative traits using RFLP linkage maps', *Genetics*, 121 (1), 185–99.

Lello, L., et al. (2018), 'Accurate genomic prediction of human height', *Genetics*, 210 (2), 477–97.

Lesch, K.-P., et al. (1996), 'Association of anxiety-related traits with a polymorphism in the serotonin transporter regulatory region', *Science*, 274 (5292), 1527–31.

Liu, X., Li, Y. I., and Pritchard, J. K. (2019), 'Trans effects on gene expression can drive omnigenic inheritance', *Cell*, 177 (4), 1022–34.

Lozovsky, E. R., et al. (2009), 'Stepwise acquisition of pyrimethamine resistance in the malaria parasite', *Proc Natl Acad Sci U S A*, 106 (29), 12025–30.

Lynch, M. and Hill, W. G. (1986), 'Phenotypic evolution by neutral mutation', *Evolution*, 40 (5), 915–35.

MacArthur, J., et al. (2017), 'The new NHGRI-EBI Catalog of published genome-wide association studies (GWAS Catalog)', *Nucleic Acids Res*, 45 (Database issue), D896–901.

Mackay, T. F. C. (2004), 'The genetic architecture of quantitative traits: lessons from Drosophila', *Curr Opinion Genet Dev*, 14 (3), 253–7.

Mackay, T. F. C. (2009), 'Q&A: genetic analysis of quantitative traits', *J Biol*, 8(3), 23.

Makinen, H. S., Cano, J. M., and Merila, J. (2008), 'Identifying footprints of directional and balancing selection in marine and freshwater three-spined stickleback (Gasterosteus aculeatus) populations', *Mol Ecol*, 17 (15), 3565–82.

Mallet, J. (2006), 'What does Drosophila genetics tell us about speciation?', *Trends Ecol Evol*, 21 (7), 386–93.

Marouli, E., et al. (2017), 'Rare and low-frequency coding variants alter human adult height', *Nature*, 542 (7640), 186–90.

Matsumura, S., Arlinghaus, R., and Dieckmann, U. (2012), 'Standardizing selection strengths to study selection in the wild: a critical comparison and suggestions for the future', *BioScience*, 61 (12), 1039–54.

Maynard Smith, J. (1970), 'Natural selection and the concept of a protein space', *Nature*, 222 (5232), 563–4.

Mayr, E. (1942), *Systematics and the Origin of Species* (New York: Columbia University Press).

Moore, A. J. and Kukuk, P. F. (2002), 'Quantitative genetic analysis of natural populations', *Nat Rev Genet*, 3 (12), 971–8.

Morrissey, M. B. and Sakrejda, K. (2013), 'Unification of regression-based methods for the analysis of natural selection', *Evolution*, 67 (7), 2094–100.

Ogbunugafor, C. B. and Hartl, D. L. (2016), 'A new take on John Maynard Smith's concept of protein space for understanding molecular evolution', *PLoS Comput Biol*, 12 (10), e1005046.

Park, D. J., et al. (2012), 'Sequence-based association and selection scans identify drug resistance loci in the Plasmodium falciparum malaria parasite', *Proc Natl Acad Sci U S A*, 109 (32), 13052–7.

Park, J. H., et al. (2011), 'Distribution of allele frequencies and effect sizes and their interrelationships for common genetic susceptibility variants', *Proc Natl Acad Sci U S A*, 108 (44), 18026–31.

Peichel, C. L. and Marques, D. A. (2017), 'The genetic and molecular architecture of phenotypic diversity in sticklebacks', *Philos Trans R Soc Lond B Biol Sci*, 372 (1713), 20150486.

Pfeifer, S. P., et al. (2018), 'The evolutionary history of Nebraska deer mice: local adaptation in the face of strong gene flow', *Mol Biol Evol*, 35 (4), 792–806.

Piperno, D. R., et al. (2009), 'Starch grain and phytolith evidence for early ninth millennium B.P. maize from the Central Balsas River Valley, Mexico', *Proc Natl Acad Sci U S A*, 106 (13), 5019–24.

Presgraves, D. C. (2003), 'A fine-scale genetic analysis of hybrid incompatibilities in Drosophila', *Genetics*, 163 (3), 955–72.

Robinson, M.R., Wray, N.R., and Visscher, P.M. (2014), 'Explaining additional genetic variation in complex traits', *Trends Genet*, 30 (4), 124–32.

Roda, F., et al. (2017), 'Genomic evidence of gene flow during reinforcement in Texas Phlox', *Mol Ecol*, 26 (8), 2317–30.

Roiser, J. P., et al. (2005), 'Association of a functional polymorphism in the serotonin transporter gene with abnormal emotional processing in Ecstasy users', *Am J Psychiatry*, 162 (3), 609–12.

Sackton, T. B. and Hartl, D. L. (2016), 'Genotypic context and epistasis in individuals and populations', *Cell*, 166 (2), 269–87.

Sanjak, J. S., et al. (2018), 'Evidence of directional and stabilizing selection in contemporary humans', *Proc Natl Acad Sci U S A*, 115 (1), 151–6.

Sham, P. C. and Purcell, S. M. (2014), 'Statistical power and significance testing in large-scale genetic studies', *Nat Rev Genet*, 15 (5), 335–46.

Siepielski, A. M., DiBattista, J. D., and Carlson, S. M. (2009), 'It's about time: the temporal dynamics of phenotypic selection in the wild', *Ecol Lett*, 12 (11), 1261–76.

Storey, J. D. and Tibshirani, R. (2003), 'Statistical significance for genomewide studies', *Proc Natl Acad Sci U S A*, 100 (16), 9440–5.

Studer, A. J., Wang, H., and Doebley, J. F. (2017), 'Selection during maize domestication targeted a gene network controlling plant and inflorescence architecture', *Genetics*, 207 (2), 755–65.

Suni, S. S. and Hopkins, R. (2018), 'The relationship between postmating reproductive isolation and reinforcement in Phlox', *Evolution*, 72 (7), 1387–98.

Tachida, H. (1991), 'A study on a nearly neutral model in finite populations', *Genetics*, 128 (1), 183–92.

Tao, Y. and Hartl, D. L. (2003), 'Genetic dissection of hybrid incompatibilities between Drosophila simulans and Drosophila mauritiana. III. Heterogeneous accumulation of hybrid incompatibilities, degree of dominance and implications for Haldane's rule', *Evolution*, 57 (11), 2580–98.

Tao, Y., et al. (2003a), 'Genetic dissection of hybrid incompatibilities between Drosophila simulans and Drosophila mauritiana. I. Differential accumulation of hybrid male sterility effects on the X and autosomes', *Genetics*, 164 (4), 1383–97.

Tao, Y., et al. (2003b), 'Genetic dissection of hybrid incompatibilities between Drosophila simulans and Drosophila mauritiana. II. Mapping hybrid male sterility loci on the third chromosome', *Genetics*, 164 (4), 1399–418.

Turelli, M. (1984), 'Heritable genetic variation via mutation-selection balance: Lerch's zeta meets the abdominal bristle', *Theor Popul Biol*, 25 (2), 138–93.

Turelli, M. (1988), 'Phenotypic evolution, constant covariances, and the maintenance of additive variance', *Evolution*, 42 (6), 1342–7.

Visscher, P. M., et al. (2017), '10 Years of GWAS discovery: biology, function, and translation', *Am J Hum Genet*, 101 (1), 5–22.

Weinreich, D. M., et al. (2006), 'Darwinian evolution can follow only very few mutational paths to fitter proteins', *Science*, 312 (5770), 111–14.

Wilson, A. and Poissant, J. (2016), Quantitative genetics in natural populations, in R. M. Kliman (ed.) *Encyclopedia of Evolutionary Biology* (Oxford: Academic Press), 361–71.

Wilson, D. J. (2019), 'The harmonic mean *p*-value for combining dependent tests', *Proc Natl Acad Sci U S A*, 116 (4), 1195–200.

Wood, A. R., et al. (2014), 'Defining the role of common variation in the genomic and biological architecture of adult human height', *Nat Genet*, 46 (11), 1173–86.

Wray, N. R., et al. (2018), 'Common disease is more complex than implied by the core gene omnigenic model', *Cell*, 173 (7), 1573–80.

Yang, J., et al. (2011), 'Genome partitioning of genetic variation for complex traits using common SNPs', *Nat Genet*, 43 (6), 519–25.

Yap, C. X., et al. (2018), 'Dissection of genetic variation and evidence for pleiotropy in male pattern baldness', *Nat Commun*, 9 (1), 5407.

Zeng, P., et al. (2015), 'Statistical analysis for genome-wide association study', *J Biomed Res*, 29 (4), 285–97.

Zeng, Z. -B. (1994), 'Precision mapping of quantitative trait loci', *Genetics*, 136 (4), 1457–68.

Index